REVIEWS in MINERALOGY

Volume 8

KINETICS of GEO-CHEMICAL PROCESSES

A. C. LASAGA & R. J. KIRKPATRICK, Editors

The Authors: Antonio C. Lasaga

Department of Geosciences
Pennsylvania State University
University Park, Pennsylvania 16802

Robert A. Berner

Department of Geology and Geophysics
Yale University
P. O. Box 6666
New Haven, Connecticut 06511

George W. Fisher

Department of Earth and Planetary Sciences
Johns Hopkins University
Baltimore, Maryland 21218

David E. Anderson and R. James Kirkpatrick

Department of Geology
245 Natural History Building
University of Illinois
Urbana, Illinois 61801

Series Editor: Paul H. Ribbe

Department of Geological Sciences
Virginia Polytechnic Institute & State University
Blacksburg, Virginia 24060

MINERALOGICAL SOCIETY OF AMERICA

PRINTED BY

BookCrafters, Inc.

Chelsea, Michigan 48118

REVIEWS IN MINERALOGY

ISSN 0275-0279

VOLUME 8: KINETICS OF GEOCHEMICAL PROCESSES

ISBN 0-939950-08-1

ADDITIONAL COPIES

Additional copies of this volume as well as those listed below may be obtained at moderate cost from

Mineralogical Society of America
2000 Florida Avenue, NW
Washington, D.C. 20009

Vol. 1:	SULFIDE MINERALOGY, P.H. Ribbe, Editor	(1974)
Vol. 2:	FELDSPAR MINERALOGY, P.H. Ribbe, Editor	(1975)
Vol. 3:	OXIDE MINERALS, Douglas Rumble III, Editor	(1976)
Vol. 4:	MINERALOGY AND GEOLOGY OF NATURAL ZEOLITES, F.A. Mumpton, Editor (1977)	(1977)
Vol. 5:	ORTHOSILICATES, P.H. Ribbe, Editor	(1980)
Vol. 6:	MARINE MINERALS, Roger G. Burns, Editor	(1979)
Vol. 7:	PYROXENES, C.T. Prewitt, Editor	(1980)

KINETICS of GEOCHEMICAL PROCESSES

FOREWORD

In 1974 the Mineralogical Society of America began publishing paperback books in conjunction with short courses presented at the time of its annual meetings. The purpose of these courses and their accompanying "Short Course Notes" has been to assemble cogent and concise reviews of the literature and current research in particular disciplines or subject-areas in mineralogy, petrology and crystallography. To date, seven groups of minerals have been featured: sulfides, feldspars, oxides, zeolites, orthosilicates, pyroxenes, and minerals of the marine environment. Titles are listed on page ii (opposite): those volumes and this one are available at moderate cost from M.S.A. at the address indicated.

In 1980 the Council of the Mineralogical Society of America changed the title of the serial publications proceeding from its short courses from "Short Course Notes" to "Reviews in Mineralogy" in order to more accurately reflect the content of these volumes. They are all basically designed as 'reference textbooks' for postgraduates and are used as such at numerous universities.

This volume, *Kinetics of Geochemical Processes*, is the first in the "Reviews in Mineralogy" series to be process-oriented. It was the brainchild of David R. Wones, who overheard Tony Lasaga and Jim Kirkpatrick talking about a book they were preparing to write on the subject. They were persuaded to convene a short course at Airlie House near Warrenton, Virginia on May 22-24, 1981, which was received with considerable enthusiasm.

The Mineralogical Society is grateful to the many editors and authors who have contributed to the success of this series of publications. It is hoped that they will continue to serve the geological community well. Ideas for future short courses and publications are cordially invited.

Paul H. Ribbe
Series Editor
Blacksburg, VA

PREFACE and ACKNOWLEDGMENTS

Geochemistry is a science that is based on an understanding of chemical processes in the earth. One of the principal tools available to the chemist for understanding systems at equilibrium is thermodynamics. The awareness and application of thermodynamic techniques has increased at a very fast pace in geosciences; in fact, one may be so bold as to say that thermodynamics in geology has reached the "mature" stage, although much future thermodynamic research is certainly needed. However, the natural processes in the earth are often sluggish enough that a particular system may not reach equilibrium. This observation is being supported constantly by new experimental and field data available to the geochemist e.g. the non-applicability of the phase rule in some assemblages, the compositional inhomogeneities of mineral grains, the partial reaction rims surrounding original minerals, the lack of isotopic equilibration or the absence of minerals (e.g. dolomite), which should be present according to thermodynamics.

The need to apply kinetics has produced a large number of papers dealing with kinetics in geochemistry. As an initial response to this growing field, a conference on geochemical transport and kinetics was conducted at Airlie House, VA, in 1973, sponsored by the Carnegie Institution of Washington. The papers there dealt with several kinetic topics including diffusion, exsolution, metasomatism and metamorphic layering. Since 1973 the number of kinetic papers has continued to increase greatly. Therefore, the time is ripe for a Short Course in Kinetics, which brings together the fundamentals needed to explain field observations using kinetic data. It is hoped that this book may serve, not only as a reference for researchers dealing with the rates of geochemical processes, but also as a text in courses on geochemical kinetics. One of us has found this need of a text in teaching a graduate course on geochemical kinetics at Harvard and at Penn State during the past several years. Finally, it is our hope that the book may itself further even more research into the rates of geochemical processes and into the quantification of geochemical observations.

The book is organized with a rough temperature gradient in mind, i.e. low temperature kinetics at the beginning and igneous kinetics at the end (no prejudices are intended with this scheme!). However, the topics in each chapter are general enough that they can be applied often to any geochemical domain: sedimentary, metamorphic or igneous. The theory of kinetics operates at two complementary levels: the phenomenological and the atomistic. The former relies on macroscopic variables (e.g. temperature or concentrations) to describe the rates of reactions or the rates of transport; the latter relates the rates to the basic forces operating between the particular atomic or molecular species of any system. This book deals with both descriptions of the kinetics of geochemical processes. Chapter one sets the framework for the phenomenological theory of reaction rates. If any geochemical reaction is to be described quantitatively, the rate law must be experimentally obtained in a kinetically sound manner and the reaction mechanism must be understood. This applies to heterogeneous fluid-rock reactions such as those occurring during metamorphism, hydrothermal alteration or weathering as well as to homogeneous reactions. Chapter 2 extends the theory to the global kinetics of geochemical cycles. This enables the kinetic concepts of stability and feedback to be applied to the cycling of elements in the many reservoirs of the earth. Chapter 3 applies the phenomenological treatment of chapter 1 to diagenesis and weathering. The rate of dissolution of minerals as well as the chemical evolution of pore waters are discussed. The atomistic basis of rates of reaction, transition state theory, is introduced in Chapter 4. Transition state theory can be applied to relate the rate constants of geochemical reactions to the atomic processes taking place. This includes not only homogeneous reactions but also reactions that occur at the surface of minerals. Chapter 5 discusses the theory of irreversible thermodynamics and its application to petrology. The use of the second law of thermodynamics along with the expressions for the rate of entropy production in a system have been used successfully since 1935 to describe kinetic phenomena. The chapter applies the concepts to the growth of minerals during metamorphism as well as to the formation of differentiated layers (banding) in petrology. Chapter 6 describes the phenomenological theory of diffusion both in aqueous solutions and in minerals. In particular, the multicomponent nature of diffusion and its consequence in natural systems is elaborated. Chapter 7 provides the atomistic basis for the rates of reactions in minerals. Understanding of the rates of diffusion, conduction, order-disorder reactions or exsolution in minerals depends on proper description of the defects in the various mineral structures. Chapter 8 provides the kinetic theory of crystal nucleation and growth. While many of the concepts in the chapter can be applied to aqueous systems, the emphasis is on igneous processes occurring during crystallization of a melt. To fully understand both the mineral composition as well as the texture of igneous rocks, the processes whereby new crystals form and grow must be quantified by using kinetic theory. Due to space and time limitations (kinetics!) some topics have not been covered in detail. In particular, the mathematical solution of diffusion or conduction equations is discussed very well by Crank in his book, Mathematics of Diffusion, and so is not covered to a great extent here. The treatment of fluid flow (e.g. convection) is also not covered in the text.

We would like to thank the various authors for sending the manuscripts in time and overcoming their own activation energies with suitable catalysts. A. Lasaga would also like to thank two of his graduate students, Gregory E. Muncill and Randy Cygan, for their extensive help in the technical preparation of the book, and Mary Frank for her excellent typing of the preliminary manuscripts.

P.H. Ribbe, as series editor of *Reviews in Mineralogy*, was responsible for preparing camera-ready copy. We are grateful to Margie Strickler for typing the text of this volume, and to Ramonda Haycocks, Rachel Elliott, and Sharon Chiang for their assistance with additional typing, drafting and editorial work, all carried out at the Department of Geological Sciences, Virginia Polytechnic Institute and State University.

TABLE of CONTENTS

Page

Copyright and Additional Copies ii

Foreword iii

Preface and Acknowledgments iv

CHAPTER 1. RATE LAWS OF CHEMICAL REACTIONS Antonio C. Lasaga

INTRODUCTION 1

OVERALL AND ELEMENTARY REACTIONS 2

STEADY STATE 7

SEQUENTIAL REACTIONS, PARALLEL REACTIONS, AND RATE-DETERMINING STEP. . 11

EXPERIMENTAL DETERMINATION OF RATE LAWS. 13

Initial rate method 13
Integration of phenomenological rate laws -- method of isolation 18
(Pseudo) first-order and second-order reactions 19
Relaxation methods 29

TEMPERATURE DEPENDENCE OF RATE CONSTANTS 30

HETEROGENEOUS KINETICS, CATALYSIS AND SURFACE ADSORPTION 35

PRINCIPLE OF DETAILED BALANCING. 48

REACTION MECHANISM: SIMPLE CHAIN REACTION 57

REACTION PROGRESS VARIABLE, MASS TRANSFER AND STOICHIOMETRY. 61

CONCLUSION 66

ACKNOWLEDGMENTS 66

REFERENCES: CHAPTER 1 67

CHAPTER 2. DYNAMIC TREATMENT OF GEOCHEMICAL CYCLES: GLOBAL KINETICS Antonio C. Lasaga

INTRODUCTION 69

LINEAR CYCLES. 70

SULFUR CYCLE 86

TREATMENT AND CONSEQUENCES OF NON-LINEAR CYCLES. 90

CONCLUSION 104

CHAPTER 2 APPENDIX -- MATRIX FUNDAMENTALS. 106

REFERENCES: CHAPTER 2 110

CHAPTER 3. KINETICS OF WEATHERING AND DIAGENESIS Robert A. Berner

INTRODUCTION 111

SOME FUNDAMENTAL CONCEPTS. 112

Crystallization 112
Dissolution 117
Organic matter decomposition 118

WEATHERING 120

DIAGENESIS 127

CHAPTER 3, Kinetics of weathering and diagenesis, continued | Page

CONCLUSION 131

ACKNOWLEDGMENTS. 132

REFERENCES: CHAPTER 3 133

CHAPTER 4. TRANSITION STATE THEORY — Antonio C. Lasaga

INTRODUCTION 135

REVIEW OF PARTITION FUNCTIONS. 137

DERIVATION OF THE RATE CONSTANT. 140

ARRHENIUS PARAMETERS AND ENTROPY OF ACTIVATION 143

IMPORTANT CONSEQUENCES OF TRANSITION STATE THEORY. 145

Fundamental frequency 145
Relation between rates and ΔG 145
Implication of transition state theory on ionic strength effects 148
Compensation law 150

MISCONCEPTIONS AND LIMITATIONS OF TRANSITION STATE THEORY. 150

Meaning of $K^\dagger$ 151
The value of $\gamma^\dagger$ 151
Elementary versus overall reactions 151

APPLICATION OF TRANSITION STATE THEORY 153

Diffusion in liquids 156
Solid state diffusion 160
Energetics of surface reactions 164

CONCLUSION 168

ACKNOWLEDGMENTS. 168

REFERENCES: CHAPTER 4 169

CHAPTER 5. IRREVERSIBLE THERMODYNAMICS IN PETROLOGY
George W. Fisher & Antonio C. Lasaga

INTRODUCTION 171

THE FUNDAMENTAL KINETIC EQUATIONS. 172

Application of the kinetic equations 174

THE DISSIPATION FUNCTION AND ITS ROLE IN COMPLEX PROCESSES 178

Further applications 185

IRREVERSIBLE THERMODYNAMICS, CHEMICAL REACTIONS AND PATTERN FORMATION IN PETROLOGY 186

Linear phenomenological laws for chemical reactions 187
Transport and non-linear irreversible processes 191

CONCLUSION 206

REFERENCES: CHAPTER 5 208

Page

CHAPTER 6. DIFFUSION IN ELECTROLYTE MIXTURES David E. Anderson

INTRODUCTION 211

NOTATION USED IN CHAPTER 6 212

GENERAL STRUCTURE OF DIFFUSION THEORY FOR MOLECULAR COMPONENTS 214

MEASURABLE QUANTITIES. 222

REFERENCE FRAMES 230

TRANSFORMATION AND CALCULATION OF L-COEFFICIENTS 234

THE DIAGONALIZATION OF $[D^V]$. 239

IONIC FLUXES 244

ESTIMATION OF DIFFUSION COEFFICIENTS 249

COMMENTS 254

ACKNOWLEDGMENTS. 256

REFERENCES: CHAPTER 6 257

CHAPTER 7. THE ATOMISTIC BASIS OF KINETICS: DEFECTS IN MINERALS Antonio C. Lasaga

INTRODUCTION 261

POINT DEFECTS. 262

Terminology and concentration of point defects in crystals 263
Vacancy pairs 267
Impure crystals 270
Migration energies 273

POINT DEFECTS AND DIFFUSION. 274

ESTIMATION OF Δh_m, Δh_s, Q_D, AND D_o 284

Compensation law 290

NON-STOICHIOMETRY DEFECTS AND DIFFUSION. 293

ELECTRONIC DEFECTS 299

DEFECT CALCULATIONS IN MINERALS. 301

Short-range potential 304
Polarization energy 305

LARGE DEVIATIONS FROM STOICHIOMETRY. 309

EXTENDED DEFECTS: DISLOCATIONS AND PLANAR DEFECTS 309

SUMMARY. 315

ACKNOWLEDGMENTS. 316

REFERENCES: CHAPTER 7 317

Page

CHAPTER 8. KINETICS OF CRYSTALLIZATION OF IGNEOUS ROCKS
R. James Kirkpatrick

INTRODUCTION 321

NUCLEATION 322

Theory 322

Steady state nucleation 322
Transient homogeneous nucleation 329
Heterogeneous nucleation 330
Solid state nucleation 332

Experimental nucleation data in simple silicate compositions 332

The system Li_2O-SiO_2 332
The system Na_2O-CaO-SiO_2 338

More complex compositions 341
Nucleation delay in undercooled samples 345

CRYSTAL GROWTH 349

Theory 349

Rate controlling processes 349
Interface controlled growth 350
rate laws 350
growth mechanisms 351
Computer simulation of crystal growth 352
Diffusion controlled growth 359
Stability of planar crystal surfaces 363

Crystal growth data 368

Rates of crystal growth 368

BULK CRYSTALLIZATION OF IGNEOUS ROCKS. 376

PROGRAMMED COOLING EXPERIMENTS 387

APPENDIX: THE RATE OF CRYSTALLIZATION OF NUCLEATION AND GROWTH CONTROLLED REACTIONS 392

REFERENCES: CHAPTER 8 396

Chapter 1

RATE LAWS of CHEMICAL REACTIONS

Antonio C. Lasaga

"Old geochemists never die, they merely reach equilibrium."

INTRODUCTION

In the last two decades, geochemists have produced a large body of data on the properties of materials at equilibrium. The laws of thermodynamics state precisely how to use these data to obtain other related thermodynamic quantities. Although progress has been made in understanding the application of thermodynamics, the application of kinetics to geological problems is still embryonic. For many geochemical systems, kinetics is the controlling factor in deciding their fate and evolution. Thermodynamics in these cases identifies a point toward which the reactions are aiming; in contrast, sometimes the kinetic path may not even seem to do that much. Of course, it is interesting to note that geology has always stressed the element of *time* as a central concept in describing the earth. In this respect the description of time-dependent phenomena (*e.g.*, kinetics) is even more akin to geology than thermodynamics (time independent).

The study of kinetics is inherently more difficult that that of thermodynamics, because time-dependent processes are *path* dependent. For example, as Benson (1960) aptly explained, a body of water on top of a hill may be described thermodynamically in terms of its composition, pressure, and temperature. At a later time, this same body of water may find its way to a lake below. The thermodynamic description of the body of water in the lake is again well defined. However, if we try to describe the transition -- the water in process of flowing from the hilltop -- we see that it may depend on almost innumerable factors: on the outlets, on the contour of the hillside, on the structural stability of the contour, and on the numerous subterranean channels through the hillside that may exist and permit seepage. Furthermore, if someone has bored a hole under the hilltop, it will take careful experimental investigation to uncover this additional factor which affects the flow. The complexity of geochemical kinetics may lead one at times to wonder whether the hilltop has in fact been punctured with numerous and tortuous holes!

The emphasis in this chapter will be on the general *phenomenological* methods used to characterize the approach of chemical systems to equilibrium and their use in geochemistry. The phenomenological approach tries to relate the evolution of a system to *macroscopically* observable quantities such as composition, temperature, pressure, volume, and time. It expresses the results in terms of "rate constants" and the relevant macroscopic parameters. This approach is necessarily the first step in describing the kinetics of a system. Ultimately, however, the studies must extend to molecular mechanisms for explanations of the rate laws. This is necessary if we are to be able to extend our knowledge to other systems and predict the kinetic behavior of new reactions.

The phenomenological treatment of kinetics is important in analyzing a variety of geochemical processes. These processes include rates of chemical reactions in natural waters, adsorption of solutes onto surfaces, leaching, dissolution and weathering of minerals, rates of isotopic exchange in hydrothermal solutions, and reactions at grain boundaries in metamorphic reactions. A substantial effort has been made recently to decipher the rate laws of mineral leaching and dissolution. Hence, several of the examples will be based on this process. However, it should be stressed that the concepts outlined here have broad utility.

An important distinction in earth sciences must be made between *homogeneous* and *heterogeneous* reactions. If a reaction occurs wholly within a single macroscopic phase, then the reaction is said to be homogeneous. If, on the other hand, the reaction occurs at the interface between two phases, the reaction is said to be heterogeneous. We will encounter examples of both below.

OVERALL AND ELEMENTARY REACTIONS

The first step in understanding the kinetics of complex processes is to distinguish *overall* reactions from *elementary* reactions. An elementary reaction describes a reaction that actually occurs *as written* at the molecular level. Therefore, the reaction $H^+ + HCO_3^- \rightarrow H_2CO_3$ is an elementary reaction because it represents the molecular collision between a proton and a bicarbonate ion to produce H_2CO_3.

On the other hand, an overall reaction represents the net result of a series of elementary reactions. For example, the reduction of sulfate in sediments by the bacterial decomposition of organic matter, $2CH_2O + SO_4^{2-} \rightarrow H_2S + 2HCO_3^-$, is an overall reaction. This reaction is the net result of a series of processes which include the breakdown of complex organic molecules and the various reactions of SO_4^{2-} with bacterial enzymes, each of which comprises several elementary reactions. In this section, the differences between the kinetic treatment of overall and elementary reactions will be discussed.

If we write a general overall reaction as

$$aA + bB + \ldots \rightarrow pP + qQ + \ldots, \tag{1}$$

the rate of the reaction can be written as

$$\frac{1}{a}\frac{dC_A}{dt} = \frac{1}{b}\frac{dC_B}{dt} = \ldots = -\frac{1}{p}\frac{dC_P}{dt} = -\frac{1}{q}\frac{dC_Q}{dt} \ldots \tag{2}$$

$$= k\, C_A^{n_a}\, C_B^{n_b} \ldots C_P^{n_p}\, C_Q^{n_q}$$

where the C represents some units of concentration and the n's can be any real number. In this section, the rate will refer to the forward rate of a reaction exclusively, *i.e.*, the rate at which the reaction proceeds from left to right as written. In later sections, both forward and reverse rates will be considered. The form of the rate law in equation (2) (*i.e.*, product of concentration terms) has been validated in the majority of kinetic studies. The exact rate law may be more complicated than that in equation (2), but in most cases certain simplifications allow the rate law to reduce to the form in equation (2). Examples of the general rate law are given below. The units of the rate constant k will depend on the form of the rate law. Therefore, for the expression in equation (2), k has units of concentrations to the power $(n_a + n_b + \ldots + n_p + n_q - 1)$ per unit time. In the case that the reaction is heterogeneous, one or more of the "concentrations" in equation (2) should refer to the *specific area* (*i.e.*, area/unit volume of solution) of the solids involved. In heterogeneous kinetics, it is very important to measure the total areas of the reactive solids, either directly by BET and related methods or indirectly by constraining grain size and shape within narrow confines. The need to use *concentration* units in writing rate laws distinguishes kinetics from thermodynamics. Whereas the "thermodynamic concentration," or activity, determines the equilibrium between thermodynamic components, the *spatial* concentration (*e.g.*, moles/cm^3) of the colliding molecules determine the molecular collision rates, and hence the rates of reactions.

In the case of complex reactions with several steps between reactants and final products, an important requirement implicit in equation (2) is that the various equalities on the left side of equation (2) be valid. For example, for every b molecules of B used up, p molecules of P must appear *simultaneously*. This requirement may fail if some of the intermediate steps are not fully adjusted to steady state. Another way of stating this requirement is that the kinetic reaction (1) be *stoichiometrically* true.

The rate law of complex reactions can depend on products as well as reactants; hence, equation (2) includes terms involving C_P and C_Q. On the other hand, the rate law of elementary reactions will never depend on the concentration of products (recall that only forward reactions are considered here). The sum of exponents, $n_a + n_b + \ldots$, in equation (2) is termed the "order" of the overall reaction. For example, Fenton's reaction for the oxidation of Fe^{2+} by H_2O_2,

$$H_2O_2(aq) + 2Fe^{2+} + 2H^+ \rightarrow 2Fe^{3+} + 2H_2O(\ell) \qquad (3)$$

might be expressed by the following general rate law:

$$-\frac{dC_{H_2O_2}}{dt} = k\, C^n_{H_2O_2}\, C^m_{Fe^{2+}}\, C^l_{H^+}\, C^r_{Fe^{3+}}\, C^s_{H_2O} \ .$$

Experimental data (Benson, 1960) show that, in fact, the rate law is given by

$$-\frac{dC_{H_2O_2}}{dt} = k\, C_{Fe^{2+}}\, C_{H_2O_2} \ . \qquad (4)$$

By our definition, this is a second-order reaction.

In thermodynamics the energy changes associated with any particular reaction, *e.g.*, ΔH, ΔS, ΔG, are path independent. Once ΔG is known for one reaction, it can be added or subtracted in any other scheme where the reaction appears. For example, from a knowledge of ΔG_1 and ΔG_2 for the reactions

$$3FeO + \tfrac{1}{2}O_2 \rightarrow Fe_3O_4 \qquad \Delta G_1$$

$$2Fe_3O_4 + \tfrac{1}{2}O_2 \rightarrow 3Fe_2O_3 \qquad \Delta G_2$$

we can immediately find ΔG of the reaction

$$6FeO + \tfrac{3}{2}O_2 \rightarrow 3Fe_2O_3; \qquad \Delta G_3 = 2\Delta G_1 + \Delta G_2$$

To evaluate ΔG_3 it is not necessary to know *how* ΔG_1 and ΔG_2 were obtained as long as the thermodynamic variables (including f_{O_2}) are the same for all three reactions.

In kinetics, unfortunately, the critical parameters do not have unique general values, because state functions are lacking. For example, kinetically the oxidation of FeO to Fe_2O_3 may proceed with the formation of an intermediate Fe_3O_4 layer coating the original FeO crystal. On the other hand, FeO in an aqueous medium may dissolve and form FeOOH which ultimately dehydrates to give Fe_2O_3. While thermodynamics is invariant to the path from FeO to Fe_2O_3, the kinetics of this transformation will be substantially different in the two cases posed above. A consequence of this dependence on path is that a particular rate of oxidation of FeO, an overall reaction, cannot be applied indiscriminately to every situation where oxidation of FeO is occurring. A major difference between elementary and overall reactions is that the rates of elementary reactions can be used without regard to path. In this sense, elementary reactions are the closest analogues of state functions in kinetics.

The description of an overall reaction in terms of each elementary reaction involved is termed the *reaction mechanism*. For example, Fenton's reaction [equation (3)] is an *overall* reaction. By this we simply mean that, at the molecular level, it is quite improbable that a dissolved H_2O_2 molecule, *two* ferrous ions and *two* hydronium ions would come together in the same region of solution at the same time and react to yield two ferric ions and two water molecules. Rather, the path may be something like this (Benson, 1960, p. 596):

Initiation + Termination

$$Fe^{2+} + HOOH \rightleftarrows Fe^{3+} + OH^- + OH \qquad (5a)$$

$$Fe^{3+} + HOOH \rightleftarrows Fe^{2+} + HO_2 + H^+ \qquad (5b)$$

$$\text{Chain propagation} \begin{cases} HO + HOOH \rightarrow H_2O + HO_2 & (5c) \\ Fe^{3+} + HO_2 \rightleftarrows Fe^{2+} + O_2 + H^+ & (5d) \\ Fe^{2+} + HOOH \rightleftarrows Fe^{3+} + OH^- + OH & (5e) \end{cases}$$

$$\text{Termination} \quad Fe^{2+} + OH \rightarrow Fe^{3+} + OH^- \qquad (5f)$$

(We have ignored the possibly extensive complexing of Fe^{2+} and Fe^{3+}.) These reactions are *elementary* reactions; that is, they mechanistically describe the stepwise transformations at the molecular level. Skirting the relation of the mechanisms in equations (5) to the overall rate law, the rate constants associated with one elementary reaction in equations (5) can be used in *any* other process where the particular elementary reaction occurs in the reaction mechanism. We obviously cannot do that with overall reactions because the rate depends on the mechanism, and the *same* overall reaction can occur via quite different paths in different circumstances.

The mechanism for Fenton's reaction points out another difference between overall and elementary reactions. In the mechanism there appear new *intermediate* species such as the radicals, OH and HO_2. These species did not occur in the overall reaction shown in equation (3). It is usual for overall reactions containing many steps to proceed via new species, which although metastable or of negligible thermodynamic activity, are crucial in determining the overall rate. This role of intermediate species, also called *reactive intermediates*, is absent in the case of elementary reactions which depict exactly the molecular species involved in the reaction. As another example, consider the exchange of sulfur isotopes between sulfates and sulfides in hydrothermal solutions, a process of great importance in understanding the evolution of ore deposits:

$$^{34}SO_4^{2-} + H_2{}^{32}S \rightleftarrows {}^{32}SO_4^{2-} + H_2{}^{34}S \;. \qquad (6)$$

Reaction (6) is an overall reaction since the isotope swap does not occur upon a chance collision between sulfate and sulfide species. In fact, the reactive intermediate in this case seems to be the thiosulfate molecule or ion (Ohmoto and Lasaga, 1981).

Another important simplicity associated with elementary reactions stems from the obvious nature of their rate law. The rate of *any* elementary reaction is always proportional to the concentration of each reactant. For example,

$$A + A \rightarrow P \qquad \text{Rate} = \frac{dC_P}{dt} = k\, C_A^2$$

or

$$A + B \rightarrow P \qquad \text{Rate} = \frac{dC_P}{dt} = k\, C_A\, C_B, \text{ and so on.}$$

The probability that two or three species collide is proportional to the concentrations of each reactant. Since elementary reactions occur only via these molecular collisions, the simplicity of the rate law is to be expected. As can be seen in the examples to follow [equations (7) and (8)], the rates of overall reactions *cannot* be simply predicted from the stoichiometries of the reactants!

The application of molecular or statistical mechanical theories, *e.g.*, transition state theory, in kinetics refers exclusively to individual elementary reactions. This important difference between elementary and overall reactions has caused much confusion in the literature. In terms of the activated complex of transition state theory (see Chapter 4), the molecularity (or the *order* of an elementary reaction) is equal to the number of molecules of reactants that are used to form the activated complex. Molecularity has meaning only for elementary reactions. One cannot apply transition state theory directly to an overall reaction, such as equation (3) or equation (6), because transition state theory is a statistical mechanical (*i.e.*, molecular) theory which only applies to each elementary reaction individually. In fact, doing so will lead to meaningless conclusions as to activation enthalpies and entropies. Insofar as the rate of the overall reaction is related to the rates of the elementary reactions comprising the reaction mechanism, transition state theory can be of general use in the kinetics of overall reactions. We will have occasion to return to this important point in Chapter 4.

STEADY STATE

Having introduced the difference between overall and elementary reactions and the role of reactive intermediates in a reaction mechanism, one more important concept is needed which is often essential in relating the details of the mechanism to the expression for the overall reaction rate. This is the concept of *steady state for reaction intermediates*. Let us illustrate the use of steady state in the case of the ozone decomposition reaction. The overall reaction is

$$2O_3 \rightarrow 3O_2 \tag{7}$$

and the overall reaction rate of the forward reaction is given by (Weston and Schwartz, 1972)

$$\frac{1}{3}\frac{d[O_2]}{dt} = -\frac{1}{2}\frac{d[O_3]}{dt} = \frac{k[O_3]^2}{[O_2]} \ . \tag{8}$$

To eliminate cumbersome notation in what follows, [ ] will stand for concentration. In this case the apparent overall order is 2 - 1 = 1. But it is clear that reaction (7) is not an elementary reaction, since, in particular, a product (O_2) can influence the rate of reaction. How does a product influence the rate? The answer lies in the reaction mechanism. A suggested mechanism for reaction (7) is

$$O_3 \underset{k_{-1}}{\overset{k_1}{\rightleftharpoons}} O_2 + O \qquad \text{fast reverse reaction} \tag{9a}$$

$$O + O_3 \xrightarrow{k_2} 2O_2 \qquad \text{slow} \tag{9b}$$

The reverse reaction in (9b) is extremely slow and is hence ignored. The reactions in equation (9) *are* elementary reactions. Hence, the total rate of change of the concentration of oxygen atoms can be readily evaluated by adding the contributions from each reaction in (9):

$$\frac{d[O]}{dt} = k_1[O_3] - k_{-1}[O_2][O] - k_2[O][O_3] \ . \tag{10}$$

Note that if the values of k_1, k_{-1}, or k_2 are obtained, these values can be used in any other overall reaction where the particular elementary reaction occurs in the reaction mechanism.

The ozone reaction mechanism introduces a reactive intermediate, O, which is not present in the overall reaction (7). It often happens that the concentration of an intermediate (especially radicals and atoms) reaches a steady state value after a transient period, which is short compared to reaction times. In this case, the rate of formation of the intermediate is balanced by the rate of annihilation. This state of dynamic equilibrium is termed *steady state*. Note that it is not a true thermodynamic equilibrium. Applying this concept to the reactive oxygen atoms in equation (10) after the initial transient period,

$$\frac{d[O]}{dt} = 0 = k_1[O_3] - k_{-1}[O_2][O] - k_2[O][O_3] \ .$$

Using this steady state equation to solve for the concentration of oxygen atoms yields

$$[O]_{st} = \frac{k_1[O_3]}{k_{-1}[O_2] + k_2[O_3]} \ . \qquad (11)$$

A similar rate equation can be written for the product, O_2, from equations (9):

$$\frac{d[O_2]}{dt} = k_1[O_3] - k_{-1}[O_2][O] + 2k_2[O][O_3] \ .$$

Assuming steady state for oxygen atoms so that equation (11) can be used for [O] in the previous equation:

$$\frac{d[O_2]}{dt} = k_1[O_3] - \frac{k_{-1}[O_2]k_1[O_3]}{k_{-1}[O_2] + k_2[O_3]} + \frac{2k_2k_1[O_3]^2}{k_{-1}[O_2] + k_2[O_3]} \ . \qquad (12)$$

Equation (12) is the final expression for the rate of the overall reaction. However, equation (12) does not resemble the measured rate law, equation (8). The resolution between equations (8) and (12) will be shown to stem from the simplification afforded by the relatively slow rate of reaction (9b). In fact, it is the slowness of reaction (9b) relative to reaction (9a), which makes a steady state possible. If k_2 is very slow ("rate determining"), or more precisely, if the concentration of O_2 and O_3 and the relative values of k_{-1} and k_2 are such that

$$k_{-1}[O_2] >> k_2[O_3] \ , \qquad (13)$$

then the first two terms in equation (12) will nearly cancel. This reduces equation (12) to

$$\frac{d[O_2]}{dt} \simeq \frac{2k_1k_2[O_3]^2}{k_{-1}[O_2]} \ . \qquad (14)$$

Comparing equation (14) with equation (8), we see that indeed the mechanism fits the overall rate. In fact, some new insight is gained, since equating equations (8) and (14) yields

$$k = \frac{2}{3}\frac{k_1k_2}{k_{-1}} \ . \qquad (15)$$

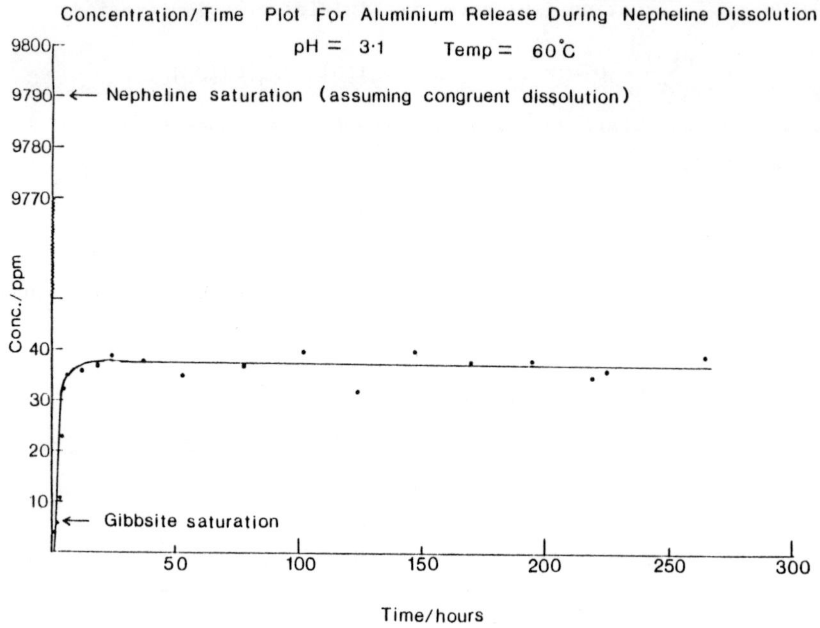

Figure 1. Concentration-time plot for aluminum dissolution from nepheline at pH = 3.1 and T = 60°C. Arrows show that Al^{3+} concentration in equilibrium with nepheline and gibbsite under experimental conditions (Tole, pers. comm.).

Thus, the overall rate constant is related to all three elementary rate constants. If two of the elementary rate constants are known, then equation (15) can be used, along with the experimentally derived k, to obtain the value of the third elementary rate constant.

A geologically common occurrence of steady state is found in the alteration of minerals by aqueous solution. Typically, the intermediate species are the aqueous species. For example, the dissolution of nepheline may lead to the precipitation of gibbsite, where the overall reaction is (using the sodium end-member for simplicity)

$$\underset{\textit{nepheline}}{NaAlSiO_4} + H^+ + 3H_2O \rightarrow \underset{\textit{gibbsite}}{Al(OH)_3} + Na^+ + H_4SiO_4(aq) \ .$$

However, the reaction proceeds by at least two steps:

$$NaAlSiO_4(s) + 4H^+ \rightarrow Na^+ + Al^{3+} + H_4SiO_4(aq)$$

and

$$Al^{3+} + 3H_2O \rightarrow Al(OH)_3(s) + 3H^+$$

where we have assumed a pH low enough that Al^{3+} is a dominant species. Therefore, the Al^{3+} concentration rises initially due to the first step but eventually reaches a plateau as the second step begins to utilize

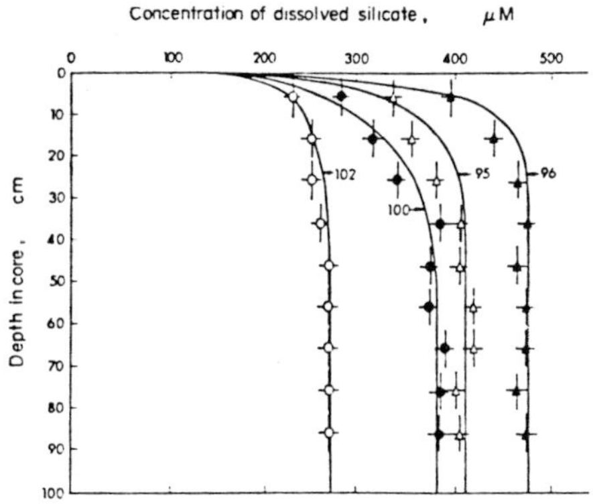

Figure 2. Concentration of dissolved silica versus depth profiles in sedimentary cores. Station number is given beside each array of points. At ambient temperatures, 0°C, the solubility of quartz is 44 μM while that of amorphous silica is 1103 μM. Data from Hurd (1973).

Al^{3+} as shown by Figure 1. Note that the steady state concentration of Al^{3+} is *not* an equilibrium value because the concentration is determined by the relative rates of dissolution of nepheline and precipitation of gibbsite. On the other hand, the equilibrium concentration is determined by the relative rates of precipitation *and* dissolution of gibbsite (or nepheline). The equilibrium concentration of Al^{3+} with both gibbsite and nepheline under experimental conditions is given in the figure. It is also important to realize that the achievement of steady state by the Al^{3+} intermediate ion is also required for the maintenance of stoichiometry in the overall reaction; *i.e.*, we need to obtain one gibbsite molecule per nepheline molecule dissolved.

The silica content of interstitial waters in marine sediments provides another example of a steady state reaction. Marine organisms secrete biogenic silica, which is buried in sediments. As this amorphous silica dissolves, the silica concentration in solution rises to a steady state level (see Fig. 2), which is intermediate between the solubility of amorphous silica (110 ppm at 25°C) and that of quartz (6 ppm at 25°C) (Wollast, 1974; Iler, 1980).

SEQUENTIAL REACTIONS, PARALLEL REACTIONS, AND RATE-DETERMINING STEP

In the ozone example, the *slow* second step determines the form of the reaction rate; hence, this rate (k_2) is called the *rate-determining step* and, in fact, the overall rate constant k [equation (15)] is directly proportional to k_2. We would stress, however, that the concept of the "rate-determining step" being the *slowest* or the "bottleneck" of the reaction mechanism is not always true. It will be true in the case of a *sequential* mechanism such as in equation (9), *i.e.*, where all steps must be traversed in sequence in yielding the products. On the other hand, in the case of parallel reactions the opposite is true; *i.e.*, the *fastest* rate is the rate-determining step. Thus, in the following mechanism

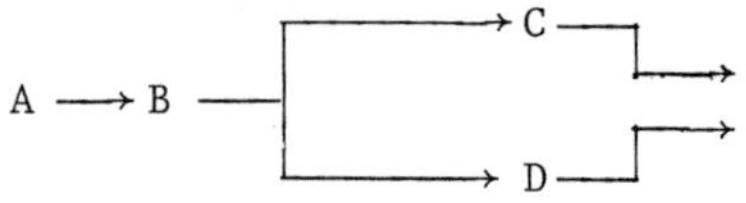

(16)

if path A → B → C → P is faster than A → B → D → P, then the rate law for A → P will be determined by the sequence B → C → P. This is an important but sometimes overlooked point in understanding complex geochemical kinetics. For example, the kinetics of oxygen isotope exchange between water and rock during alteration may involve both (a) bulk diffusion of oxygen isotopes in the mineral crystal lattices and (b) dissolution and recrystallization of an isotopically modified mineral. The rate of approach to isotopic equilibrium in this case is controlled by the *fastest* of the two parallel processes. Depending on the conditions (P, T, grain size, etc.) *either* of the two processes has been found to determine the rate.

Parallel reactions also occur in the kinetics of oxidation-reduction reactions in solution. The reaction (Sykes, 1966, p. 131 -- where Fe* refers to a radiotracer, ^{55}Fe)

$$Fe^{2+} + Fe*^{3+} \xrightarrow{k_1} Fe*^{2+} + Fe^{3+} \qquad (17a)$$

$$k_1 = 0.87 \text{ l/mole/sec}$$

can be speeded by the *parallel* reaction

$$Fe^{2+} + Fe*Cl^{2+} \xrightarrow{k_2} Fe*^{2+} + FeCl^{2+} \qquad (17b)$$

$$k_2 = 5.4 \text{ l/mole/sec}$$

In this case, the addition of a Cl^- ligand to the solution reduces the Coulomb repulsion between Fe^{2+} and Fe^{3+}, hence the much faster rate constant, k_2, of the electron transfer reaction (17b). Since the rate of reaction (17b) depends on the product $k_2[Fe^{2+}][Fe^*Cl^{2+}]$, the rate must vary with the total chloride content of solution. For very low values of ΣCl^- such that $k_2[Fe^*Cl^{2+}] < k_1[Fe^{*3+}]$, reaction (17a) is the fastest and the total rate, being controlled by (17a), is independent of ΣCl^-. On the other hand, as the total chloride content increases, a value may be reached where $k_2[Fe^*Cl^{2+}] > k_1[Fe^{*3+}]$. For these values of total chloride, the overall reaction is controlled by reaction (17b) and the total rate is therefore dependent on ΣCl^-. Both reactions are going on simultaneously; however, the overall rate is determined largely by only one reaction at any given time. The usual result of parallel reactions is that the net rate follows one type of rate law under certain conditions and then switches to a different rate law as conditions change. This change has been found recently for the pH dependence of the rate of sulfur isotope exchange between sulfide and sulfate in hydrothermal solutions (Ohmoto and Lasaga, 1981). The parallel reactions in the last case arise from the speciation of sulfur in solution.

EXPERIMENTAL DETERMINATION OF RATE LAWS

Before overall rates can be compared with possible reaction mechanisms, the overall rates must be known. In this section, we discuss some useful experimental methods which have been applied in geochemical kinetics.

Initial Rate Method

A very useful method to deduce the rate law of overall reactions (whether homogeneous or heterogeneous) is the initial rate method. The general principle is very simple: If the overall rate is sufficiently slow (*i.e.*, more than a few hours) so that no extensive chemical change has occurred during the measurements (or if we artificially fix or buffer the concentrations), we can vary the initial amounts of *one* reactant and find the relation between the rate and that particular reactant. For example, assume the initial rate law is written

$$\frac{dC_P}{dt} = k\, {C_A^o}^{n_a}\, {C_B^o}^{n_b}\, {C_C^o}^{n_c} \ldots \quad (18)$$

where C^o refers to the *initial* concentrations. If C_B^o, C_C^o, ... are held constant, the rate law becomes

$$\text{Rate} = \frac{dC_P}{dt} = k' \, {C_A^o}^{n_a} \tag{19}$$

where k' is some constant. Therefore, if the initial amount of A, C_A^o, is varied in a series of experiments keeping all else constant, the coefficient n_a can be deduced. The usual procedure is to take the logarithm of both sides of equation (19):

$$\log \text{Rate} = \log k' + n_a \log C_A^o \tag{20}$$

and obtain n_a from the slope of the plot of log Rate versus log C_A^o.

We can find recent uses of this method in the geochemical literature. Rickard (1975) studied the rate of pyritization of FeS by elemental sulfur. This reaction is of importance in the marine formation of pyrite deposits, since some "polymorph" of FeS usually forms when H_2S evolved from the reduction of SO_4^{2-} reacts with Fe^{2+} in solution. Berner (1971) has found that in sedimentary environments the *overall* reaction between the metastable, initially formed FeS (either "amorphous" FeS, or pyrrhotite or greigite or mackinawite) and FeS_2 can be written $FeS + S^\circ \rightarrow FeS_2$ (an oxidation-reduction reaction). Rickard employed amorphous FeS in his experiments and reacted it with an aqueous solution containing elemental sulfur in equilibrium with P_{H_2S}. The solution was stirred so that transport would not be rate limiting. The initial amount of FeS, S, P_{H_2S}, and the initial pH were varied in a set of different runs. Figures 3-6 show the results. Writing the general rate law pertinent here

$$\frac{d(FeS_2)}{dt} = k \, (A_{FeS})^{n_1} \, (A_S)^{n_2} \, (P_{H_2S})^{n_3} \, (H^+)^{n_4} \tag{21}$$

where A_S and A_{FeS} stand for the specific surface area of the solids S and FeS, respectively, we can obtain n_i from the slopes of the curves in the figures. The result is an overall rate of the form

$$\frac{d(FeS_2)}{dt} = k_1 \, A_{FeS}^2 \, A_S \, P_{H_2S} \, (H^+)^{\sim 0.1} \tag{22}$$

An unexpected result is the dependence of the rate on the square of the surface area. This may be caused by the small particle size used (0.01 μm grains).

The initial rate method was also used by Morse and Berner (1972) and by Morse (1974; 1978). They applied the method to the dissolution

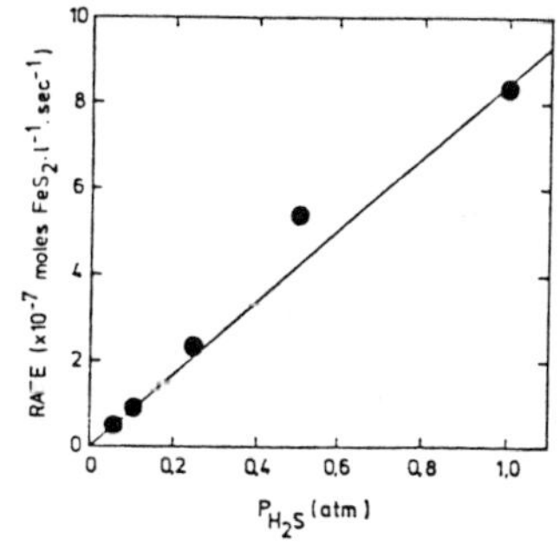

Figure 3. Initial rate of pyrite formation versus H_2S partial pressure at 40°C. The linear plot through the origin demonstrates the first-order rate dependence on P_{H_2S}. (From Rickard, 1975.)

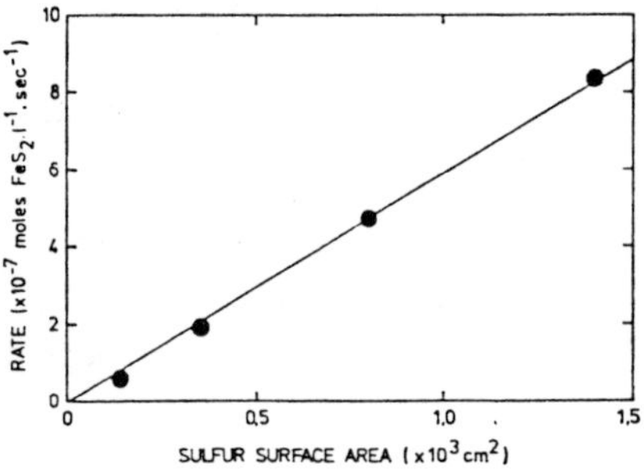

Figure 4. Initial rate versus initial sulfur surface area at 40°C. A surface area of 1.4 x 10^3 cm is equivalent to 1 g sulfur. The linear plot through the origin demonstrates the first-order rate dependence on sulfur surface area. (From Rickard, 1975.)

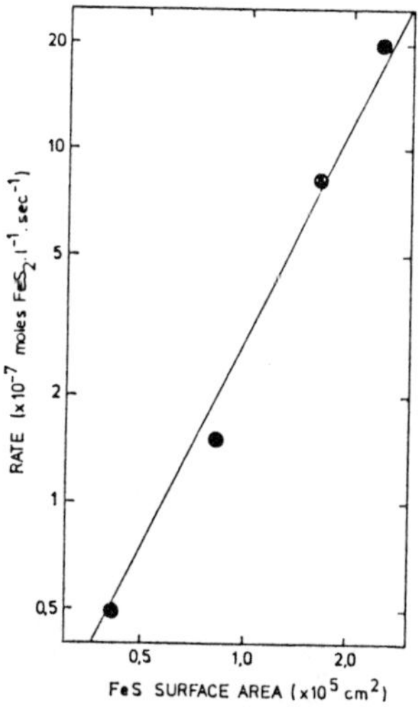

Figure 5. Log-log plot of initial rate versus initial FeS surface area at 40°C. A surface area of 1.6 x 10^5 cm^2 is equivalent to 0.36 g FeS. The line is drawn with a slope of 2, demonstrating the second-order rate dependence on FeS surface area. (From Rickard, 1975.)

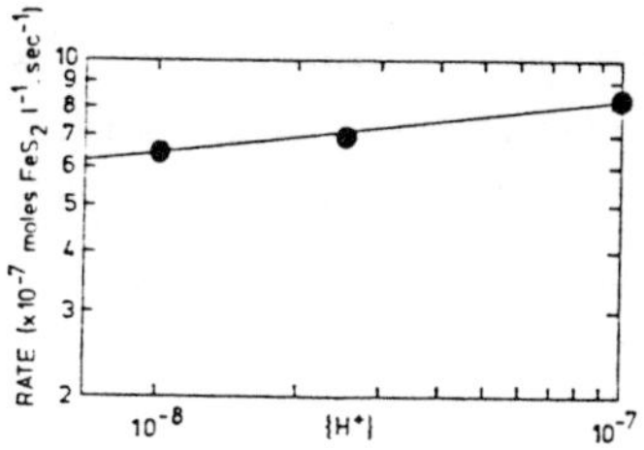

Figure 6. Log-log of initial rate versus $[H^+]$ for pH 7 to 8 at 40°C. The line is drawn with a slope of 0.1, indicating the virtual independence of the rate on pH when P_{H_2S} is taken as a rate-determining parameter. (From Rickard, 1975.)

kinetics of calcite and aragonite. These rates are critical in determining the origin of the lysocline, which is the depth in the oceans where the rate of carbonate dissolution rapidly increases, and of the carbonate compensation depth, below which carbonate sediments are lacking, in the oceans. We can write the overall reaction as

$$CaCO_3 + 2H^+ \rightarrow Ca^{2+} + H_2O + CO_2(g) \quad . \qquad (23)$$

In their experiments, Morse and Berner maintained P_{CO_2} and temperature constant. The pH was also held constant by the constant addition of acid to the solution with the use of a pH stat. A pH stat is basically an automated pH meter which can drive an acid-containing or base-con-

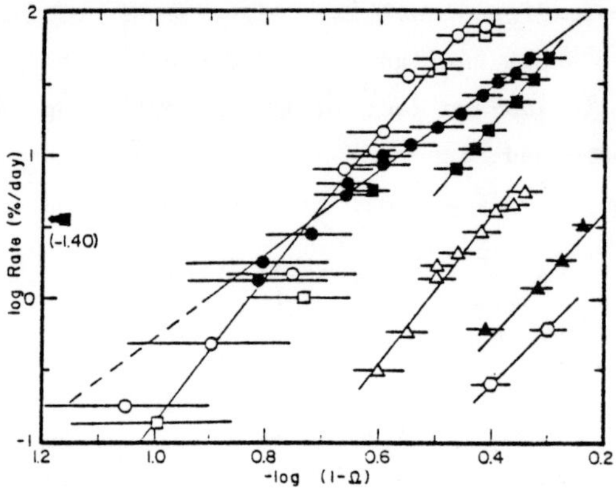

Figure 7. The log of the rates of dissolution in percent per day versus the log of (1-Ω). Open circles = whole Indian Ocean sediment dissolved in Atlantic deep-sea water, open squares = whole Indian Ocean sediment dissolved in deep-sea sediment pore water, open triangles = the greater than 62 micron size fraction of the Indian Ocean sediment dissolved in Atlantic deep-sea water, filled circle = whole Pacific Ocean sediment dissolved in Atlantic deep-sea water, filled triangle = the 125 to 500 micron size fraction of Pacific Ocean sediment dissolved in Atlantic deep-sea water, filled square = whole Atlantic Ocean sediment dissolved in Long Island Sound sea water (Morse and Berner, 1972), open hexagon = 150 to 500 micron foraminifera dissolved in the Pacific Ocean water column. Error bars are based on an uncertainty in pH of ± 0.01. Data from Morse (1978).

taining syringe if the pH deviates from a specified value. Since many rates in solution vary with the pH, use of a pH stat simplifies the extraction of kinetic data. In fact, the rate of dissolution can be followed by knowing how much acid was added and the stoichiometry of reaction (23). Note that if pH, $m_{Ca^{2+}}$, and P_{CO_2} are fixed, then the product $m_{Ca^{2+}}\ m_{CO_3^{2-}}$ is constant since

$$m_{Ca^{2+}}\ m_{CO_3^{2-}} = \frac{m_{Ca^{2+}}\ K\ P_{CO_2}}{a^2_{H^+}} ,$$

where K is the equilibrium constant for the reaction (corrected for $\gamma_{CO_3^{2-}}$)

$$CO_2(g) + H_2O = 2H^+ + CO_3^{2-} .$$

The experiments, therefore, yield the initial rate, defined as the number of Ca^{2+} ions released per unit time, $dm_{Ca^{2+}}/dt$, as a function of the fixed *supersaturation ratio* S, where

$$S = \frac{m_{Ca^{2+}}\ m_{CO_3^{2-}}}{K'_{sp}}$$

and K'_{sp} is the apparent solubility constant of calcite or aragonite in sea water.

Typical results for dissolution experiments on carbonates in natural waters are shown in Figure 7. The experiments used deep-sea sediment from the Indian and Pacific Oceans containing variable amounts of calcium carbonate and either deep-sea water or deep-sea-sediment pore water, which includes any natural inhibitors such as phosphate ions. This plot shows log rate versus $-\log(1-\Omega)$ for the different runs. Note that in the case of mineral dissolution (or precipitation) the rate is related to the corresponding saturation product, Ω, which can be obtained from S and is defined as

$$\Omega = \frac{a_{Ca^{2+}} \, a_{CO_3^{2-}}}{K_{sp}}$$

where K_{sp} is the true solubility product of $CaCO_3$ at the given P and T conditions. Therefore, the rate of dissolution of calcium carbonate at constant surface area, P and T, a reaction which involves Ca^{2+} and CO_3^{2-} as products, is not written as

$$\text{Rate} = k \, m_{Ca^{2+}}^{n_1} \, m_{CO_3^{2-}}^{n_2}$$

where n_1 and n_2 are *different* real numbers. Rather, the rate is a function of the product $m_{Ca^{2+}} m_{CO_3^{2-}}$ (*i.e.*, $n_1 = n_2$). Furthermore, the overall rate in Figure 7 is made a function of $1-\Omega$ rather than Ω. The rationale beind this expression stems from crystal growth theory and will be further discussed in the section on the principle of detailed balancing. The need for difference terms $(1-\Omega)$ rather than simple products (Ω) stems from the fact that near equilibrium *both* dissolution and precipitation reactions are simultaneously, quantitatively important.

The log-lot plot in Figure 7 yields straight lines; hence, from the slopes of the lines the dependence of the rate, R, on $1-\Omega$ can be deduced to be $R = k(1-\Omega)^n$, where k is some constant and n is the slope. The data in Figure 7 yield

Ocean	R (% per day)	
Indian	$10^{4.3}(1-\Omega)^{5.2}$	
Atlantic	$10^{3.1}(1-\Omega)^{4.5}$	(24)
Pacific	$10^{2.7}(1-\Omega)^{3.0}$	

The reasons for the wide variations in the exponent in equation (24) are not fully known. The variations may be related to different particle size distributions and/or different surface histories. Note that the rate equals zero when $\Omega = 1$, as it should. The large size of the exponents of $(1-\Omega)$ in equation (24) certainly suggests that the dissolution mechanism is not simple. In fact, it seems that for calcite, as well as other minerals, dissolution kinetics are intimately involved with complex surface phenomena (see section on *Heterogeneous Kinetics*).

Integration of Phenomenological Rate Laws -- Method of Isolation

Often one can obtain a general rate law from data on the rate of reaction as a function of time, rather than just the *initial* rate as in part (a). This is especially the case if there is no way of unambiguously defining an initial rate -- for example, when plots of concentration versus time exhibit pronounced curvature. In most cases of interest to geochemists, reactions will be sufficiently complex that the interpretation of the concentration versus time data must still assume the "constancy" of one or more variables. This is the basis of the *Method of Isolation*. For example, let us take a reaction whose stoichiometry is given by $A + B \rightarrow C + D$. Suppose that the true rate obeys the equation

$$-\frac{dC_A}{dt} = k\ C_A\ C_B\ , \qquad (25)$$

i.e., a second-order rate law. If we now choose the initial concentration of B, C°_B, to be much greater than C°_A, then we can rewrite equation (25) as:

$$-\frac{dC_A}{dt} \simeq k\ C^\circ_B\ C_A = k'\ C_A \qquad (26)$$

where $k' \equiv k\ C^\circ_B$. Equation (26) is a first-order equation; our choice of initial conditions has thus reduced the experimental order of our system. The reaction is now referred to as a *pseudo-first-order* reaction.

With suitable constraints, the kinetic behavior of geochemical reactions can be reduced to that exhibited by simpler reactions. The simplest possible case is that of a *zero-order* reaction. Such a reaction obeys the equation

$$-\frac{dC}{dt} = k \qquad (27)$$

which integrates to

$$C = C_o - kt \tag{28}$$

where C_o is the initial concentration. Zero-order kinetics yield straight lines, if concentration is plotted versus time. Equation (28) can only hold for t such that $C \geq 0$. The reduction of the sulfate ion by the bacteria *Desulfovibrio* in marine environments seems to obey a zero-order rate law with respect to sulfate if the concentration of SO_4^{2-} is greater than 2 mM (Berner, 1971; Goldhaber and Kaplan, 1974). Thus if the amount of decomposable organic matter, the temperature, and other variables are kept constant (except SO_4^{2-}), the concentration of sulfate in the interstitial waters of sediments would obey equation (28). The rate of leaching of alkalis from single crystals of alkali feldspar also follows a zero-order rate law (Tole and Lasaga, pers. comm.; Fung and Sanipelli, 1980).

(Pseudo) First-Order and Second-Order Reactions

A first-order reaction obeys the equation

$$-\frac{dC_A}{dt} = k\, C_A \tag{29}$$

Rate constants for first-order reactions have units of time^{-1}. This equation can readily be integrated to obtain

$$C_A = C_A^\circ\, e^{-kt} \tag{30}$$

or

$$ln\, C_A = ln\, C_A^\circ - kt \tag{31}$$

The time required to consume half of the original amount of A, termed the half-life or the reaction, is given by [setting $C_A = \frac{1}{2}\, C_A^\circ$ in (31)]

$$t_{1/2} = \frac{0.693}{k} \tag{32}$$

Equation (31) shows that a plot of $ln\, C_A(t)$ versus time would yield a straight line with slope equal to -k. Such a plot is the best way to prove that a reaction is a first-order reaction.

First-order reactions (either true or pseudo) are quite important in kinetics. As an example of a true (elementary) first-order reaction, we can take the case of radioactive decay. Thus the reaction $^{14}C \rightarrow {}^{14}N + e$ has a rate law given by $\frac{d[C^{14}]}{dt} = -k[C^{14}]$, where $k = 1.22 \times 10^{-4}\ yr^{-1}$.

An example of a pseudo-first-order reaction is the reduction of sulfate referred to earlier, which may be written in simple terms as:

$$2CH_2O + SO_4^{2-} \rightarrow 2HCO_3^- + H_2S \ .$$

If the concentration of SO_4^{2-} is less than 2 mM, then the rate of disappearance of sulfate follows a first-order rate law.

Another example of a pseudo-first-order reaction is the oxidation of sulfide in solution by hydrogen peroxide (a method proposed to eliminate unwanted odors in sewage systems). The reaction can be written as

$$H_2S + H_2O_2 \rightarrow 2H_2O + S \text{ (colloid)} \ . \tag{33}$$

Under acidic conditions, however, polysulfides and sulfates are also obtained as reaction products (Hoffmann, 1977). Hoffmann measured the EMF of a solution containing H_2S and H_2O_2 with the Ag/Ag_2S electrode. The measured EMF obeys the Nernst equation for Ag/Ag_2S, *i.e.*,

$$\text{EMF} = E° - \frac{RT}{nF} \ln a_{S^{2-}} \ . \tag{34}$$

If constant ionic strength is maintained (so that $\gamma_{S^{2-}}$ is fixed), then equation (34) can be rewritten as

$$\text{Eh} = E° - \frac{RT}{nF} \ln \gamma_{S^{2-}} - \frac{RT}{nF} \ln m_{S^{2-}} \tag{35}$$

$$\text{Eh} = E°' - \frac{RT}{nF} \ln m_{S^{2-}}$$

If the reaction is carried out with excess H_2O_2 (*i.e.*, $m_{H_2O_2} >>$ total sulfide) at constant H^+ and temperature, the kinetics become pseudo-first-order, *i.e.*,

$$\frac{d[\Sigma S^{2-}]}{dt} = -k[\Sigma S^{2-}] \tag{36}$$

where $[\Sigma S^{2-}]$ stands for total sulfide in the solution, *i.e.*,

$$[\Sigma S^{2-}] = m_{H_2S} + m_{HS^-} + m_{S^{2-}} \tag{37}$$

Of course, at constant pH, the rates $\frac{dm_{H_2S}}{dt}$, $\frac{dm_{HS^-}}{dt}$, $\frac{dm_{S^{2-}}}{dt}$ are all proportional to each other. This follows from the equilibrium between the sulfide species which can be expressed as:

$$\frac{m_{H^+} \, m_{HS^-}}{m_{H_2S}} = K_1$$

$$\frac{m_{H^+} \, m_{S^{2-}}}{m_{HS^-}} = K_2 \ ,$$

K_1 and K_2 being effective equilibrium constants (at constant ionic strength).

Taking the derivative of the two previous equations yields

$$\frac{dm_{H_2S}}{dt} = \left(\frac{m_{H^+}}{K_1}\right) \frac{dm_{HS^-}}{dt} = \left(\frac{m^2_{H^+}}{K_1 K_2}\right) \frac{dm_{S^{2-}}}{dt} \tag{38}$$

Therefore, $d[\Sigma S^{2-}]/dt$ can be related to $dm_{S^{2-}}/dt$; from equation (37) we have

$$\frac{d[\Sigma S^{2-}]}{dt} = \frac{dm_{H_2S}}{dt} + \frac{dm_{HS^-}}{dt} + \frac{dm_{S^{2-}}}{dt} .$$

Using (38) in the last equation

$$\frac{d[\Sigma S^{2-}]}{dt} = \left(\frac{m^2_{H^+}}{K_1 K_2} + \frac{m_{H^+}}{K_2} + 1\right) \frac{dm_{S^{2-}}}{dt} . \tag{39}$$

At constant temperature, pH and ionic strength, the term in parentheses in equation (39) will be a constant. In this case, if our rate expression follows equation (36), we expect also

$$\frac{dm_{S^{2-}}}{dt} = -k\ m_{S^{2-}} \qquad \left(\text{or} \quad \frac{dm_{H_2S}}{dt} = -k\ m_{H_2S}\right) \tag{40}$$

with the same k (the constant in equation (39) cancels from both sides). Hence, using equation (31) to solve equation (40)

$$ln\ m_{S^{2-}} = ln\ m^\circ_{S^{2-}} - kt$$

and from equation (35):

$$Eh = E^{\circ\prime} - \frac{RT}{nf} ln\ m^\circ_{S^{2-}} + \frac{RT}{nF} kt . \tag{41}$$

Equation (41) indicates that a plot of measured Eh versus time should yield a straight line with positive slope equal to (RT/nF)k. Figure 8 shows that this is found to be the case. The rate constant, k, can then be easily evaluated.

The full rate law of a complex reaction can be derived from data on pseudo-first-order rate constants. For example, any variation in the rate of reaction (33) due to variation in the concentrations of H_2O_2 or H^+ is embedded in the experimental rate constant, k, of equation (41). We need, therefore, to compare k obtained for different

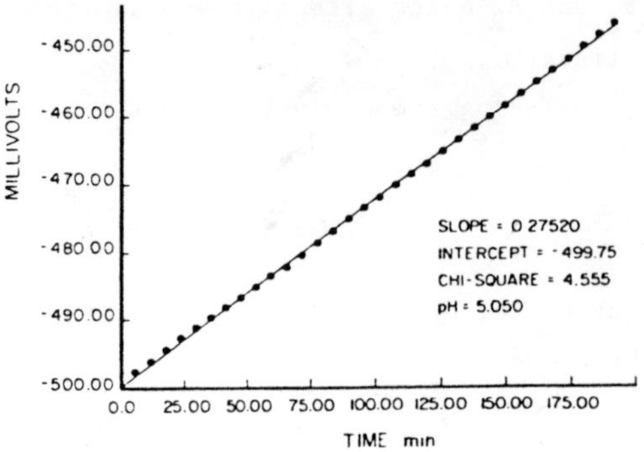

Figure 8. Plot of EMF against time for H_2S-H_2O_2 reaction at pH 5.05 and 25.0°C, where $[S^{-2}]_0 = 1.5 \times 10^{-3}$ M and $[H_2O_2]_0 = 3.0 \times 10^{-2}$ (From Hoffman, 1977.)

initial values of H_2O_2 or H^+. At this point, the method proceeds identically to the procedure outlined in the section on the initial rate method. To obtain the rate law, the pseudo-first-order rate constant, k, obtained from equation (41) is written in terms of the true rate constant, k_r,

$$k = k_r \, m^a_{H_2O_2} \, m^b_{H^+} \tag{42}$$

where the constants a and b are the reaction orders of (33) with respect to H_2O_2 and H^+, respectively. A plot of log k versus log $m_{H_2O_2}$ (at constant pH) is shown in Figure 9 (Hoffmann, 1977). The coefficient a is the slope of the straight line and is equal to 1. A plot of log k versus

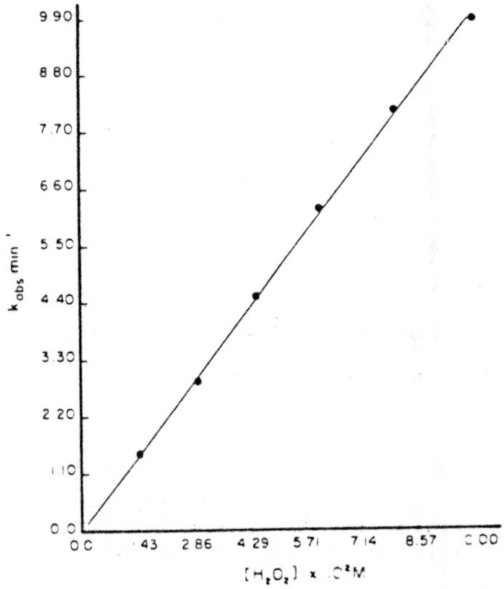

Figure 9. Plot of k_{obs} versus $[H_2O_2]$ where $[S^{-2}]_0 = 1.5 \times 10^{-3}$ M (From Hoffman, 1977.)

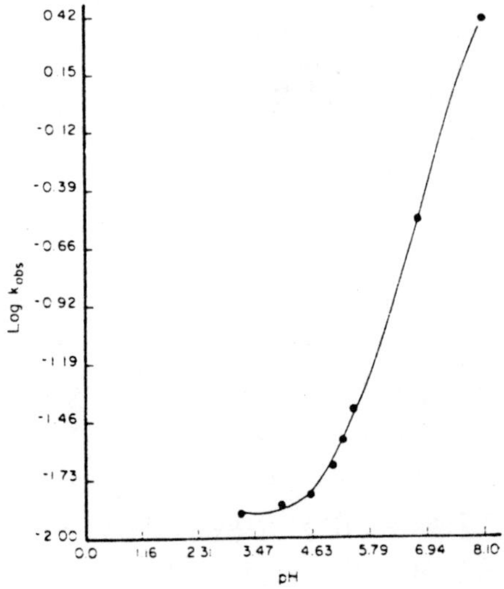

Figure 10. Plot of log k_{obs} versus pH. (From Hoffman, 1977.)

pH (Fig. 10) shows that the pH dependence of the rate is complex. The data in Figure 10 show a near horizontal line at low pH and a straight line with a slope of ∿0.8 at higher pH (5-8). This suggests the existence of *two* different reaction mechanisms operating simultaneously. The rate law for one mechanism is

$$\frac{d[\Sigma S^{2-}]}{dt} = -k'_r \, m_{H_2O_2} \, [\Sigma S^{2-}] \; ,$$

whereas the rate law for the other is

$$\frac{d[\Sigma S^{2-}]}{dt} = -k''_r \, a_{H^+}^{-0.8} \, m_{H_2O_2} \, [\Sigma S^{2-}] \; .$$

The first mechanism (low pH) reflects reaction between H_2S and H_2O_2. The second mechanism reflects reaction between HS^- and H_2O_2, since m_{HS^-} is proportional to $[\Sigma S^{2-}]/a_{H^+}$ for pH < 7 (taking the exponent, 0.8, to be close to 1). At higher pH (pH > 7), the rate should be independent of pH ($m_{HS^-} \sim [\Sigma S^{2-}]$); the data, thus, may suggest additional species playing a role at higher pH.

The hydration of carbon dioxide offers another example of the treatment of pseudo-first-order reaction kinetics. This reaction is important both to the geochemistry of CO_2 (*e.g.*, rate of equilibration of rain water with the atmosphere and in soils) and to the understanding of the physiology of dehydration of carbonic acid in our blood. Most of the dissolved CO_2 in solution is in the form of hydrated "CO_2"(aq), *i.e.*, as molecular CO_2; less than 1% is truly as hydrated H_2CO_3 (Cotton and Wilkinson, 1971). H_2CO_3 can dissociate very quickly to HCO_3^- or CO_3^{2-}; however, the reaction between CO_2(aq) and H_2CO_3 is sluggish. The rate of this important step

$$CO_2(aq) + H_2O(\ell) \rightarrow H_2CO_3(aq) \qquad (43)$$

follows a true (elementary) first-order law:

$$\frac{dm_{CO_2(aq)}}{dt} = -k \, m_{CO_2(aq)} \; . \qquad (44)$$

The rate constant $k = 2.0 \times 10^{-3}$ sec^{-1} at 0°C (Jones *et al.*, 1964). The half-life [equation (32)] is then around six minutes. Reaction (43) is slow enough to follow visually in an interesting experiment (Jones *et al.*, 1964). A yellow 0.1 M CO_2 solution with an acid-base indicator

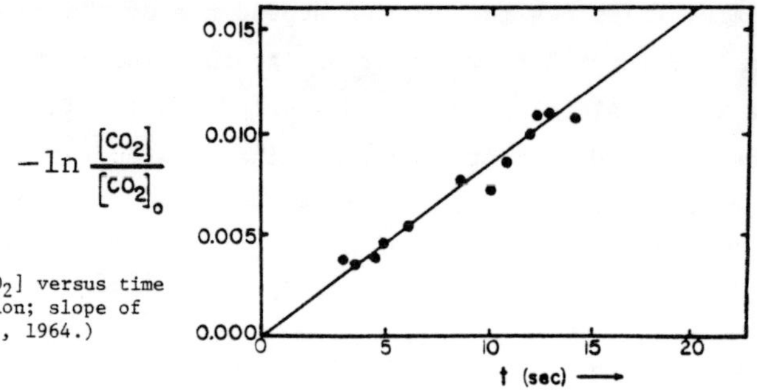

Figure 11. Plot of -log $[CO_2]$ versus time for the CO_2 hydration reaction; slope of line = k. (From Jones *et al.*, 1964.)

(bromothymol blue) turns blue (basic) upon addition of a small amount of 0.4 M NaOH solution. The solution remains blue for a minute or so and then becomes yellow again. This delay in the color change arises because, while OH^- reacts readily with whatever H_2CO_3 and HCO_3^- is present to give CO_3^{2-}, the abundant $CO_2(aq)$ must be hydrated to H_2CO_3 before further reaction can occur, and this is a slow step. Therefore, the pH rises at first (blue color) until more H_2CO_3 can be generated to react with the OH^-. Because reaction (43) is an elementary reaction, the first-order rate expression [equation (44)] holds regardless of the pH and so the reaction can be investigated even if the pH changes with time (unlike the previous case). The results are shown in Figure 11. In support of the first-order kinetics, a plot of *ln* c versus time yields a straight line. We should point out that the rate law (44) applies only if pH $\leqslant$ 9 because at higher pH's there is interference from the competing elementary reaction

$$CO_2(aq) + OH^- \rightarrow HCO_3^- \ . \qquad (45)$$

Two important traits that distinguish kinetics from thermodynamics are shown by the above two examples. First, there is the importance of establishing what the actual molecular species in solution are. Since kinetics depends on the paths of reaction, we must know exactly what species react at the molecular level. Thermodynamically, $CO_2(aq)$ and $H_2CO_3(aq)$ may be treated as being the same component in solution (usually labeled as H_2CO_3 for both), and we write equilibrium constants between total CO_2 (*i.e.*, $CO_2(aq) + H_2CO_3$) and HCO_3^- or CO_3^{2-}. However, in kinetics there is a difference of several orders of magnitude between the rates of reaction of H_2CO_3 and $CO_2(aq)$ with OH^-. To treat the CO_2 system kinetically, therefore, we must differentiate between $CO_2(aq)$ and

H_2CO_3 and consider the actual molecular processes involved in the chemical reaction.

The second point is the existence of several parallel pathways for the reactions discussed previously. While for a thermodynamic treatment one needs to state merely the initial and the final states to obtain ΔG, ΔH, or ΔS uniquely, in kinetics that is not sufficient. In the reaction between sulfides and H_2O_2, the rate is the result of two parallel reactions, one the reaction with H_2S and the other the reaction with HS^-. In the CO_2 reaction there are two *parallel* (and competing) reactions:

$$CO_2(aq) + H_2O(\ell) \rightarrow H_2CO_3(aq) \xrightarrow{OH^-} HCO_3^- + H_2O(\ell) \tag{46a}$$

and

$$CO_2(aq) + OH^- \rightarrow HCO_3^- . \tag{46b}$$

Even though the net result is the same, *i.e.*, the conversion of one $CO_2(aq)$ to HCO_3^-, the rates and rate laws for the *elementary* reactions in (46a) and (46b) are quite different. For example, reaction (46b) depends on pH whereas reaction (46a) is pH independent. It is important to emphasize that in the case of parallel reaction kinetics, the more rapid one determines the overall rate. This is why we need to keep the pH below 9 (*i.e.*, low m_{OH^-}), thereby slowing reaction (46b), in order to study the hydration reaction (46a).

When both forward and reverse first-order reactions can occur at fast enough rates that equilibrium *or* steady state may be achieved, their first-order kinetics must be written as

$$\frac{dC_A}{dt} = k\ (C_S - C_A) \tag{47}$$

where C_S is the equilibrium or steady state concentration of A. The solution to the first-order equation (47) is given by

$$ln\ \frac{C_S - C_A}{C_S - C_A^\circ} = -kt \tag{48}$$

where C_A° is the initial concentration of A. For example, the rate of dissolution of silica from biogenic opal is found to obey equation (47); here, C_S is the solubility of silica, and the rate constant k incorporates the area terms of the hetereogeneous reaction. Therefore, on a logarithmic plot of $(C_S - C_A)/C_S$ versus t (assuming $C_A^\circ = 0$) a straight line with negative slope of -k should be obtained (see Fig. 12).

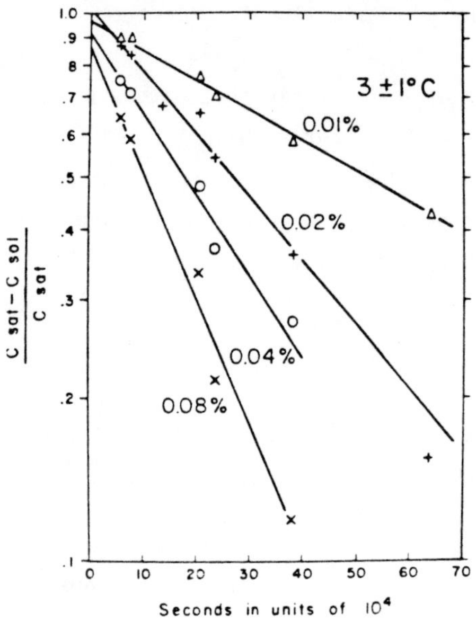

Figure 12. Rate of solution of biogenic opal in sea water at 3°C as a function of the percent opal suspended. (From Hurd, 1972.)

Equation (48) shows that the time needed to closely achieve equilibrium or steady state will be given by 1/k. Therefore, the time evolution of (47) behaves as shown schematically in Figure 13, achieving a plateau at times long enough that $t > 1/k$.

Another common rate law is that obtained in a *second-order* reaction. Thus, for the *elementary* reaction, A + B → products, the rate law is second order:

$$-\frac{dC_A}{dt} = k\ C_A\ C_B\ . \qquad (49)$$

If C_A° and C_B° are the initial concentrations and if A and B are not involved in other side reactions, the stoichiometry of the reaction requires that

$$C_A^\circ - C_A = C_B^\circ - C_B\ .$$

Using this last equation to solve for C_B and inserting in equation (49)

$$-\frac{dC_A}{dt} = k\ C_A(C_A - C_A^\circ + C_B^\circ)\ . \qquad (50)$$

Equation (50) is integrated to give

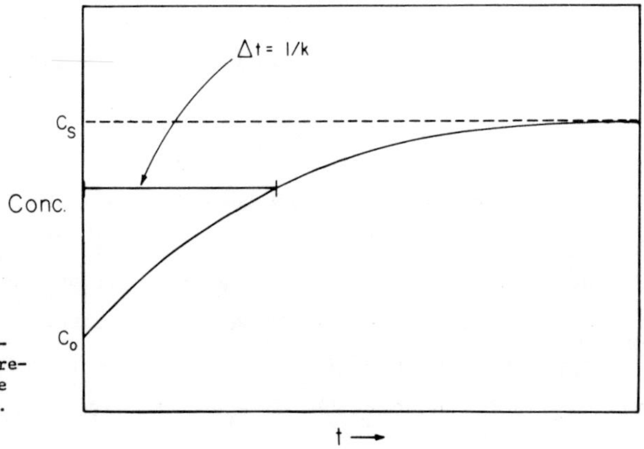

Figure 13. Sketch of the concentration versus time plot predicted by equation (48). Note the time scale depends on 1/k.

$$\ln \frac{C_A^\circ \, C_B}{C_B^\circ \, C_A} = (C_B^\circ - C_A^\circ)\, kt \; . \qquad (51)$$

Equation (51) allows us to quantitize the method of isolation described earlier. If we allow C_B° to be much greater than C_A°, then $C_B/C_B^\circ \sim 1$ and $C_B^\circ - C_A^\circ \sim C_B^\circ$. Inserting these approximations reduces equation (51) to equation (31), validating our pseudo-first-order treatment of the kinetics under these conditions.

A classical example of a second-order elementary reaction is the ligand replacement reaction, so common in aqueous solutions. Actually, there are two extreme mechanistic possibilities for such a reaction. In the S_N1 mechanism (S_N1 for substitution, nucleophilic, unimolecular), the complex dissociates first, loses the ligand to be replaced, and then the vacancy in the coordinate shell is filled by the new ligand. We can write such a path as

$$M\,L_sX^{n+} \xrightarrow{k} M\,L_s^{(n+1)+} + X^- \qquad \text{slow} \qquad (52)$$

$$M\,L_x^{(n+1)+} + Y^- \longrightarrow M\,L_sY^{n+} \qquad \text{fast} \qquad (53)$$

If equation (52) is slow (rate-determining), then the reaction rate obeys a first-order law

$$-\frac{dC_{ML_sX}}{dt} = k\, C_{ML_sX} \; .$$

In an S_N2 mechanism, the new ligand attacks the original complex directly, forming a new complex, which subsequently liberates the displaced ligand. In this case the reaction path becomes

$$M\,L_sX^{n+} + Y^- \xrightarrow{k} M\,L_sX\,Y^{(n-1)+} \qquad \text{slow} \qquad (54a)$$

$$M\,L_sX\,Y^{(n-1)+} \longrightarrow M\,L_sY^{n+} + X^- \qquad \text{fast} \qquad (54b)$$

and the rate law obeys second-order kinetics

$$-\frac{dC_{ML_sX}}{dt} = k\, C_{ML_sX}\, C_Y \; .$$

The complex in equation (54a) may sometimes be an ion pair rather than a complex based on covalent bonds.

Many of the metal octahedral complexes react via an S_N1 mechanism with very short half-lives (usually $10^{-5} - 10^{-9}$ sec -- see Cotton and

Table 1

Summary of Simple Rate Laws

Zeroth Order:

$$\frac{dC}{dt} = k \qquad C = C^\circ + kt$$

$$\frac{dC}{dt} = -k \qquad C = C^\circ - kt \qquad t < \frac{C^\circ}{k}$$

First Order:

$$\frac{dC}{dt} = -k\,C \qquad C = C^\circ\, e^{-kt} \qquad \ln C = \ln C^\circ - kt$$

$$\frac{dC}{dt} = k\,(C_S - C) \qquad \frac{C_S - C}{C_S - C^\circ} = e^{-kt}$$

$$\ln\left(\frac{C_S - C}{C_S - C^\circ}\right) = -kt$$

Higher Order:

$$\frac{dC}{dt} = -k\,C^n \qquad \frac{1}{C^{n-1}} - \frac{1}{C^{\circ\, n-1}} = (n-1)\,kt$$

$$C < C_S \qquad \frac{dC}{dt} = k\,(C_S - C)^n \qquad \frac{1}{(C_S-C)^{n-1}} - \frac{1}{(C_S-C^\circ)^{n-1}} = (n-1)\,kt$$

$$C > C_S \qquad \frac{dC}{dt} = -k\,(C - C_S)^n \qquad \frac{1}{(C-C_S)^{n-1}} - \frac{1}{(C^\circ-C_S)^{n-1}} = (n-1)\,kt$$

Wilkinson, 1967). In general, the rates of these reactions are much faster than other relevant geochemical reactions in solution.

The final example deals with the integration of high-order kinetic rate laws. In some instances, such as the calcite dissolution discussed earlier, complex phenomena at surfaces give rise to rate laws, which have the form

$$\frac{dC_A}{dt} = k(C_S - C_A)^n \qquad C_A < C_S \tag{55}$$

where C_S is, as before, either the equilibrium or the steady state concentration of species A. The exponent n is any positive number. The requirement that C_A be less than C_S ensures that the term $(C_S-C_A)^n$ is well defined for all positive n (*e.g.*, $(C_S-C_A)^{1.5}$ is imaginary if $C_S - C_A < 0$). If $C_A > C_S$ an analogous formula can be written (see Table 1). The solution of equation (55) is

$$\frac{1}{(C_S-C_A)^{n-1}} - \frac{1}{(C_S-C_A^\circ)^{n-1}} = (n-1)\ kt \qquad (n\neq 1) \tag{56}$$

where C_A° is the initial concentration of A. C_S can be obtained from the limiting value of the concentration at long times (*i.e.*, the "plateau" region). Then equation (56) can be used with different values of n, to check the linearity of a plot of $(1/C_S-C_A)^{n-1}$ versus t. The value of n, if any, that yields a straight line is then the order of the reaction in (55). If no linearity is obtained, then the simple form assumed in equation (55) is not valid. One interesting result of solution (56) is that for all reactions with n *less* than 1, the concentration, C_A, will reach its final value, C_S, in a *finite* amount of time. Thus, $C_A = C_S$ when

$$t = \frac{1}{k\ (1-n)}\ (C_S-C_A^\circ)^{1-n} \qquad n < 1\ .$$

For values of n equal to or greater than one it takes an infinite amount of time to reach the steady state value C_S. Nonetheless, in these latter cases, one can compute the time it takes to be within a few percent of the steady state value and 1/k was used in this context earlier (first-order reactions). Table 1 summarizes the kinetic equations of this section.

Relaxation Methods

The experimental method of studying fast reactions such as those discussed in the last section introduces some ideas that we will find useful. The essential element involves studying the response of a reaction to a displacement from equilibrium, *i.e.*, the so-called relaxation method. For example, consider the ionization of a weak acid,

$$HA + H_2O \underset{k_{-1}}{\overset{k_1}{\rightleftharpoons}} H_3O^+ + A^-\ .$$

If C_{H^+} is displaced from the equilibrium value, the rate of change of the pH will satisfy

$$\frac{dC_{H^+}}{dt} = k_1\ C_{HA} - k_{-1}\ C_{H^+}\ C_{A^-}\ .$$

If there are no other sources of H^+ (ignoring H_2O ionization), then $C_{A^-} = C_{H^+}$. Labeling C_{H^+} as x and the total concentration of A as

($a = C_{HA} + C_{A^-}$), then we can rewrite the previous equation as

$$\frac{dx}{dt} = k_1(a-x) - k_{-1}\, x^2 \ . \tag{57}$$

At equilibrium $dx/dt = 0$; hence, if the equilibrium value of the pH is labeled as x_e, and using equation (57), then

$$k_1(a-x_e) = k_{-1}\, x_e^2 \ . \tag{58}$$

Defining the perturbation as $\Delta x = x - x_e$, we have

$$\frac{dx}{dt} = \frac{d\Delta x}{dt} = k_1(a-x) - k_{-1}\, x^2$$

$$= k_1(a-x_e-\Delta x) - k_{-1}(x_e+\Delta x)^2$$

$$\frac{d\Delta x}{dt} = k_1(a-x_e) - k_{-1}\, x_e^2 - k_1\Delta x - 2k_{-1}\, x_e\Delta x - k_{-1}\Delta x^2 \ .$$

Using equation (58) and ignoring the small Δx^2 term

$$\frac{d\Delta x}{dt} = -(k_1 + 2k_{-1}x_e)\ \Delta x \ .$$

This equation is a pseudo-first-order equation with solution

$$\Delta x = \Delta x_o e^{-(k_1+2k_{-1}x_e)t} = \Delta x_o e^{-t/\tau} \tag{59}$$

if Δx_o is the initial displacement from equilibrium. The quantity $\tau = (k_1+2k_{-1}x_e)^{-1}$ is called the *relaxation time*. If we have a method for following rapid changes in some property of the reaction system, we can use equation (59) to measure extremely rapid reaction rates. For example, this method (with conductance measured in experiments) has been applied to fast reactions such as

$$H^+ + SO_4^{2-} \longrightarrow HSO_4^- \qquad k = 10^{11}\ \text{l/mole/sec} \tag{60a}$$

$$H^+ + OH^- \longrightarrow H_2O \qquad k = 1.5 \times 10^{11}\ \text{l/mole/sec} \tag{60b}$$

(Moore, 1964). The concept of a relaxation time will reappear in a general manner in the chapter on geochemical cycles.

TEMPERATURE DEPENDENCE OF RATE CONSTANTS

In any treatment of reaction rates, temperature is one of the most important variables. For example, rate constants can change by several orders of magnitude as the temperature is varied by 100°C. The reason for this drastic variation stems from the exponential dependence on temperature of the reaction rate constant, which is often assumed to follow the classic equation proposed by Arrhenius in 1889:

$$k = A\,e^{-\frac{E_a}{RT}} \,. \tag{61}$$

In the *Arrhenius equation* A is the *pre-exponential factor* (not to be confused with surface area) and E_a is the *activation energy*. While equation (61) is a phenomenological (*i.e.*, not derived fully from molecular theory) equation, detailed molecular theories often yield a temperature-dependent pre-exponential factor (*cf.* Moore, 1964); *i.e.*,

$$k = A(T)\,e^{-\frac{E_a}{RT}} \,. \tag{62}$$

Furthermore, the activation energies of overall reactions, E_a, may themselves depend on temperature. Due to the uncertainties attached to measured activation energies, it is generally very difficult to ascertain experimentally the weak temperature dependence of the pre-exponential factor implicit in equation (62). In any case, we can always define the energy of activation by the equation:

$$E_a = -R\,\frac{d\,ln\,k}{d(1/T)} \,.$$

Therefore, E_a can be obtained from the slope of a plot of *ln* k versus 1/T. However, curvature in this plot would give rise to temperature-dependent activation energies.

The existence of a temperature term of the Arrhenius type [equation (61)] is to be expected. Because chemical reactions involve crossing over certain *potential barriers* (*e.g.*, breaking bonds or distorting molecular structures), the exponential term in equation (61) results from application of Boltzmann's law. The probability of a molecule having an energy E* is proportional to exp(-E*/RT); therefore, integrating this probability from the minimum energy required to reach the maximum in the potential surface, $E^* = E_a$, to $E^* = \infty$ yields a term proportional to $\exp(-E_a/RT)$. As Moelwyn-Hughes put it, energy among molecules is like money among men: The poor are numerous, the rich few.

The energy of activation of an overall reaction is really the composite of several activation energies from the elementary reactions comprising the reaction mechanism. For example, in the ozone reaction

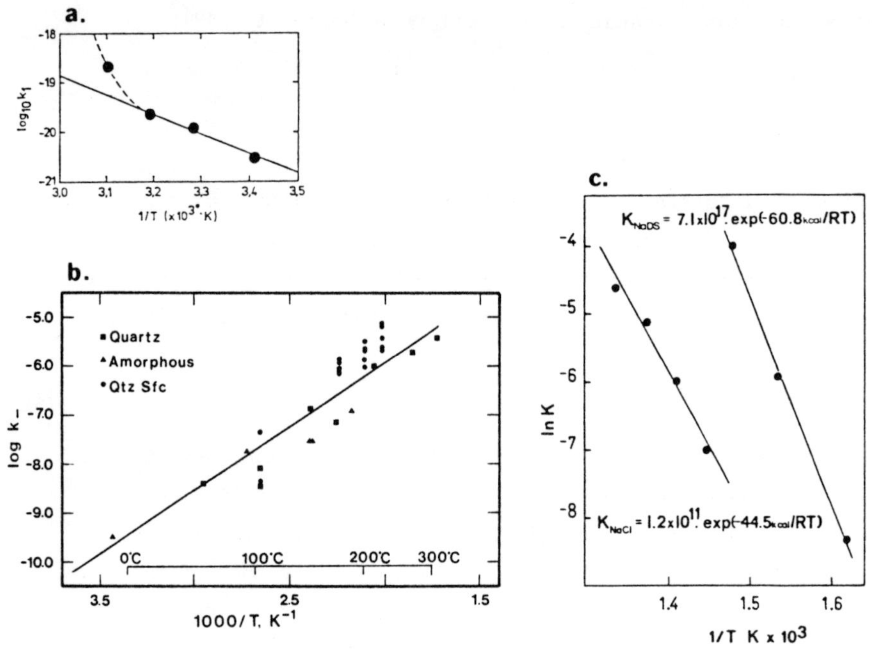

Figure 14. (a) Arrhenius plot of the reciprocal absolute temperature (1/T) versus the natural logarithm of the apparent rate constant. The divergence of the point at 1/T = 3.1 is believed to be real and to reflect the increasing dominance of the solid-solid reaction at higher temperatures. (From Rickard, 1975.)
(b) Arrhenius plot of experimentally determined preciptation rate constants for silica-water reactions. Qtz sfc indicates values calculated from the dissolution rate of the disturbed layer on quartz grains. The line is a least-squares fit of the points shown where each was weighted by the number of concentration-time determinations for each experiment. (From Rimstidt and Barnes, 1980.)
(c) Arrhenius plot of *ln* K versus 1/T(K) for analcite + quartz $\rightleftharpoons$ albite + H_2O reaction in NaCl and NaDS. Equations to the linear regression curves are given adjacent to the two sets of data. (From Matthews, 1980.)

(7) the overall activation energy for the rate constant k depends on three activation energies for elementary reactions, as shown by equation (15). If equilibrium reactions are involved [see equation (65) below] in the mechanism, then the overall activation energy may depend also on the standard enthalpies of the particular reactions. We will not dwell on the fundamentals of the temperature terms in equation (61) or equation (62) in this section but concentrate on the experimental results and leave the fundamentals for the chapter on transition state theory.

Figures 14a-c show some geochemical applications of the Arrhenius rate law. Generally, reasonably straight lines are obtained in *ln* k versus 1/T plots, upholding the validity of equation (61). (The minor curvature predicted from (62) is within experimental error in all cases.) Table 2 gives some typical activation energies including some

Table 2

Activation Energies of Geochemical Reactions (Kcal/mole)[a]

[a]In some cases the temperature range studied is given.

	E_a	Reference
$H + D_2 \rightarrow HD + H$	6.5	Benson (1960)
$O + O_3 \rightarrow 2\ O_2$	5.6	Benson (1960)
$[Co(NH_3)_5Cl]^{2+} + OH^- \rightarrow [Co(NH_3)_5(OH)]^{2+} + Cl^-$	27.2	Moelwyn-Hughes (1971)
$Fe^{2+} + {*Fe}^{3+} \rightarrow Fe^{3+} + {*Fe}^{2+}$	16.8	Ibid
$C_2H_4 + H_2 \rightarrow C_2H_6$	43	Eyring et al. (1950)
$CO(g) + Cl_2(g) \rightarrow COCl(g) + Cl(g)$	53	Benson (1960)
$I + I \rightarrow I_2$ in CCl_4 solvent	3.2	Benson (1960)
$2\ HI \rightarrow H_2 + I_2$	44	Moore (1964)
$2\ HI \rightarrow H_2 + I_2$ (gold)	25	Moore (1964)
Racemization of amino acids	18-23 (25-136°C)	Bada (1972)
Geochemical Reactions:		
$NaAlSi_2O_6 \cdot H_2O + SiO_2 \rightarrow NaAlSi_3O_8 + H_2O$ analcite quartz albite	44.5 (300-500°C)	Matthews (1980)
Porcelanite → Quartz (chert) (Hydrothermal)	23.2 (300-500°C)	Ernst and Calvert (1969)
Calcite + Mg^{2+} → Dolomite + Ca^{2+}	49 (252-295°C)	Katz and Matthews (1977)
FeS(am) + S° → FeS_2 (pyrite)	17	Rickard (1975)
$H_4SiO_4 \rightarrow SiO_2(qtz) + 2\ H_2O$	11.9 (0-300°C)	Rimstidt and Barnes (1980)
$^{32}SO_4 + H_2{}^{34}S \rightarrow {}^{34}SO_4 + H_2{}^{32}S$	18 pH<3 30 4<pH<7 48 7<pH	Ohmoto and Lasaga (1981)

Table 2

(Continued)

	E_a	Reference
dissolution:		
Diopside dissolution	12-36	Berner (1980)
Enstatite dissolution	12	Ibid
Augite dissolution	19	Ibid
$Mg_2SiO_4 + 4H^+ \rightarrow 2Mg^{2+} + H_4SiO_4$ forsterite	9.1	Grandstaff (1980)
$Mg_{.77}Fe_{.23}SiO_3$ dissolution (incongruent) orthopyroxene	10.5 (1-42°C)	Grandstaff (1977)
$SiO_2(qtz) + 2H_2O \rightarrow H_4SiO_4$	16-18 (0-300°C)	Rimstidt and Barnes (1980)
$SiO_2(am) + 2H_2O \rightarrow H_4SiO_4$	14.5-15.5 (0-300°C)	Ibid
$CaCO_3$ (calcite) $\rightarrow Ca^{2+} + CO_3^{2-}$	8.4 (5-50°C)	Sjoberg (1976)
Diffusion in Minerals	20-120	(see Chapter 7)

of the reactions discussed earlier and other geochemical reactions. The activation energies span quite a range from less than 1 Kcal/mole for reactions usually involving individual atoms or radicals, to more than 100 Kcal/mole, when a covalent bond is fully broken.

The size of the activation energy, besides providing the very important temperature dependence of the rate of reaction, often offers an important clue regarding the reaction *mechanism*. For example, diffusion-controlled reactions in fluid media have rather low activation energies (E_a less than 5 Kcal/mole). An example is the $I + I \longrightarrow I_2$ reaction in Table 2. Thus, a high experimental E_a would rule out a solution-transport mechanism. This is applicable, for instance, to calcite or silica dissolution kinetics, where E_a = 8.4 Kcal/mole and 16 Kcal/mole, respectively, which are higher than the diffusion activation energy. If the dissolution rate is not controlled by transport away from the surface, the reactions at the surface (which are responsible for the higher activation energy) must be rate-controlling.

This type of dissolution is termed "surface-controlled" (see Chapter 3 by Berner).

It is interesting that most of the activation energies for a wide variety of mineral-solution alteration processes whether sedimentary or metamorphic (Table 2) seem to lie in the range 10-20 Kcal/mole. This range, while distinctly higher than that based on transport control, is also less than the range expected from breaking bonds in crystals, for example, bulk diffusion activation energies ∿20-80 Kcal/mole (see Chapter 7). Apparently, the catalytic effects of adsorption on surfaces (see next section) or the role of surface defects reduces the activation energies to this intermediate energy range.

Chemists have developed the art of predicting activation energies for elementary reactions in the gas phase very successfully (*cf.* Benson, 1976). This is not the case with heterogeneous reactions or reactions in fluid phase, which are of most interest to the geochemist. Nonetheless, some theoretical progress is being made in this direction (see Moelwyn-Hughes, 1971, Chapter 5). At any rate, besides using the activation energy to correct for the temperature change of a rate constant, geochemists should begin to link the experimentally derived E_a to the molecular processes taking place. (See Ohmoto and Lasaga (1981) for an example.)

HETEROGENEOUS KINETICS, CATALYSIS AND SURFACE ADSORPTION

Most igneous, metamorphic and sedimentary reactions are, at least in part, *heterogeneous*. The study of heterogeneous kinetics and surface chemistry will be very important in geochemistry in the coming years. For example, crystal growth from melts (see Chapter 8 by Kirkpatrick) depends on reactions at the crystal-melt interface; the breakdown of a previous mineral and the growth of a new mineral in regional metamorphism (see Chapter 5 by Fisher and Lasaga) depends on either the heterogeneous reaction with a pervasive grain boundary fluid or the reaction between the two minerals at a common boundary; finally, the chemical weathering of minerals via dissolution depends on the reactions of fluid species at the surface of the mineral (see Chapter 3 by Berner). Two important aspects which are necessary to understand all these geochemical heterogeneous kinetics are *catalysis* and *adsorption*.

These will be discussed in this section. Other aspects specific to some of the processes above are discussed further in the respective chapters quoted.

An important effect of surfaces is the reduction of the energy of activation. This point is the essence behind the use of *catalysts* in kinetics. Catalysts permit the system to follow a reaction path having a much lower activation energy. Consequently, catalysts are very useful in industrial reactions. The most commonly encountered cases of catalysis involve the presence of an active solid in a contact with a solution. For example, metals are useful catalysts for many ionic reactions because metals contain highly mobile electrons -- they are good conductors. These electrons can act as "intermediaries" in ionic reactions. Of more importance geochemically, semiconductors such as metal oxides and sulfides can also act as good catalysts for redox reactions. Redox reactions involve a transfer of electrons among the reactants. This transfer must occur as the reactants approach each other closely in a homogeneous (*e.g.*, aqueous) reaction. On the other hand, a reductant can transfer an electron to the surface of a semiconductor; the semiconductor can then transport this excess electron to another locality on the surface, where an oxidant can receive the electron. This latter heterogeneous process catalyzes the redox reaction; because the electrons of semiconductors can be thermally excited at relatively low activation energies (10-20 Kcal/mole) to give electronic conduction, the energy of activation of the redox reaction is substantially reduced by the heterogeneous pathway.

The process of catalysis as well as most other heterogeneous reactions depends ultimately on the kinetics and energetics of the various *adsorption* reactions. Therefore, adsorption and its relation to geochemistry is the main topic of this section. Most heterogeneous reactions proceed by the following sequence of steps:

(1) Adsorption of fluid species onto surface.

(2) Reaction of adsorbed species among themselves or with the surface atoms.

(3) Desorption of product species.

There are two different types of adsorption: *physical adsorption* and *chemisorption*. Physical adsorption is caused by intermolecular or van der Waal's forces. During physical adsorption the heat liberated is generally in the region 2-6 Kcal/mole of species adsorbed. On the other hand, chemisorption involves the formation of chemical or ionic bonds between the surface atoms and the adsorbed species. For chemisorption the *heat of adsorption* is usually greater than 20 Kcal/mole. It is precisely this heat of adsorption which is responsible for lowering the activation energy and catalyzing chemical reactions in fluids. For example, suppose a heterogeneous reaction has the following mechanism:

$$A + S \rightleftarrows A{\cdot}S \qquad K$$

$$A{\cdot}S \xrightarrow{k_1} B{\cdot}S \qquad \text{slow}$$

$$B{\cdot}S \xrightarrow{k_2} B + S \qquad \text{fast}$$

S stands for an available surface site and A·S and B·S are adsorbed species. B could be a molecularly rearranged A or a species leached from the solid (leaving A behind). The overall rate of production of B is given by

$$\frac{dC_B}{dt} = k_2[B{\cdot}S] = k_1[A{\cdot}S]$$

where steady state is assumed for B·S in the last equation. If equilibrium is assumed between A and A·S, then $[A{\cdot}S] = K[A][S]$. Thus,

$$\frac{dC_B}{dt} = K\ k_1[A][S]\ .$$

The activation energy of this reaction is then

$$E_a = \Delta H_{ad} + E_1 \qquad (63)$$

where ΔH_{ad} is the heat of adsorption of the A molecule and E_1 is the activation energy of the surface reaction. Equation (63) shows that if ΔH_{ad} is negative (strong adsorption), the adsorption will catalyze the reaction.

Figure 15 illustrates the variety of possible adsorption types for a diatomic molecule A_2. Another distinction between physical adsorption and chemisorption is the rate at which adsorption occurs. Although physical adsorption is very fast and requires almost no activation energy, chemisorption requires an activation and is much slower. Some details of the energetics and potential surfaces of

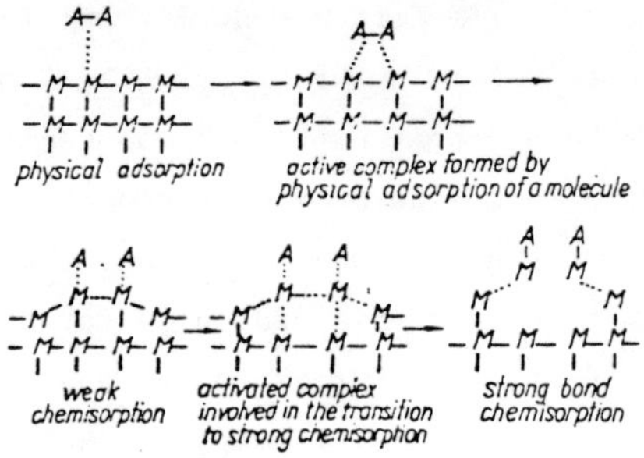

Figure 15. Schematic illustration of the possible processes involved in the formation of a strong chemisorption bond and dissociation of molecules on adsorption.

adsorption are discussed further in Chapter 4 on transition state theory. Quite often, it is the rate of adsorption or the rate of desorption which controls the overall rate of a heterogeneous reaction.

Whenever solids are involved in a reaction occurring in a fluid, the process of adsorption (or absorption) may play an important role in the form of the rate law. An illustration of the principles involved is found in the reaction:

$$CO(g) + 1/2\ O_2(g) \rightarrow CO_2(g)\ .$$

In the presence of a solid catalyst, for example crushed fused quartz (Bodenstein and Othmer, 1945), the reaction is catalyzed and is found to obey the rate law:

$$\frac{dP_{CO_2}}{dt} = k\,\frac{P_{O_2}}{P_{CO}}\ . \tag{64}$$

The explanation for the rate law in equation (64) must rely on a detailed treatment of the adsorption process (Wagner, 1970) as follows. Labeling the surface sites with the symbol S (*i.e.*, A·S represents adsorbed A on the surfaces), the kinetic path is given by the following mechanism:

$$S + O_2 \rightleftarrows O_2{\cdot}S \qquad K_1 \tag{65a}$$

$$S + CO \rightleftarrows CO{\cdot}S \qquad K_2 \tag{65b}$$

$$O_2{\cdot}S + CO{\cdot}S \xrightarrow{k_3} CO_2{\cdot}S + O{\cdot}S \tag{65c}$$

$$O{\cdot}S + CO{\cdot}S \xrightarrow{k_4} CO_2{\cdot}S \tag{65d}$$

$$CO_2{\cdot}S \xrightarrow{k_5} CO_2 + S\ . \tag{65e}$$

Using [ ] to denote concentration, and assuming that there is rapid reaction between adsorbed CO, adsorbed O_2, and the corresponding molecules in the gas [equations (65a) and (65b)], the number of adsorbed species at equilibrium is given by

$$[O_2 \cdot S] = K_1 \; [O_2(g)][S] \tag{66}$$

$$[CO \cdot S] = K_2 \; [CO(g)][S] \; . \tag{67}$$

The equilibrium constants, K_1 and K_2, are given by

$$K_1 = k_1/k_{-1} \qquad\qquad K_2 = k_2/k_{-2}$$

where k_i refer to the individual rate constants of the adsorption and desorption reactions in (65a) and (65b). Note that equations (66) and (67) assume reactions (65c) and (65d) are so slow that the equilibrium in (65a) and (65b) is not perturbed into a steady state for species $CO \cdot S$ and $O_2 \cdot S$ (*i.e.*, k_3 and k_4 do not enter in determining $[CO \cdot S]$ and $[O_2 \cdot S]$). [S] in equations (66) and (67) refers to the concentration of surface sites free of adsorbed species. If the total concentration of surface sites is given by $[S°]$, then

$$[S] = [S°] - [CO \cdot S] - [O_2 \cdot S] \; , \tag{68}$$

ignoring $CO_2 \cdot S$ as minor. Inserting equations (66) and (67) into equation (68):

$$[S] = [S°] - K_2[CO(g)][S] - K_1[O_2(g)][S] \; ,$$

and solving for [S],

$$[S] = \frac{[S°]}{1 + K_1[O_2] + K_2[CO]} \; . \tag{69}$$

Therefore, the fraction of surface sites occupied by CO and O_2, X_{CO} and X_{O_2}, will be given by

$$X_{O_2} = \frac{[O_2 \cdot S]}{[S°]} = \frac{K_1[O_2(g)]}{1 + K_1[O_2(g)] + K_2[CO(g)]} \tag{70}$$

and

$$X_{CO} = \frac{[CO \cdot S]}{[S°]} = \frac{K_2[CO(g)]}{1 + K_1[O_2(g)] + K_2[CO(g)]} \; . \tag{71}$$

Equations (70) and (71) are the adsorption equations for the reactant molecules. Note that in this case the molecules must compete for the available surface sites. To obtain the rate law, we now apply the *steady state* condition to the reaction intermediates, O·S and CO_2·S. From the sequence in equations (65c) - (65d)

$$\frac{d[O\cdot S]}{dt} = k_3[O_2\cdot S][CO\cdot S] - k_4[O\cdot S][CO\cdot S] = 0 \ .$$

Hence, at steady state,

$$[O\cdot S]_{st} = \frac{k_3}{k_4}[O_2\cdot S] \ . \qquad (72)$$

Similarly, for [CO_2·S] we find that at steady state:

$$[CO_2\cdot S]_{st} = \frac{k_4[O\cdot S][CO\cdot S] + k_3[O_2\cdot S][CO\cdot S]}{k_5} \ .$$

Inserting equation (72) for [O·S] in the last equation yields

$$[CO_2\cdot S]_{st} = \frac{2k_3[O_2\cdot S][CO\cdot S]}{k_5} \ . \qquad (73)$$

Finally, from equation (65e) the rate of production of CO_2 can be written as

$$\frac{d[CO_2(g)]}{dt} = k_5[CO_2\cdot S] \ .$$

Inserting the steady state expression for [CO_2·S] [equation (73)] and the expressions for [O_2·S] and [CO·S] from equations (70) and (71) into the last equation yields

$$\frac{d[CO_2(g)]}{dt} = \frac{2k_3K_1K_2[O_2(g)][CO(g)]}{(1 + K_1[O_2(g)] + K_2[CO(g)])^2} \ .$$

CO is very strongly adsorbed on quartz, probably forming CO_3^{2-} on the surface. Therefore, $K_2[CO(g)] \gg 1 + K_1[O_2(g)]$. Using this information, we can simplify the last equation to obtain

$$\frac{d[CO_2(g)]}{dt} = \frac{2k_3K_1[O_2(g)]}{K_2[CO(g)]} \ . \qquad (74)$$

Equation (74) agrees with the observed experimental law and relates the experimental rate constant to k_3, K_1, and K_2. It is important to note that the peculiar form of the rate law arises from the competition of CO and O_2 for the available surface sites. Furthermore, once gaseous species are adsorbed, the solid surface of the quartz offers a new path for the reactants. Specifically, the rate constant in equation (74) dictates that the activation energy of the overall reaction is given by

$$E_a = E_3 + \Delta H_1 - \Delta H_2$$

where E_3 is the activation energy of reaction (65c) and ΔH_1 and ΔH_2 are the standard enthalpies of reactions (65a) and (65b). Any catalytic effect arises from the low value of E_3, which offsets any enthalpy differences ($\Delta H_1 - \Delta H_2$), relative to the activation energy of the homogeneous reaction in the fluid phase. Note that since the gas species compete for available sites, the energy of activation does not have the simple dependence on ΔH, given by equation (63).

A treatment similar to that carried out here for the formation of CO_2 is quite widely used in handling an amazing variety of heterogeneous reactions [*e.g.*, see Wagner (1970) and Darken and Turkdogan (1970)] and should be applicable to geochemical heterogeneous reactions.

The homogeneous hydrogenation of ethylene (C_2H_4), which has an activation energy of 43 Kcal/mole (Table 2), offers another example of surface catalysis. In the presence of metal catalysts such as Fe, Ni, or Pt, the activation energy is reduced to 10 Kcal/mole (Eyring *et al.*, 1950).

The effects of competition for adsorption sites has been introduced here by equations (70) and (71). In the case of just one adsorbate, the rate of adsorption (in moles of A adsorbed/cm^3/sec) A + S $\xrightarrow{k_1}$ A·S is given by $k_1 C_A (1-\Theta_A)$, where C_A is the concentration of A in solution and Θ_A is the fraction of available surface sites, which are occupied by species A. The rate of desorption is given by $k_{-1}\Theta_A$ (in moles of A desorbed/cm^3/sec). Therefore at equilibrium

$$k_1 C_A (1-\Theta_A) = k_{-1}\Theta_A \tag{75}$$

or

$$\Theta_A = \frac{KC_A}{1 + KC_A} \tag{76}$$

where $K = k_1/k_{-1}$. Equation (76) is the *Langmuir isotherm*. The Langmuir adsorption isotherm predicts a linear relation between Θ_A and C_A for low concentrations of A. As C_A increases, however, the surface sites become more saturated with adsorbed A and Θ_A approaches 1. The linear relation between Θ_A and C_A has been used often in modeling adsorption in natural waters. For example, in the early diagenesis of sediments, diffusion of a species in the pore waters can be significantly reduced by the adsorption of the particular species. This reduction in diffusion relative to advective mass transfer is important for cations, which can adsorb onto clays. Some of the adsorption coefficients, K, have been measured for several cations in some sediments (*cf.* Berner, 1980).

There are other adsorption isotherms, which improve on the Langmuir isotherm. For example, the successful BET isotherm accounts for physical adsorption from the submonolayer to the multilayer stage and is central in the measurement of surface area (*cf.* Ponec *et al.*, 1967; Adamson, 1976). For many treatments, however, the Langmuir isotherm is a suitable starting point for a description of the kinetics.

If a molecule dissociates into two entities upon chemisorption (see Fig. 15), the adsorption reaction can be written as

$$A_2 + \underset{}{\overset{|}{S}}\!-\!\!-\!\overset{|}{S} \xrightarrow{k_1} \overset{A}{\overset{|}{S}}\!-\!\!-\!\overset{A}{\overset{|}{S}} \longrightarrow 2\ \overset{A}{\overset{|}{S}} \qquad (77)$$

Hence, the rate of adsorption is $k_1C_{A_2}C_{S_2}$ (in moles of A_2 adsorbed/cm^3/sec), where C_{S_2} is the concentration of free *pairs* of adsorption sites. If there are [S°] total sites per unit area, and a fraction Θ_A of them are occupied by A, then there are $(1-\Theta_A)[S°]$ free sites. For each free site, the probability of a neighbor being free also is $1 - \Theta_A$. Hence, if each free site has z neighbors, the number of free pairs of adsorption sites per unit area is given by

$$C_{S_2} = \frac{z}{2}[S°](1-\Theta_A)^2 ,$$

and division by two corrects for each S_2 pair being counted twice. This indicates that C_{S_2} is proportional to $(1-\Theta_{A_2})^2$; as a consequence the adsorption rate can be written as $k_1'C_{A_2}(1-\Theta_A)^2$. The rate of desorption [reverse of (76)] requires collision of two A·S species and so is given by $k_{-1}\Theta_A^2$. At equilibrium, $k_1'(C_{A_2}(1-\Theta_A)^2 = k_{-1}\Theta_A^2$, or

$$\Theta_A = \frac{(KC_{A_2})^{1/2}}{1 + (KC_{A_2})^{1/2}} \qquad (78)$$

where $K = k_1'/k_{-1}$. Equation (78) shows that the number of adsorbed species, if dissociation occurs, depends on $C_{A_2}^{1/2}$. Recent data obtained of the influence of organics on the dissolution rate of olivine

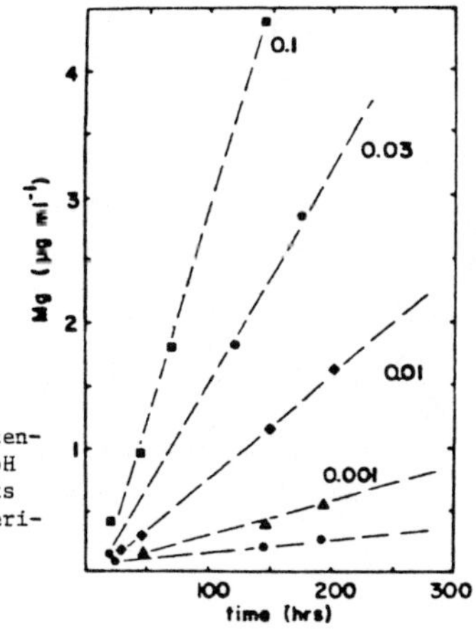

Figure 16. Dissolution rate as a function of phthalate concentration (data for Mg only) in 0.1 M KCl 'matrix' solution at pH 4.5 and 25°C. Phthalate concentration (molar) in experiments given besides lines fitted to data. Unlabelled line is experiment conducted at pH 4.5 without added phthalate. (From Grandstaff, 1980.)

(Grandstaff, 1980) provides an example of the use of equation (78). Figure 16 shows the *increase* in the rate of dissolution of Mg^{2+} from forsterite due to increased amounts of phthalate in solution. The increase in the rate is probably due to adsorption of the organic ligand on the surface of the olivine, followed by desorption of a Mg^{2+}-phthalate complex. When a plot of log Rate versus log $C_{phthalate}$ is made, the dissolution rate (*i.e.*, the slope of the lines in Fig. 16) is found to vary as $k\ C^{1/2}_{phthalate}$. This dependence on the square root of the ligand concentration is expected from equation (78) if the organic ligand adsorbs onto *two* sites (and if $KC_L^{1/2} << 1$). The phthalate ligand (a dicarboxyl ligand) is bidentate (each COO^- can adsorb onto a site) and so this adsorption behavior is reasonable. Note that, in this case, the ligand does not fully dissociate and so equation (78) is only approximate. Nonetheless, the overall kinetic behavior can be explained on the basis of the adsorption isotherm in equation (78).

A general phenomenon of great importance in the description of weathering and hydrothermal alteration is the hydrolysis of minerals (see Garrels, 1957; Helgeson, 1971). While recent experiments (Petrovic *et al.*, 1976) suggest that the depth of leached layers in some minerals may be minimal, the pH dependence of dissolution rates indicates that H^+ reaction at the surface is rate-controlling. The chemisorption of H^+ (as H_3O^+) is therefore a critical step in the weathering and alteration of minerals. For albite the overall reaction may be written as $NaAlSi_3O_8 + H^+ \rightarrow HAlSi_3O_8 + Na^+$ with subsequent dissolution of the protonated structure. The reaction may be viewed as chemisorption of

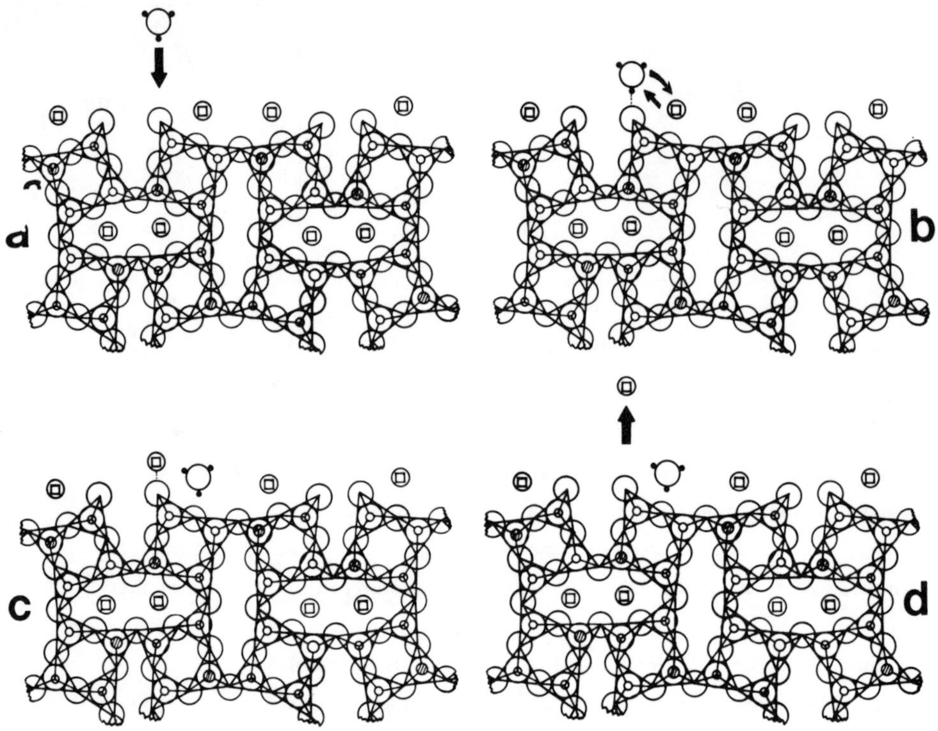

Figure 17. Diagram illustrating the possible kinetic steps in the hydrolysis of feldspar.
a) Hydrogen ion approaches surface of feldspar
b) Hydrogen ion is adsorbed onto surface of feldspar.
c) Nearby Na+ ion in feldspar surface exchanges place with the adsorbed H^+ ion to produce an adsorbed Na^+ ion.
d) Na^+ ion is desorbed leaving behind a locally protonated feldspar surface.
Square-in-circle = Na^+; large circle = O^{2-}; shaded circle = H^+.

H^+ followed by reaction with a surface Na^+ to produce adsorbed Na^+ and then detachment of the Na^+ ion (see Fig. 17). One of the puzzling results of the kinetic work on alteration of minerals is the complex dependence of the rates on the pH. Often the rate is proportional to a fractional power of the hydrogen ion activity, $a_{H^+}^{1/n}$, where $0.5 < 1/n < 1$. To understand the complex rate law, the details of the adsorption process during hydrolysis must be understood.

Before discussing feldspar alteration mechanism in detail, we should note that a normal surface is not a featureless plane but a topographically complex ensemble such as shown in Figure 18. Therefore, on a non-uniform surface, a complete range of sites and corresponding activation energies are available for adsorption and desorption processes. The non-uniform nature of the surface is responsible

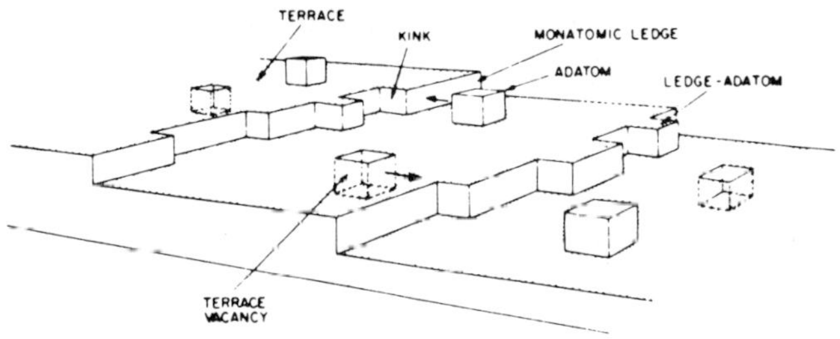

Figure 18. Schematic diagram of terrace-ledge-kink model of surfaces.

for the geochemically important effect of poisons or inhibitors. For example, very small amounts of phosphate ions in solution (in the μM range) will inhibit the dissolution rate of calcite in sea water by order of magnitude (see Chapter 3). Since dissolution is favored at a few active sites (see Fig. 18), a substance (such as phosphate) which adsorbs strongly onto these active sites will diminish the rate significantly. Furthermore, it takes little phosphate to saturate the relatively few active sites.

The variety of sites on a surface also gives rise to different activation energies for the surface reactions. Recall that the adsorption energy is responsible for the reduction in the activation energy for reaction. However, if adsorption is very strong on a given site (thereby favoring the surface reaction), it may be also difficult to desorb the products of the reaction. For example, adsorption of H^+ onto a kink in Figure 18 is very favorable (high number of bonds formed), but the desorption of Na^+ from the same site, after the Na^+-H^+ exchange, will also be much impeded. Each type of site will lead to a particular free energy of activation for reaction, ΔG_r, and for desorption of products, ΔG_d (the rate of adsorption of reactants is ignored here). As we vary the sites, a decrease in ΔG_r generally leads to an increase in ΔG_d (for the reasons just given) and vice versa. If either ΔG_r or ΔG_d is made very large, the rate of the overall reaction will be small. It is expected, therefore, that there are some sites with optimal values for ΔG_r and ΔG_d which maximize the overall rate (*i.e.*, reaction + desorption). Since the reactions on the different sites are parallel reactions, the fastest overall reaction will control the rate, as discussed earlier. Therefore, it is important to obtain the conditions for a maximum overall rate. In the vicinity of these optimal conditions, the increase in ΔG_d accompanying a decrease in ΔG_r will obey an equation of the form:

$$d\Delta G_r = -n\ d\Delta G_d \tag{79}$$

where n is some arbitrary (positive) constant.

Let us now examine the kinetic paths involved in the hydrolysis reaction as depicted in Figure 17. The following approach to non-uniform surfaces is based on an interesting paper presented by Halsey (1949). First, assume that the H^+ adsorption follows the Langmuir isotherm. Because Na^+ is also present, the sorption-desorption equivalent of equation (75) becomes

$$k_1 C_{H^+} (1 - \Theta_{H^+} - \Theta_{Na^+}) = k_{-1}\Theta_{H^+} \tag{80}$$

where $1 - \Theta_{H^+} - \Theta_{Na^+}$ is the fraction of free sites. Solving for Θ_{H^+}:

$$\Theta_{H^+} = \frac{KC_{H^+}}{1 + KC_{H^+}} (1-\Theta_{Na^+}) \ . \tag{81}$$

Equation (81) is the modified Langmuir isotherm. In addition, *steady state* is assumed for Θ_{Na^+}; therefore,

$$\frac{d\Theta_{Na^+}}{dt} = \frac{kT}{h}\left[\Theta_{H^+}\, e^{-\frac{\Delta G_r}{RT}} - \Theta_{Na^+}\, e^{-\frac{\Delta G_d}{RT}}\right] = 0 \ . \tag{82}$$

The first term in equation (82) refers to reaction of H^+ with the surface, producing adsorbed Na^+, and the second term is the removal rate of adsorbed Na^+. Note that we are using transition state theory in writing the rate constants in (82) as

$$k_i = \frac{kT}{h} e^{-\frac{\Delta G^\dagger}{RT}}$$

where $\Delta G^\dagger$ is the free energy change in forming the activated complex (see Chapter 4). By inserting the Langmuir isotherm for Θ_{H^+} into (82) we obtain the following expression for Θ_{Na^+}:

$$\Theta_{Na^+} = \frac{\dfrac{KC_{H^+}}{1 + KC_{H^+}}\, e^{-\frac{\Delta G_r}{RT}}}{\dfrac{KC_{H^+}}{1 + KC_{H^+}}\, e^{-\frac{\Delta G_r}{RT}} + e^{-\frac{\Delta G_d}{RT}}} \ . \tag{83}$$

But since the overall rate is given by the desorption of Na^+

$$\text{Rate} = \frac{dC_{Na^+}}{dt} = \frac{kT}{h}\, \Theta_{Na^+}\, e^{-\frac{\Delta G_d}{RT}} \;,$$

we can insert equation (83) for Θ_{Na^+} at steady state to obtain:

$$\text{Rate} = \frac{kT}{h} \frac{A' e^{-\frac{\Delta G_r}{RT}}\, e^{-\frac{\Delta G_d}{RT}}}{A' e^{-\frac{\Delta G_r}{RT}} + e^{-\frac{\Delta G_d}{RT}}} \qquad (84)$$

where we defined $A' \equiv KC_{H^+}/(1+KC_{H^+})$. Finally, using equation (79), it is clear that

$$\Delta G_r + n\Delta G_d = \Delta G_o \;, \qquad (85)$$

where ΔG_o is a *constant*. Using (85) to evaluate ΔG_r in equation (84), the final equation for the overall rate is obtained:

$$\text{Rate} = \frac{kT}{h} \frac{A\, e^{(n-1)\frac{\Delta G_d}{RT}}}{A\, e^{n\frac{\Delta G_d}{RT}} + e^{-\frac{\Delta G_d}{RT}}} \qquad (86)$$

where now $$A \equiv A' e^{-\frac{\Delta G_o}{RT}} = \frac{KC_{H^+}}{1 + KC_{H^+}}\, e^{-\frac{\Delta G_o}{RT}} \;. \qquad (87)$$

To obtain the maximum rate, we set $d\,\text{Rate}/d\Delta G_d = 0$ and solve for ΔG_d. After recasting, the result is

$$A\, e^{(n+1)\frac{\Delta G_d}{RT}} = n \;. \qquad (88)$$

Equation (88) can be rewritten

$$e^{\frac{\Delta G_d}{RT}} = \left(\frac{n}{A}\right)^{\frac{1}{n+1}} \;.$$

Inserting this optimal value of ΔG_d into the rate expression (86) we obtain

$$\text{Rate}_{max} = \frac{dC_{Na^+}}{dt} = \text{constant} \times A^{\frac{1}{n+1}}$$

or $$\text{Rate}_{max} = \frac{dC_{Na^+}}{dt} = \text{constant} \times \left(\frac{KC_{H^+}}{1 + KC_{H^+}}\right)^{\frac{1}{n+1}} . \qquad (89)$$

Equation (89) is our final result. It shows that the dependence of the dissolution rate on pH will follow

$$\text{Rate} \ \alpha \ a_{H^+}^{\frac{1}{n+1}} \qquad (90)$$

if $KC_{H^+} < 1$. The exponent in equation (90) can take any value between 0 and 1. If n = 1, then we obtain a dependence on $a_{H^+}^{1/2}$. It is interesting to note that dissolution studies on minerals have yielded pH dependencies in the range

$$\text{Rate} \ \alpha \ a_{H^+}^{0.5-1} .$$

For example, Grandstaff (1977) reports that the rate of dissolution of orthopyroxene is proportional to $a_{H^+}^{1/2}$. Similarly, nepheline dissolution seems to obey a $a_{H^+}^{1/2}$ rate law (Tole and Lasaga, pers. comm.) and the release of silica during diopside dissolution is proportional to $a_{H^+}^{0.7}$ (Berner *et al.*, 1980) in the pH range 2-6 for both cases. While this model of heterogeneous reaction is certainly simplified, it shows how the non-trivial pH dependence of mineral rock interactions (*e.g.*, fractional exponents) can indeed be explained from a treatment of the surface chemistry.

Further work in deciphering the chemistry of weathering, as well as the rate of metamorphic reactions and hydrothermal alteration, must utilize and develop the theory of surface chemistry and adsorption as it relates to minerals.

PRINCIPLE OF DETAILED BALANCING

Even though rate constants are certainly path-dependent and activation energies may be changed drastically by catalysts, there is a very direct and important link between the kinetic rate constants and thermodynamic equilibrium constants. The principle of detailed balancing states that at equilibrium, the *rates* of forward and reverse microscopic processes are *equal* for every *elementary* reaction. Equilibrium, therefore, is associated not with a cessation of all processes but rather with a dynamic balance, where all the rates in a given direction have counterbalancing rates in the opposite direction. The fundamental explanation for this principle goes back to the theory of *microscopic reversibility*, which forms a central part of the theory of irreversible thermodynamics to be treated in a later chapter. Our interest here, however, is the application of this principle.

In a given chemical system there are many elementary reactions going on simultaneously. If only certain reactions or species have balanced rates, the result is a steady state condition as discussed earlier. However, for equilibrium to be reached, *all* elementary processes must have equal forward and reverse rates and all species, not just reactive intermediates, must be at steady state. The kinetic behavior of systems near equilibrium will be shown to be linear and hence is always simpler to characterize than that for systems very far from equilibrium. The latter systems (even near steady state) can exhibit dramatic (non-linear) changes stemming from minor fluctuations in concentration or temperature (see Chapter 5).

The application of the principle of detailed balancing is best understood with the case of two opposite first-order reactions:

$$A \underset{k_-}{\overset{k_+}{\rightleftharpoons}} P \ . \tag{91}$$

The principle of detailed balancing states that at equilibrium the concentrations of A and P, $[A]_e$ and $[P]_e$ respectively, are such that rates in both directions of reaction (91) are equal

$$k_+[A]_e = k_-[P]_e \ . \tag{92}$$

We can rewrite equation (92) utilizing the equilibrium constant of reaction (91) as

$$\frac{k_+}{k_-} = \frac{[P]_e}{[A]_e} = K_{eq} \ . \tag{93}$$

Equation (93) is our link between kinetics and thermodynamics! It allows us, for example, to obtain a value for k_- from only a kinetic measurement of k_+ and knowledge of thermodynamics (*e.g.*, K_{eq}).

An obvious extension of the correlation between the equilibrium constant and the ratio of forward and reverse rate constants applies to *heterogeneous* reactions. If we write the rate of dissolution of a mineral during alteration of a rock by water as proportional to the surface area of the mineral in contact with solution, then it must also follow that the rate of precipitation or formation of the mineral is proportional to the same surface area. This is a consequence of the well-known fact that, unless the particles are very tiny, the equilibrium constant of a mineral-water reaction is independent of the surface area of the mineral.

A geochemical example of the application of the principle of detailed balancing has been offered by Rimstidt and Barnes' (1980) work on silica dissolution and precipitation kinetics. Rimstidt and Barnes worked on both the dissolution and precipitation reactions

$$SiO_2(s) + 2H_2O(l) \underset{k_+}{\overset{k_-}{\rightleftarrows}} H_4SiO_4(aq) \tag{94}$$

and found the rate law for the precipitation to be first order:

$$-\frac{dm_{H_4SiO_4}}{dt} = \frac{A}{M} k_- \, m_{H_4SiO_4} \tag{95}$$

where A = total surface area of silica (they used both amorphous silica and quartz) and M is the mass of water in the system. Analysis of the experimental data yielded values of the *precipitation* rate constant k_-. Suppose the rate of *dissolution* of quartz or amorphous silica was needed. One may be tempted to carry out a new series of experiments with undersaturated solutions. However, the principle of detailed balancing makes these latter experiments redundant.

The equilibrium constant for the solubility of SiO_2 (quartz, amorphous silica, etc.), reaction (94), is given by

$$K_{eq} = \frac{a_{H_4SiO_4(aq)}}{a_{SiO_2(s)} a^2_{H_2O}} \approx a_{H_4SiO_4(aq)} \, . \tag{96}$$

If the precipitation rate obeys equation (95), then based on equation (96) the *dissolution* rate of silica must obey the equation

$$\frac{dm_{H_4SiO_4}}{dt} = \frac{A}{M} k_+ \, . \tag{97}$$

Combining equations (95) and (97) at equilibrium

$$\frac{dm_{H_4SiO_4}}{dt} = \frac{A}{M} k_+ - \frac{A}{M} k_- \, m^{eq}_{H_4SiO_4} = 0$$

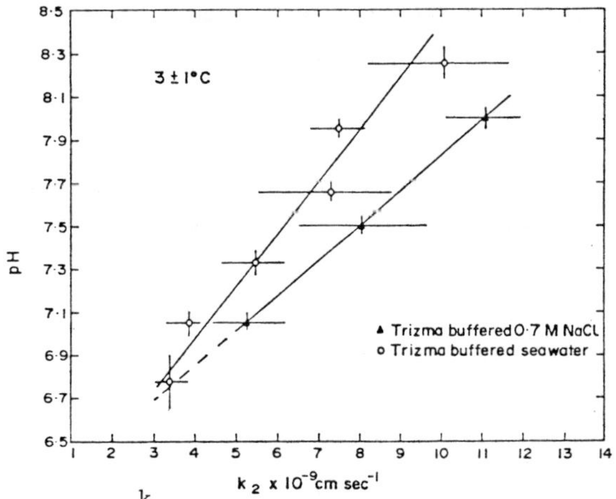

Figure 19. Heterogeneous solution rate constant of acid-cleaned biogenic opal as a function of pH at 3±1°C. Note lesser values of k_2 in sea water at increasing pH values relative to the 0.7 M NaCl solutions. Data from Hurd (1973).

$$m^{eq}_{H_4SiO_4} = \frac{k_+}{k_-} \; .$$

Hence, using equation (80) we obtain

$$k_+ = k_- \frac{K_{eq}}{\gamma_{H_4SiO_4}} \qquad (98)$$

which is similar to equation (93). Rimstidt and Barnes (1980), therefore, were able to obtain dissolution rate constants, k_+, for different silica species as a function of temperature (25°C - 300°C) based on the temperature data for k_- (see Fig. 14) and on data for the solubility of the various silica species as a function of temperature. No new experiments were required.

Use of equation (98) can be extended further. For example, the solubility of silica at 25°C is independent of the pH for pH values less than circa 9.6. Rimstidt and Barnes found k_- to be independent of pH in the slightly acid to neutral pH region. Therefore, from equation (98), it follows that the dissolution rate, k_+, is also pH-independent in this region. However, Wollast (1974) and Hurd (1973) have found that for more basic pH conditions (pH > 7.5, *i.e.*, relevant to sea water) the dissolution rate is pH-dependent [see Fig. 19; also Stober (1967)]. Since K_{eq} is constant up to pH ∿9.6 (Iler, 1980), it also follows from equation (98) that the pH dependence of both k_+ and k_- is identical in the region $7.5 < pH \lesssim 9.6$. It is interesting to note that the rate constants begin to increase with increasing pH *before* the solubility of silica does the same. It may be that the dissociated forms of silica, or the polynuclear species, which are not thermodynamically important until pH $\gtrsim$ 9.6, can act as *kinetic* reactive intermediates (which require only minor concentration) in speeding k_+ and k_- at lower pH's. Once the pH is high enough for the reaction to proceed faster via these reactive intermediates, the mechanism changes and becomes pH-dependent (just as the concentration of the intermediates is pH-dependent). For pH above 9.6, K_{eq} increases with pH and so equation (98) dictates that the dissolution rate constant, k_+, will increase with pH at a higher rate that the precipitation rate constant, k_-.

For some general reaction such as

$$A + B + \ldots \longrightarrow P + Q + \ldots \tag{99}$$

the rates of forward and back reactions may be written $k_+f_+(C_A, C_B, \ldots, C_P, C_Q\ldots)$ and $k_-f_-(C_A, C_B, \ldots, C_P, C_Q\ldots)$ where k_+ and k_- are both rate constants and the functions f_+ and f_- can be any general functions of the concentrations. Earlier sections have commented on the methods available for obtaining the form of f_+ and f_-. Regardless of the form of f_+ and f_-, at equilibrium the rates are still equal and so

$$\frac{k_+}{k_-} = \frac{f_-(C_A, C_B, \ldots, C_P, C_Q\ldots)}{f_+(C_A, C_B, \ldots, C_P, C_Q\ldots)} . \tag{100}$$

Therefore, the ratio of the functions f_-/f_+ is a constant at equilibrium and must be equal to some equilibrium constant, K, though in some cases K is not necessarily the equilibrium constant for reaction (99) as written (see below). In fact, a powerful use of detailed balancing is the predicting of rate laws; if the rate law is properly written for a forward reaction, the equilibrium constant expression [*e.g.* (100)] will allow us not only to obtain k_- but also to constrain the *rate law* for the reverse reaction. For example, the growth of garnet from chlorite under prograde metamorphic conditions and in the presence of a grain boundary fluid can be written as

$$\underset{\text{Mg-chlorite}}{Mg_5Al_2Si_3O_{10}(OH)_8} + 4H^+_{(aq)} \rightleftarrows \underset{\text{pyrope}}{Mg_3Al_2Si_3O_{12}} + 2Mg^{2+}_{(aq)} + 6H_2O ,$$

if chlorite has the composition shown and only pyrope garnet (no Fe^{2+}) is considered for this illustration. Suppose the formation of garnet is found to have a rate law such as

$$\text{Forward Rate} = k \frac{A_{garnet}}{V} a_{H^+}^2$$

where (A_{garnet}/V) is the specific area of the garnet and k is a rate constant (which depends only on P and T). Then, if this rate law holds near equilibrium, the rate of chloritization of garnet under retrograde conditions would have to obey

$$\text{Reverse Rate} = k' \frac{A_{garnet}}{V} \frac{a_{Mg^{2+}}^2}{a_{H^+}^2} .$$

This latter rate law must hold, since at equilibrium where the two rates are equal,

$$k \frac{A_{garnet}}{V} a_{H^+}^2 = k' \frac{A_{garnet}}{V} \frac{a_{Mg^{2+}}^2}{a_{H^+}^2}$$

or
$$\frac{a_{Mg^{2+}}^2}{a_{H^+}^4} = \frac{k}{k'} = K .$$

The last expression is the equilibrium constant for the overall reaction and must therefore be maintained at equilibrium. Note that the area of the garnet was chosen (arbitrarily in this illustration) as rate-limiting. If, instead, the reaction at the chlorite surface is rate-limiting, then *both* forward and reverse rates would now contain the term $A_{chlorite}$ instead.

Note that if there are several alternative (parallel) paths leading from the initial to the final products, at equilibrium the net rates must cancel and so

$$k_{+1}f_{+1} + k_{+2}f_{+2} + k_{+3}f_{+3} + \ldots = k_{-1}f_{-1} + k_{-2}f_{-2} + k_{-3}f_{-3} + \ldots \quad (101)$$

But, as discussed earlier, the principle of microscopic reversibility goes much further than (101), since it requires that *each* individual path be at equilibrium, *i.e.*,

$$k_{+1}f_{+1} = k_{-1}f_{-1} \; ; \; k_{+2}f_{+2} = k_{-2}f_{-2} \text{ etc.}$$

There are two important caveats in the use of the principle of detailed balancing. First, if our reactions are *overall* reactions, we must be careful in formulating the rate law. An example may be the disproportionation of zerovalent sulfur (polysulfides) in aqueous solution to thiosulfate and bisulfide. We can write the reaction as

$$4S_nS^{2-} + (4n-4)OH^- + (4-n)H_2O \underset{k_-}{\overset{k_+}{\rightleftarrows}} n\ S_2O_3^{2-} + (2n+4)HS^- \qquad (102)$$

where S° is present in solution in the form of polysulfide ions S_nS^{2-}. Rate studies up to 240°C (Giggenbach, 1974) yield a complex rate law of the form

$$-\frac{dm_{S_nS^{2-}}}{dt} = k_+ m_{S_nS^{2-}} m_{HS^-}^{-2} m_{OH^-} - k_- m_{S_2O_3^{2-}} m_{HS^-} m_{OH^-}^{-2} \qquad (103)$$

where $m_{S_nS^{2-}}$ stands for the molality of polysulfide ions. Therefore, the ratio k_+/k_- is given at equilibrium by

$$\frac{k_+}{k_-} = \frac{m_{S_2O_3^{2-}} m_{HS^-}^3}{m_{S_nS^{2-}} m_{OH^-}^3} \ . \qquad (104)$$

Up to now the polysulfide species has been written as S_nS^{2-} without specifying the value of n. The rate law [equation (103)] and equation (104) suggest that in fact the pentasulfide ion (n = 4) is the main active polysulfide species in the kinetics of (102) under the experimental conditions, because the equilibrium constant for the reaction

$$S_4S^{2-} + 3OH^- \rightleftarrows S_2O_3^{2-} + 3HS^- \qquad K_{eq} \qquad (105)$$

satisfies the equation

$$\frac{k_+}{k_-} = K_{eq} \ . \qquad (106)$$

Equation (104) is the detailed balance relation for this case. Since both k_+ and k_- were measured independently by Giggenbach (1974), the experimental data were used to confirm the validity of equation (104). Note, however, that we must use the precise reaction (105) and not the general reaction (102) in obtaining K_{eq}. As a by-product, the principle of detailed balancing also succeeded in determining the dominant active species in the polysulfide reaction under the given conditions.

A second caveat on the use of detailed balancing involves the use of "effective" or concentration equilibrium constants [*e.g.*, (104)] rather than true equilibrium constants, which are based on activities.

The reason for this lies in the use of concentrations in rate expressions.

Another example of the principle of detailed balancing is found in calcite dissolution. In the kinetic treatment of calcite dissolution, rate laws are usually written as proportional to $(1-\Omega)^n$ where Ω is the saturation ratio defined earlier and n is some positive constant [see equation (24)]. In general, the net rate measured is the difference between the dissolution rate and the precipitation rate. The dissolution rate can be written as

$$-\frac{dm_{calcite}}{dt} = \frac{A}{V} k_+ \qquad \text{(dissolution)} \qquad (107)$$

where A/V is the specific area (area/solution volume) of the calcite. The precipitation rate may be some function of the ion activity product in solution, *i.e.*,

$$\frac{dm_{calcite}}{dt} = \frac{A}{V} k'_- m^n_{Ca^{2+}} m^n_{CO_3^{2-}} \qquad \text{(precipitation)} \qquad (108)$$

where n is some constant. Combining equations (107) and (108), the net rate of calcite dissolution is

$$-\frac{dm_{calcite}}{dt} = \frac{A}{V} (k_+ - k'_- m^n_{Ca^{2+}} m^n_{CO_3^{2-}}) .$$

Inserting the activity coefficients into k'_-, we can rewrite the above as

$$-\frac{dm_{calcite}}{dt} = \frac{A}{V} (k_+ - k_- a^n_{Ca^{2+}} a^n_{CO_3^{2-}}) . \qquad (109)$$

But by detailed balancing at equilibrium $dm_{calcite}/dt = 0$; thus,

$$\frac{k_+}{k_-} = a^n_{Ca^{2+}} a^n_{CO_3^{2-}} = K^n_{sp} \qquad (110)$$

where K_{sp} is the solubility product of calcite. Solving (110) for k_+ and inserting in (109), we have

$$-\frac{dm_{calcite}}{dt} = \frac{A}{V} k_- (K^n_{sp} - a^n_{Ca^{2+}} a^n_{CO_3^{2-}})$$

$$= \frac{A}{V} k_- K^n_{sp} (1-\Omega^n) . \qquad (111)$$

Note that equation (111) yields a rate law, very similar to the empirical $(1-\Omega)^n$ law, although not quite the same. In fact, in studying calcite dissolution, Sjoberg (1976) used a rate law identical to (111); in his case n had the value 1/2. The use of the same exponent, n, for both $a_{Ca^{2+}}$ and $a_{CO_3^{2-}}$ in equations (108) and (109) arises from the necessity to incorporate one Ca^{2+} ion for each CO_3^{2-} ion in the calcite structure; otherwise, a charge imbalance would result. Now, if some other ions such as Mg^{2+} are also being incorporated into calcite, the rate dependence on $a_{Ca^{2+}}$ in equation (108) or (109) may be slightly different from the dependence on $a_{CO_3^{2-}}$.

Finally, it may be that the dissolution rate itself depends on the solution composition, *i.e.*,

$$\frac{dm_{calcite}}{dt} = \frac{A}{V} k_+ a^m_{Ca^{2+}} a^m_{CO_3^{2-}} \qquad \text{(dissolution)} \qquad (112)$$

In this case equation (109) becomes

$$\frac{dm_{calcite}}{dt} = \frac{A}{V} (k_+ a^m_{Ca^{2+}} a^m_{CO_3^{2-}} - k_- a^n_{Ca^{2+}} a^n_{CO_3^{2-}}) \ . \qquad (113)$$

Applying the principle of detailed balancing to (113)

$$\frac{k_+}{k_-} = a^{n-m}_{Ca^{2+}} a^{n-m}_{CO_3^{2-}} = K^{n-m}_{sp} \ .$$

Therefore, equation (113) can be rewritten as

$$-\frac{dm_{calcite}}{dt} = \frac{A}{V} k_- K^n_{sp} (\Omega^m - \Omega^n) \ . \qquad (114)$$

To determine the correct rate law to use, experiments must be carried out *far* from equilibrium so that only one of the two terms in (107) and (108) is dominant.

The principle of detailed balancing also relates the activation energies of overall reactions. If one writes the Arrhenius law for the temperature dependence of the forward and reverse rate constants

$$k_+ = A_+ e^{-E_+/RT} \qquad (115)$$

$$k_- = A_- e^{-E_-/RT} \qquad (116)$$

and likewise introduces the van't Hoff equation from thermodynamics

$$K_{eq} = K^\circ_{eq} e^{-\Delta H^\circ/RT} , \qquad (117)$$

it follows from equation (93) and equations (115) - (117) that

$$\frac{A_+}{A_-} = K^\circ_{eq} \quad \text{and} \quad E_+ - E_- = \Delta H^\circ \ . \qquad (118)$$

Equation (118) shows that the difference in the activation energies must equal the standard enthalpy change for the appropriate overall reaction. Our link to thermodynamics, therefore, is quite powerful indeed.

REACTION MECHANISM: SIMPLE CHAIN REACTION

Previous sections have offered examples of reaction mechanisms. In this section, we will look more closely at the steps involved in relating the reaction mechanism to overall rate laws. The analysis is based on the simple, but very important, example of a chain reaction.

A simple chain reaction can be written as

$$A \xrightarrow{k_1} B \xrightarrow{k_2} P \ . \tag{119}$$

Based on reaction (119), the rate equation can be written for each individual species

$$\frac{dC_A}{dt} = -k_1 C_A \tag{120a}$$

$$\frac{dC_B}{dt} = k_1 C_A - k_2 C_B \tag{120b}$$

$$\frac{dC_P}{dt} = k_2 C_B \tag{120c}$$

Equation (120a) can be integrated and yields [see equations (29) - (31)]

$$C_A = C_A^\circ \, e^{-k_1 t} \ . \tag{121}$$

Inserting equation (121) into (120b)

$$\frac{dC_B}{dt} = k_1 C_A^\circ \, e^{-k_1 t} - k_2 C_B$$

or

$$\frac{dC_B}{dt} + k_2 C_B = k_1 C_A^\circ \, e^{-k_1 t} \ . \tag{122}$$

The easy way to solve these types of equations is by substitution. Let

$$C_B' \equiv C_B \, e^{k_2 t} \tag{123}$$

then

$$\frac{dC_B'}{dt} = e^{k_2 t} \left(\frac{dC_B}{dt} + k_2 C_B\right) \ .$$

Using equation (122) to replace the term in parentheses

$$\frac{dC'_B}{dt} = e^{k_2 t} \; k_1 C^\circ_A \; e^{-k_1 t}$$

and integration of both sides yields

$$\int_{C'_B(0)}^{C'_B} dC'_B = \int_0^t k_1 \; C^\circ_A \; e^{(k_2 - k_1)t} \; dt$$

$$C'_B - C'_B(0) = \frac{k_1 \; C^\circ_A}{k_2 - k_1} \left(e^{(k_2-k_1)t} - 1\right) .$$

From (123) $C_B = C'_B \; e^{-k_2 t}$; therefore,

$$C_B(t) = C^\circ_B \; e^{-k_2 t} + \frac{k_1 C^\circ_A}{k_2 - k_1} \left(e^{-k_1 t} - e^{-k_2 t}\right) . \qquad (124)$$

Finally, inserting equation (124) into the right side of equation (120c) and integrating

$$C_P(t) = C^\circ_P + C^\circ_B \left[1 - e^{-k_2 t}\right] + C^\circ_A \left[1 - e^{-k_1 t} - \frac{k_1}{k_2 - k_1} \left(e^{-k_1 t} - e^{-k_2 t}\right)\right] . \qquad (125)$$

Equations (121), (124), and (125) completely determine the kinetic evolution of our system.

Let us now analyze these equations. In what cases might one be safe in assuming steady state for the intermediate species B? If the reaction A → B is slow, while B → C is fast ($k_2 >> k_1$), we expect that B will reach a steady state. The reason is that as soon as B is produced from A (slow) it is consumed to make the product P (fast). Therefore, the concentration of B cannot build up and B will reach steady state. If $k_2 >> k_1$ and $C^\circ_B = 0$, then the denominator in equation (124), $k_2 - k_1$, reduces to k_2 and since $e^{-k_1 t} >> e^{-k_2 t}$ ($k_1 << k_2$) the overall expression reduces to

$$C_B(t) = \frac{C^\circ_A \; k_1}{k_2} \; e^{-k_1 t} .$$

But using equation (121) for $C_A(t)$ we can write this as

$$C_B(t) = \frac{k_1}{k_2} \; C_A(t) . \qquad (126)$$

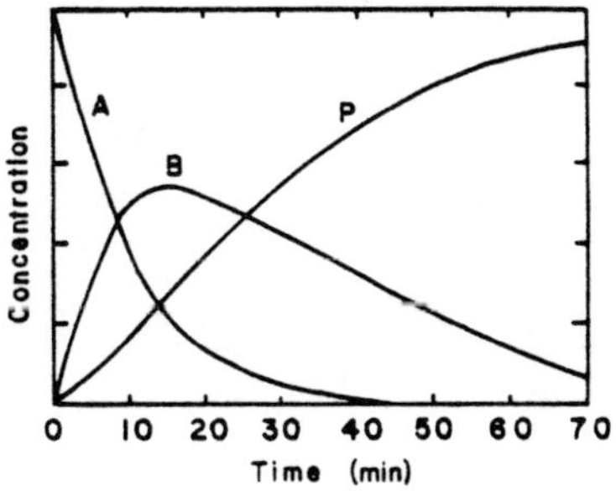

Figure 20. Concentration-time curves for substances A, B, and P.
$k_1 = 0.1\ min^{-1}$ and $k_2 = 0.05\ min^{-1}$.

Expression (126) is precisely what would be obtained if the steady state assumption was applied to species B, $dC_B/dt = 0$. Hence, equation (126) justifies the steady state assumption when $k_2 >> k_1$. Furthermore, equation (124) predicts that if $k_2 >> k_1$ steady state will be achieved within a time span of $t = 1/k_2$. Thus, if $k_1 = 1\ (yr)^{-1}$ and $k_2 = 10\ (yr)^{-1}$, steady state will ultimately be closely achieved (10 >> 1) but only after about 0.1 year has elapsed.

On the other hand, note that if $k_1 >> k_2$, then equation (124) (for $C_B^\circ = 0$) reduces to

$$C_B(t) = +C_A^\circ\ e^{-k_2 t} = C_A(t)\ e^{(k_1-k_2)t} \qquad (127)$$

which is certainly *not* a steady state situation. In fact, equation (127) simply states that all of initial A becomes B and then decomposes according to a first-order rate equation. If $k_1 \sim k_2$, then the kinetic behavior is more complex (see Fig. 20).

We should mention an important clarification. As is obvious in equation (124), dC_B/dt is *never* actually zero except as $t \to \infty$. Then, what does steady state really mean? Steady state simply means that the magnitude of dC_B/dt is very small relative to either of the two terms dC_A/dt and dC_P/dt. Put in other words, the concentration of B is much less than that of A. Therefore, even though $C_A(t)$ varies with time and

$$C_B(t) = \frac{k_1}{k_2}\ C_A(t)$$

the variation of C_B with time, at steady state, is very small since $dC_B/dt = k_1/k_2(dC_A/dt) << dC_A/dt$ if $k_1/k_2 << 1$.

The type of behavior exhibited by the intermediate species B in Figure 20 is found often in the time evolution of rock-water systems.

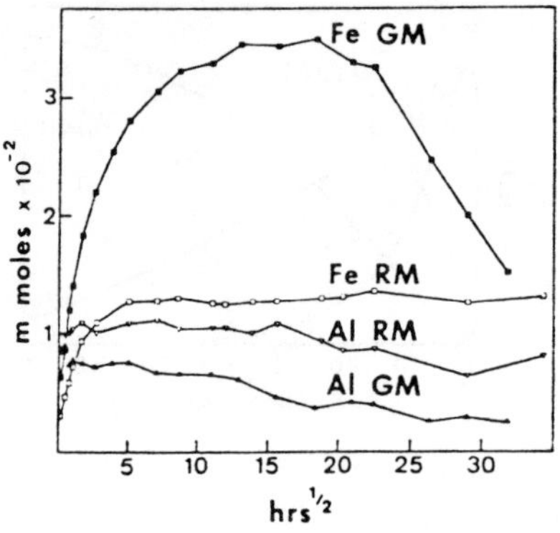

Figure 21(a). The concentration of Al and Fe in the filtrates plotted as a function of the square root of time during muscovite dissolution at 25°C, P_{CO_2} = 1 atm. Data from Lin and Clemency (1980).

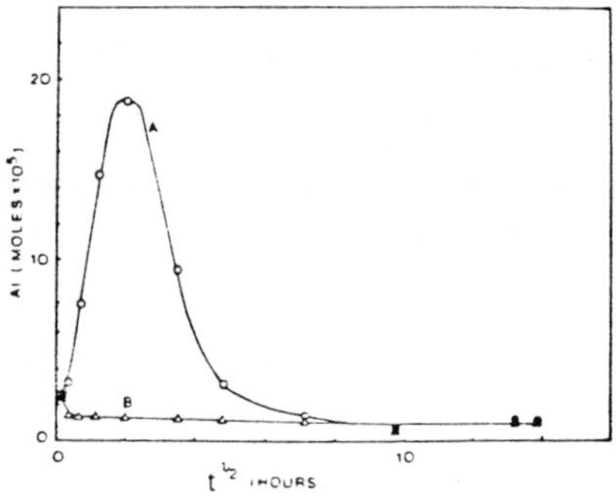

Figure 21(b). The analyzed concentration of Al as a function of time during feldspar dissolution at 25°C, pH = 4 = 5.5. Solutions (A) and (B) were filtered through 0.45 and 0.10 μm respectively. See Busenberg (1978) for details.

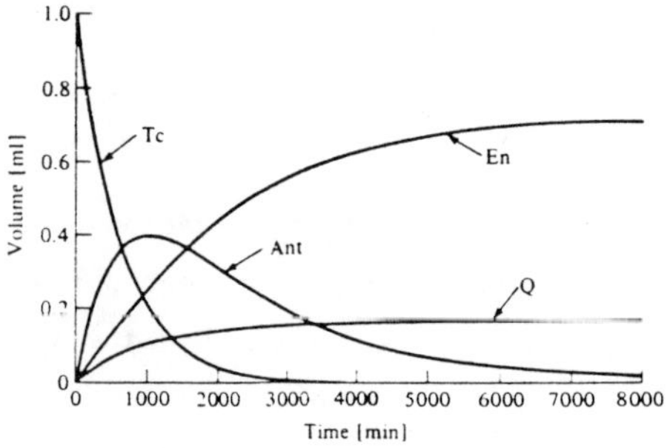

Figure 21(c). Volumes of talc, anthophyllite, enstatite, and quartz as functions of time at 830°C and 1000 bars, as calculated from measured rate constants. Note that although the stable assemblage is enstatite plus quartz, anthophyllite is the most abundant phase from 1000 to 1500 min. (From Mueller and Saxena, 1977.)

For example, Figure 21a shows this behavior for (Fe) during the dissolution of muscovite (Lin and Clemency, 1980) and Figure 21b exhibits similar results for Al^{3+} in feldspar dissolution (Busenberg, 1978).

Another example of the time evolution of the intermediate B species exhibited in Figure 20 comes from data on the $MgO-SiO_2-H_2O$ system by Greenwood (1963). Mueller and Saxena (1977) have analyzed his data from the point of view of equations (121), (124), and (125). In this case, the initial reactant is talc ($Mg_3Si_4O_{10}(OH)_2$), which is undergoing dehydration reactions. Ultimately, enstatite ($MgSiO_3$) is formed; however, the reaction proceeds as

$$Mg_3Si_4O_{10}(OH)_2 \rightarrow Mg_7Si_8O_{22}(OH)_2 \rightarrow MgSiO_3 \ .$$
talc anthophyllite enstatite

Therefore, anthophyllite is analogous to the B intermediate. Because the rates in the talc dehydration are similar ($k_1 \sim k_2$), the result is a complex evolution of the anthophyllite intermediate, which never reaches steady state as shown in Figure 21.

The stoichiometry of overall reactions was discussed earlier. Equations (121) and (125) as well as Figures 20 and 21 allow us to return to this important point. From these equations and assuming $C^\circ_B = 0$, the rate of production of products is:

$$\frac{dC_P}{dt} = k_1 \ C^\circ_A \ e^{-k_1 t} - \frac{k_1 \ C^\circ_A}{k_2 - k_1} (k_2 \ e^{-k_2 t} - k_1 \ e^{-k_1 t})$$

$$\frac{dC_P}{dt} = - \frac{dC_A}{dt} - \frac{k_1 \ C^\circ_A}{k_2 - k_1} (k_2 \ e^{-k_2 t} - k_1 \ e^{-k_1 t}) \ . \qquad (128)$$

If the *overall* reaction $A \rightarrow P$ were to occur stoichiometrically -- a central assumption in much of this chapter -- then for every A lost, one P must be formed, *i.e.*, $dC_P/dt = -dC_A/dt$. Equation (128), however, contains an additional term and, unless this term is made small, the kinetic stoichiometry of the overall reaction will not be followed. This extra term will become small relative to the first term in the case where $k_2 >> k_1$. Therefore, both the achievement of steady state and the conforming to stoichiometry occur only if $k_2 >> k_1$ and *only* when a sufficient time ($1/k_2$) has elapsed.

REACTION PROGRESS VARIABLE, MASS TRANSFER AND STOICHIOMETRY

In the previous section, the topic of stoichiometry in the kinetics of overall reactions has been discussed. For further clarification, this topic is taken up here within the context of water-rock interactions. The presence of an intermediate solid phase in the dissolution of a mineral is analogous to the B species in the previous section. This intermediate phase leads to *incongruent* dissolution, which is another terminology for the fact that the mineral does not dissolve in proportion to its stoichiometry. Recent papers (Helgeson, 1979) have employed Prigogine's (1954) concept of the reaction progress variable. In a reaction such as the dissolution of alkali feldspar:

$$KAlSi_3O_8(s) + 4H^+ + 4H_2O(\ell) \rightarrow K^+ + Al^{3+} + 3H_4SiO_4(aq) \qquad (129)$$

stoichiometric kinetics will require that

$$-\frac{dn_{KAlSi_3O_8}}{dt} = -\frac{1}{4}\frac{dn_{H^+}}{dt} = \frac{dn_{K^+}}{dt} = \frac{dn_{Al^{3+}}}{dt} = \frac{1}{3}\frac{dn_{H_4SiO_4}}{dt} \qquad (130)$$

where n_i stands for the total number of moles of species i. Equation (130) assumes a closed system with no other parallel reactions. If (130) is true, then the *progress variable* ξ can be defined unambiguously by $dn_i = \nu_i d\xi$, where ν_i is the stoichiometric coefficient of species i in the reaction [*e.g.*, $\nu_{H^+} = -4$ and $\nu_{K^+} = +1$ in (129)]. This simple concept can be applied to monitor the evolution of each component in an aqueous system if the correct stoichiometric equations are used (Helgeson, 1979). Note that in this treatment, nothing is said about the *rate* of the reactions but only that an overall reaction does occur in the manner written.

The application of the concept of progress variable to mineral-rock interactions is usually supplemented by certain assumptions of equilibrium. For example, as the data in equation (60) show, aqueous reactions between ions are normally very fast. Therefore, equilibrium can be assumed between certain aqueous species. For example, in the process of feldspar dissolution, the Al^{3+} leached into solution can itself be hydrolyzed:

$$Al^{3+} + n\ OH^- \rightleftarrows Al(OH)_n^{3-n}\ . \qquad (131)$$

If equilibrium is achieved, then

$$\frac{a_{Al(OH)_n^{3-n}}}{a_{Al^{3+}}\, a_{OH^-}^n} = K_n'\ . \qquad (132)$$

Incorporating the activity coefficients into an effective equilibrium constant, K_n,

$$\frac{m_{Al(OH)_n^{3-n}}}{m_{Al^{3+}}\, m_{OH^-}^n} = K_n\ . \qquad (133)$$

Therefore, the total aluminum concentration in solution is given by

$$m^T_{Al} = m_{Al^{3+}} + m_{Al(OH)^{2+}} + m_{Al(OH)_2^+} + m_{Al(OH)_3^0} + m_{Al(OH)_4^-}$$

$$= m_{Al^{3+}} (1 + K_1 m_{OH^-} + K_2 m^2_{OH^-} + K_3 m^3_{OH^-} + K_4 m^4_{OH^-}) \tag{134}$$

Of course, equation (134) merely states that if the total amount of aluminum dumped into solution from the feldspar is known, the speciation can be computed, once the pH (hence m_{OH^-}) is known.

The progress variable can also be extended to include precipitation of new minerals. As the aluminum concentration increases in the feldspar dissolution and the pH increases (since H^+ is used up), the first reaction is the hydrolysis of Al^{3+}. But soon the Al^{3+} and OH^- molalities reach high enough values that the solution becomes supersaturated with respect to gibbsite. Now a new reaction occurs:

$$Al^{3+} + 3OH^- \rightarrow \underset{\text{gibbsite}}{Al(OH)_3} . \tag{135}$$

At this point some kinetic input is needed. A simple approach would assume that the gibbsite precipitates *and* dissolves fast enough relative to the rate of dissolution of feldspar that equilibrium is reached, *i.e.*,

$$a_{Al^{3+}} a^3_{OH^-} = K_{gibbsite} . \tag{136}$$

Equation (136) places severe constraints on the variation of Al^{3+} and OH^- in solution. Taking the logarithm of both sides

$$\log a_{Al^{3+}} + 3 \log a_{OH^-} = \log K_{gibbsite} .$$

Differentiating the previous equation with respect to time and assuming that the activity coefficients do not vary with time:

$$\frac{1}{m_{Al^{3+}}} \frac{dm_{Al^{3+}}}{dt} + \frac{3}{m_{OH^-}} \frac{dm_{OH^-}}{dt} = 0 . \tag{137}$$

If all reactions are assumed to be in equilibrium except for the dissolution of feldspar (129), then the entire process of dissolution of $KAlSi_3O_8$ and precipitation of gibbsite can be described by the progress variable, ξ, for reaction (129). Therefore, equation (137) may be written as

$$\frac{1}{m_{Al^{3+}}} \frac{dm_{Al^{3+}}}{d\xi} + \frac{3}{m_{OH^-}} \frac{dm_{OH^-}}{d\xi} = 0 \tag{138}$$

where $d\xi/dt$ is the rate of dissolution reaction. Therefore, the variations in $m_{Al^{3+}}$ and m_{OH^-} with the progress of the dissolution reaction are not independent. These types of equations can then be used to evaluate dm_i for each species once $d\xi$ is varied. For further details of the procedure the reader is referred to Helgeson (1979).

The main point to make here about mass transfer is that only certain reactions are considered to be going irreversibly (for which the progress variable is used) and equilibrium is assumed for the rest. Therefore, one is left with basically an *equilibrium* calculation problem, each time an irreversible reaction is allowed to proceed to a small extent. The techniques of matrix algebra are employed (see Helgeson, 1979) to simplify the equilibrium calculation. In this context, the generalization of these calculations would be to actually incorporate a rate law for each reaction. To illustrate, assume that Al^{3+} does not undergo hydrolysis in solution. Then the reactions discussed above can all be included in the following scheme:

$$\underset{\text{feldspar}}{KAlSi_3O_8} + 4\,H^+ + 4\,H_2O \underset{k_{-1}}{\overset{k_1}{\rightleftharpoons}} Al^{3+} + 3\,H_4SiO_4 + K^+ \tag{139}$$

$$Al^{3+} + 3\,OH^- \underset{k_{-2}}{\overset{k_2}{\rightleftharpoons}} \underset{\text{gibbsite}}{Al(OH)_3} \tag{140}$$

$$H^+ + OH^- \underset{k_{-3}}{\overset{k_3}{\rightleftharpoons}} H_2O \ . \tag{141}$$

Note that the last reaction is an elementary reaction while the other reactions (heterogeneous) are overall reactions, as discussed earlier. The rate law of (141) is completely straightforward; the other rate laws must be obtained from experiment using the techniques of this chapter. The rate laws for (139) and (140) will be some functions, f_i, of the areas of the solids and the molalities of the aqueous species. Therefore, leaving the rate laws (*e.g.*, the terms in equation (2) as general functions, f_i, we can write an expression for the rate of change of the concentration of each aqueous species:

$$\frac{dm_{Al^{3+}}}{dt} = k_1 f_1 - k_{-1} f_{-1} - k_2 f_2 + k_{-2} f_{-2}$$

$$\frac{dm_{K^+}}{dt} = k_1 f_1 - k_{-1} f_{-1}$$

$$\frac{dm_{H_4SiO_4}}{dt} = 3k_1 f_1 - 3k_{-1} f_{-1}$$

$$\frac{dm_{H^+}}{dt} = -4k_1 f_1 + 4k_{-1} f_{-1} - k_3\, m_{H^+}\, m_{OH^-} + k_{-3}$$

$$\frac{dm_{OH^-}}{dt} = -3k_2 f_2 + 3k_{-2} f_{-2} - k_3\, m_{H^+}\, m_{OH^-} + k_{-3}$$

$$\frac{1}{V} \frac{dn_{KAlSi_3O_8}}{dt} = -k_1 f_1 + k_{-1} f_{-1}$$

$$\frac{1}{V} \frac{dn_{Al(OH)_3}}{dt} = k_2 f_2 - k_{-2} f_{-2}$$

where V is the volume of solution (assumed fixed). These equations can be integrated numerically once the initial conditions are known and once the functions f_1, f_{-1}, f_2, f_{-2} are known. Note that the heterogeneous reactions (f_1, f_{-1}, f_2, f_{-2}) will depend on area (and hence on the number of moles of solids, n_{solids}). For example, f_1 may be given by

$$f_1 = \frac{A_{feldspar}}{V} m_{H^+} .$$

In this approach, high values for the rate constants k_3 and k_{-3} would guarantee that

$$m_{H^+}\, m_{OH^-} = K_w$$

(K_w being the effective equilibrium constant), as long as

$$\frac{k_{-3}}{k_3} = K_w .$$

Similar equilibria could be imposed by high rate constants. Note also that for Al^{3+} to be in equilibrium with gibbsite in (140), the rate of precipitation of gibbsite must be faster than the rate of dissolution of feldspar. If this is not the case, then Al^{3+} will reach a steady state value (as in Fig. 1), which is not the equilibrium value.

More work should be done in this direction to incorporate the mass transfer ideas with new geochemical data on the kinetics of dissolution and precipitation.

CONCLUSION

The treatment of geochemical kinetics builds upon the powerful thermodynamic structure that has been developed during the last decades. In addressing geochemical phenomena from the kinetic point of view, however, a deeper understanding is obtained of the mechanisms and molecular pathways by which the reactions are carried out. A geochemist envisages the elemental distribution in the earth as the result of a series of transformations. These transformations involve sedimentary, metamorphic, and igneous processes and lead to the concept of geochemical cycles. As a particular mineral assemblage undergoes changes in pressure and/or temperature, disequilibrium will drive certain chemical reactions. Changes in the fluid content of the system will also drive other reactions. Whether the assemblage ever reaches equilibrium and also the path by which equilibrium is achieved depend on the kinetics. This chapter has related the rate of change of the concentration of species participating in chemical reactions to the state of the system, *i.e.*, the pressure, the temperature, and the bulk composition. Many of the concepts discussed, *e.g.*, steady state, overall reactions, reactive intermediates, activation energies,adsorption, detailed balancing, mass transfer, are essential to the proper understanding of non-equilibrium in the earth. As more careful and quantitative studies are carried out, one conclusion, which results, is that many natural systems are *not* in equilibrium. Therefore, nature has left a record of the kinetic (dynamic) history in the chemistry of the various mineral assemblages. It is the hope of the author that much more use is made of the concepts outlined here in unraveling the geochemical kinetics.

ACKNOWLEDGEMENTS

This chapter has profited greatly from the comments of Professor H. L. Barnes and D. M. Kerrick, who reviewed the manuscript. Many thanks are also due to Greg E. Muncill for technical assistance. Support from NSF Grant OCE-7909240 is gratefully acknowledged.

Adamson, A. W. (1976) *Physical Chemistry of Surfaces*, John Wiley & Sons, New York, 698 p.

Bada, J. L. (1972) Kinetics of racemization of amino acids as a function of pH. J. Amer. Chem. Soc., 94, 1371-1373.

Benson, S. W. (1960) *The Foundations of Chemical Kinetics*, McGraw-Hill Co., New York.

________ (1976) *Thermochemical Kinetics*, 2nd ed., John Wiley & Sons, Inc., New York, 320 p.

Berner, R. A. (1971) *Principles of Chemical Sedimentology*, McGraw-Hill, New York, 240 p.

________ (1980) *Early Diagenesis: A Theoretical Approach*, Princeton University Press, Princeton, N. J., 241 p.

________, Sjoberg, E. L., and Schott, J. (1980) Mechanism of pyroxene and amphibole weathering: I. Experimental studies. In *Third International Symposium on Water-Rock Interaction*, Proceedings, p. 44-45.

Boyd, Robert K. (1978) Some common oversimplifications in teaching chemical kinetics. J. Chem. Educ., 55, 84-89.

Busenberg, E. (1978) The products of the interaction of feldspars with aqueous solutions at 25°C. Geochim. Cosmochim. Acta, 42, 1679-1686.

Cotton, F. A. and Wilkinson, G. (1967) *Advanced Inorganic Chemistry*, 2nd ed., Interscience Publ., New York, 1136 p.

Darken, L. S. and Turkdogan, E. T. (1970) Adsorption and kinetics at elevated temperatures. In *Heterogeneous Kinetics at Elevated Temperatures*, Belton, G. R. and Worrell, W. L. (eds.), p. 25-95.

Ernst, W. G. and Calvert, S. E. (1969) An experimental study of the recrystallization of porcelanite and its bearing on the origin of some bedded cherts. Amer. J. Sci., 267-A, 114-133.

Eyring, H., Colburg, C. B. and Zivolinski, B. J. (1950) The activated complex in chemisorption and catalysts, in heterogeneous catalysis. Discuss. Faraday Soc., No. 8, 39-46.

Fung, P. C. and Samipelli, G. G. (1980) Repeated leaching of microcline at 25°C. In *Third International Symposium on Water-Rock Interaction*, Proceedings, Int. Assoc. Geochem. Cosmochem., p. 67-68.

Garrels, R. M. (1957) Some free energy values from geologic relations. Amer. Mineral., 42, 780-791.

Giggenbach, W. F. (1974) Kinetics of the polysulfide-thiosulfate disproportion up to 240°C. Inorg. Chem., 13, 1730-1733.

Goldhaber, M. B. and Kaplan, I. R. (1974) The sulfur cycle. In *The Sea*, Vol. 5, Goldberg (ed.), p. 569-655.

Grandstaff, D. E. (1977) Some kinetics of bronzite-orthopyroxene dissolution. Geochim. Cosmochim. Acta, 41, 1097-1103.

________ (1980) The dissolution rate of forsteritic olivine from Hawaiian beach sand. In *Third International Symposium on Water-Rock Interaction*, p. 72-74.

Halsey, G. D. (1949) Catalysis on non-uniform surfaces. J. Chem. Phys., 17, 758-761.

Helgeson, H. C. (1971) Kinetics of mass transfer among silicates and aqueous solutions. Geochim. Cosmochim. Acta, 35, 421-469.

________ (1979) Mass transfer among minerals and hydrothermal solutions. In *Geochemistry of Hydrothermal Ore Deposits*, Barnes, H. L. (ed.), p. 568-610.

Hoffman, M. R. (1977) Kinetics and mechanism of oxidation of hydrogen sulfide by hydrogen peroxide in acidic solution. Env. Sci. Tech., 11, 61-66.

Hurd, D. C. (1972) Factors affecting solution rate of biogenic opal in seawater. Earth Planet. Sci. Lett., 15, 411-417.

________ (1973) Interactions of biogenic opal. Sediment and seawater in the central equatorial Pacific. Geochim. Cosmochim. Acta, 37, 2257-2282.

Iler, R. K. (1979) *The Chemistry of Silica*, John Wiley & Sons, New York, 866 p.

Jones, P., Haggett, M. L. and Longridge, J. L. (1964) The hydration of carbon dioxide. J. Chem. Educ., 41, 610-612.

Katz, A. and Matthews, A. (1977) The dolomitization of $CaCO_3$: an experimental study at 252-295°C. Geochim. Cosmochim. Acta, 41, 297-308.

Lin, F. and Clemency, C. (1980) The kinetics of dissolution of muscovites at 25°C and 1 atm CO_2 partial pressure. In *Third International Symposium on Water-Rock Interaction*, p. 44-47.

Matthews, A. (1980) Influences of kinetics and mechanism in metamorphism: a study of albite crystallization. Geochim. Cosmochim. Acta, 44, 387-402.

Moelwyn-Hughes, E. A. (1971) *The Chemical Statics and Kinetics of Solutions,* Academic Press, New York, 540 p.

Moore, W. J. (1964) *Physical Chemistry,* Prentice-Hall, Englewood Cliffs, N. J., 844 p.

Morse, J. W. (1978) Dissolution kinetics of calcium carbonate in sea water: VI. The near-equilibrium dissolution kinetics of calcium carbonate-rich deep sea sediments. Amer. Jour. Sci., 278, 344-353.

——— and Berner, R. A. (1972) Dissolution kinetics of calcium carbonate in sea water: II A kinetic origin for the lysocline. Amer. Jour. Sci., 272, 840-851.

Mueller, R. F. and Saxena, S. K. (1977) *Chemical Petrology,* Springer-Verlag, New York, 394 p.

Ohmoto, H. and Lasaga, A. C. (1981) Kinetics of isotopic and chemical reactions between sulfates and sulfides in hydrothermal systems (submitted to Geochim. Cosmochim. Acta).

Petrovic, R., Berner, R. A. and Goldhaber, M. B. (1976) Rate control in dissolution of alkali feldspars--I. Study of residual feldspar grains by x-ray photoelectron spectroscopy. Geochim. Cosmochim. Acta, 40, 537-548.

Ponec, V., Knor, Z. and Cerny, S. (1974) *Adsorption on Solids,* CRC Press, London, 693 p.

Prigogine, I. and Defay, R. (1954) *Chemical Thermodynamics.* (English translation by Everett) Longmans, New York, 460 p.

Rickard, D. T. (1975) Kinetics and mechanism of pyrite formation at low temperatures. Amer. J. Sci., 275, 636-652.

Rimstidt, J. D. and Barnes, H. L. (1980) The kinetics of silica-water reactions. Geochim. Cosmochim. Acta, 44, 1683-1699.

Sjöberg, E. L. (1976) A fundamental equation for calcite dissolution kinetics. Geochim. Cosmochim. Acta, 40, 441-447.

Sykes, A. G. (1966) *Kinetics of Inorganic Reactions,* Pergamon Press, Oxford, 310 p.

Wagner, C. (1970) The concept of the thermodynamic activity of atomic species in heterogeneous kinetics. In *Heterogeneous Kinetics at Elevated Temperatures,* Belton, G. R. and Worrell, W. L. (eds.), p. 101-112.

Weston, R. E. and Schwartz, H. H. (1972) *Chemical Kinetics,* Prentice-Hall, Englewood Cliffs, N. J., 274 p.

Wollast, R. (1974) The silica problem. In *The Sea,* vol. 5, Goldberg (ed.), p. 359-392.

Chapter 2

DYNAMIC TREATMENT of GEOCHEMICAL CYCLES: GLOBAL KINETICS

Antonio C. Lasaga

INTRODUCTION

The earth, whether viewed globally or on a local scale, is certainly a dynamic and evolving chemical system. The latest developments in plate tectonics have strengthened further this dynamic panorama. If a geochemist is interested in describing the nature of the distribution of elements in the earth, it is only appropriate to do so within a general kinetic framework. Up to now, the kinetic treatment of geochemical cycles has been mostly limited to the use of simple concepts such as residence times or response times. However, man's strong perturbations of the geological and biological systems in the present-day earth has raised fundamental questions concerning our understanding of the basic processes in geochemical cycles. Therefore, it is essential to analyze further the stability of a geochemical cycle and its dynamical response to perturbations. The recent work on the so-called "dissipative structures" by Prigogine (1980) and coworkers has had enormous influence on the understanding of complex structures and their reorganization. These studies have found their way not only into physical chemistry and biochemistry but also into social sciences and even the study of city planning. Geochemical cycles are likely candidates to benefit precisely from the study of complex structures. In particular, the consequences of positive and negative feedback processes (*e.g.*, see Broecker, 1971) on the various elemental distributions in the atmosphere, crust, lakes, rivers, and oceans must be much more seriously and quantitatively pursued.

When faced with obtaining solutions to the questions posed above, one finds the available data mostly in terms of reservoir contents for particular species and known fluxes between the reservoirs for each of the elemental cycles (Holland, 1978). The extraction of kinetic inferences from the available data becomes a fundamental problem both in (a) deciphering the functional dependence of the observed fluxes on geochemical parameters and (b) developing a framework within which to scientifically address the consequences of the kinetic models. The generalization of the kinetic treatment of cycles is most profitably achieved when we realize that geochemical cycles are conceptually similar to complex chemical kinetic reactions. In a sense, our treatment of cycles can therefore be termed "Global Kinetics."

This chapter will begin with the results of so-called linear cycles; then the introduction of non-linear coupling between the cycles will be pursued including special attention to the possibility of multiple steady states, as well as the effects on stability.

LINEAR CYCLES

The most frequent kinetic assumption is that of first-order kinetics. Southam and Hay (1976) initiated the analysis of the linear geochemical cycle. The assumption is essentially that the flux of an element out of a reservoir is linearly and directly proportional to the concentration or the content of that element in the reservoir. In many instances, this is a reasonable approximation provided no major changes occur in the reservoir contents of the other elements. For example, other things being equal the rate of photosynthesis is found to depend linearly on CO_2 content (Garrels and Perry, 1974; Lasaga, 1980). The flux of sulfate from oceans to sediments is probably linearly dependent on the amount of sulfate in the oceans (Lasaga and Holland, 1976). Furthermore, an increase in the NaCl content of the oceans probably produces a nearly proportional increase in the rate of removal of NaCl by evaporite formation, given the same amount of evaporation to dryness in evaporite basins (see also Mackenzie, 1975; Lerman *et al.*, 1975; Pytkowicz, 1975).

The simplest case of a linear cycle consists of one reservoir with inputs and outputs, such as shown in Figure 1. If the input rate of a given component into a reservoir is constant, for example, a, and the output rate is proportional to the content of the component in the reservoir, n, then the rate of change of the amount of the component in the reservoir obeys the simple equation:

$$\frac{dn}{dt} = a - kn \; . \tag{1}$$

If at t = 0, n has the value n_0, then the solution to equation (1) becomes:

$$n = \frac{a}{k} - \left(\frac{a}{k} - n_0\right) e^{-kt} \; . \tag{2}$$

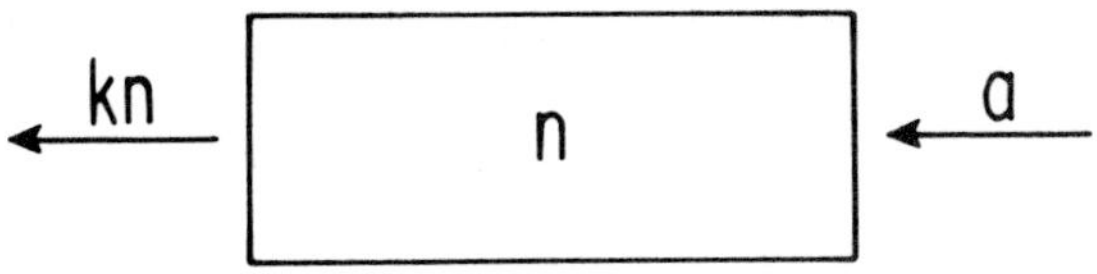

Figure 1. Simple geochemical "cycle" with only one reservoir. The output is assumed to obey first-order (linear) kinetics. n is the amount of the element in the reservoir (see text).

The initial amount, n_0, is altered to the steady state value, a/k, by the decay of the second term in equation (2), i.e., by the decay of the exponential term, exp(-kt). The time required to substantially decay the size of the second term in equation (2) is simply 1/k. Therefore, the *response time* of the system is given by 1/k. The *residence time* at steady state, on the other hand, is defined as:

$$\tau_{res} \equiv \frac{n^{steady}}{(dn/dt)_{input}} = \frac{n^{steady}}{(dn/dt)_{output}} . \tag{3}$$

The steady state limit of equation (2) is n = a/k. It is important to note that in the simple linear case [equation (1)], there is only one steady state, i.e., only one solution to the equation dn/dt = 0. The residence time defined by equation (3) becomes:

$$\tau_{res} = \frac{(a/k)}{a} = \frac{1}{k} .$$

Hence, the response time and the residence time are equal to each other in this simple case, as noted by Holland (1978). This simple example has provided a kinetic interpretation to the residence time. The interpretation is not always so simple as we shall see. Let us, however, look at another instructive variant of the one-reservoir cycle. It often happens that the source of an element has periodic variations (*e.g.*, the seasonal variations in photosynthesis). In this case, the input into the reservoir might be represented by

$$\left.\frac{dn}{dt}\right|_{input} = a + b \sin(wt) . \tag{4}$$

The overall equation for the rate of change in the amount of the element in the reservoir is $\frac{dn}{dt} = a + b \sin(wt) - kn$. If at t = 0, n has the value of n_0, then the solution is (Holland, 1978)

$$n(t) = \left(n_0 - \frac{a}{k} + \frac{bw}{k^2+w^2}\right)e^{-kt} + \frac{a}{k} + \frac{b}{k^2+w^2}\left(k \sin(wt) - w \cos(wt)\right) . \tag{5}$$

For long times ($t \gg \frac{1}{k}$) equation (5) reduces to

$$n(t) = \frac{a}{k} + \frac{b}{k^2 + w^2}\left\{k \sin(wt) - w \cos(wt)\right\} .$$

This latter can be rewritten as

$$n(t) = \frac{a}{k} + \frac{b}{(k^2 + w^2)^{1/2}} \sin(wt - \delta)$$

where $$\delta \equiv \cos^{-1}\left[\frac{k}{(k^2 + w^2)^{1/2}}\right] \quad (0 \leq \delta \leq \pi/2) .$$

Hence, the "steady state" is really a fluctuating function with a *mean* value of $\frac{a}{k}$, as in the simple case, and a *phase*, δ, which is different from the driving function, equation (4).

It is instructive to analyze two important limits of equation (5). If $w \to 0$, equation (5) reduces to equation (2). This is not surprising since $w \to 0$ implies that the b sin(wt) term is negligible (over the time period of interest) compared to the "a" term. In other words, this limit simply removes the fluctuations of the input. However, if $w \to \infty$, or more precisely if w is much greater than k or b/n_0, equation (5) also reduces to equation (2). This is an important result. Basically it states that, because the response of the cycle is so sluggish, the high frequency input oscillations are averaged out completely and the cycle is unaffected by them.

Real cycles (apart from being partly non-linear) consist of many reservoirs topologically connected by inter-reservoir fluxes. To understand the response of a general cycle to a perturbation and thereby analyze its stability, we must investigate the *collective* behavior of the reservoirs. As so often happens in nature, the collective behavior can be quite different from that of the individual components (a common example being the cooperativity of order-disorder phase transitions). If the outputs from each reservoir depend linearly on the reservoir contents, the time evolution equations become

$$\frac{dA_i}{dt} = \sum_{j=1}^{n} k_{ji} A_j - A_i \sum_{j=1}^{n} k_{ij} \qquad i = 1, \ldots, n \tag{6}$$

where A_i is the content of a particular element in reservoir i and n is the total number of reservoirs. k_{ij} are the linear rate constants and the flux of the element from reservoir i to reservoir j is given by $k_{ij} A_i$.

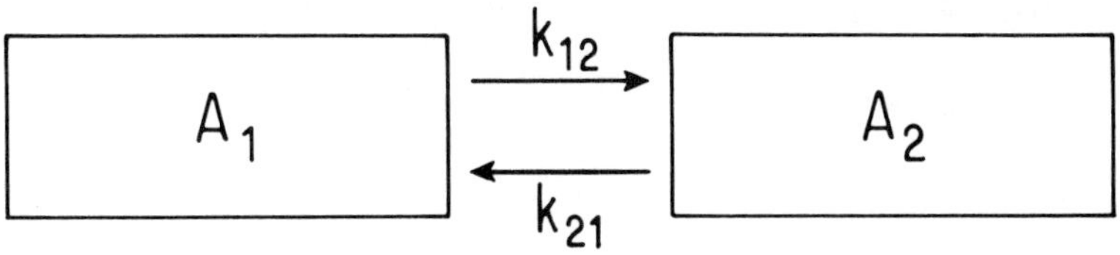

Figure 2. A two-reservoir geochemical cycle. The rate constants k_{12} and k_{21} are first-order rate constants, defined in the text. A_1 and A_2 are the amounts of the element in reservoirs 1 and 2 respectively.

The notation of Lasaga (1980) has been kept for continuity. The general solution of equation (6) and some important general conclusions are detailed in Lasaga (1980). Here, I will illustrate the concepts in Lasaga (1980) with the next simplest cycle, a two-reservoir cycle, shown in Figure 2. The advantage of this example is that most of the equations in Lasaga (1980) can be chekced out in a more direct way (i.e., without recourse to computers). The evolution of the cycle in Figure 2 depends on the following rate equations:

$$\frac{dA_1}{dt} = -k_{12} A_1 + k_{21} A_2 \tag{7a}$$

$$\frac{dA_2}{dt} = k_{12} A_1 - k_{21} A_2 \ . \tag{7b}$$

These equations can be rewritten in matrix form

$$\frac{d}{dt}\begin{pmatrix} A_1 \\ A_2 \end{pmatrix} = \begin{bmatrix} -k_{12} & k_{21} \\ k_{12} & -k_{21} \end{bmatrix} \begin{pmatrix} A_1 \\ A_2 \end{pmatrix} \ . \tag{8}$$

The use of the basic notions of matrix algebra is becoming more and more widespread in the earth sciences. A brief review of the elementary aspects is given in the Appendix. As was shown in Lasaga (1980), the solution to equation (8) has the form

$$\vec{A}(t) = a_1 e^{E_1 t} \vec{\Psi}_1 + a_2 e^{E_2 t} \vec{\Psi}_2 \tag{9}$$

where E_1, E_2 are the eigenvalues and $\vec{\Psi}_1$, $\vec{\Psi}_2$ are the eigenvectors of the matrix K defined by

$$K \begin{bmatrix} -k_{12} & k_{21} \\ k_{12} & -k_{21} \end{bmatrix} \ . \tag{10}$$

To obtain the eigenvalues of K we solve the determinantal equation

$$\begin{vmatrix} -k_{12}-E & k_{21} \\ k_{12} & -k_{21}-E \end{vmatrix} = 0 \quad \text{or} \quad (k_{12} + E)(k_{21} + E) - k_{12}k_{21} = 0 \ .$$

The solution to this quadratic in E is

$$E_1 = 0 \qquad E_2 = -(k_{12} + k_{21}) \ . \tag{11}$$

To obtain the eigenvectors we must solve the equation $K\ \vec{\Psi}_i = E_i\ \vec{\Psi}_i$ or

$$\begin{bmatrix} -k_{12} & k_{21} \\ k_{12} & -k_{21} \end{bmatrix} \begin{pmatrix} a \\ b \end{pmatrix} = E \begin{pmatrix} a \\ b \end{pmatrix}$$

If E = 0, then

$$\begin{bmatrix} -k_{12} & k_{21} \\ k_{12} & -k_{21} \end{bmatrix} \begin{pmatrix} a \\ b \end{pmatrix} = 0$$

or

$$-k_{12}\ a + k_{21}\ b = 0; \ k_{12}\ a - k_{21}\ b = 0 \ .$$

Either equation yields

$$b = \frac{k_{12}}{k_{21}}\ a \ .$$

Since a can be arbitrary, we can set a = 1:

$$\vec{\Psi}_1 = \begin{pmatrix} 1 \\ \dfrac{k_{12}}{k_{21}} \end{pmatrix}$$

Equation (9) can be used with any eigenvector; in other words, multiplying $\vec{\Psi}_1$ by any constant can easily be corrected by dividing a_1 by the same constant. Hence, whether $\vec{\Psi}_1 = (1,\ k_{12}/k_{21})$ or $(2,\ 2k_{12}/k_{21})$, *etc.*, is arbitrary. The important requirement is that $\vec{\Psi}_1$ be an eigenvector with eigenvalue, E_1 (see Appendix). Likewise, to find $\vec{\Psi}_2$ we must solve for $K\ \vec{\Psi}_2 = E_2\ \vec{\Psi}_2$:

$$\begin{bmatrix} -k_{12} & k_{21} \\ k_{12} & -k_{21} \end{bmatrix} \begin{pmatrix} a \\ b \end{pmatrix} = -(k_{12} + k_{21}) \begin{pmatrix} a \\ b \end{pmatrix}$$

or $-k_{12}\ a + k_{21}\ b = -(k_{12} + k_{21})a \ ; \quad k_{12}\ a - k_{21}\ b = -(k_{12} + k_{21})b \ .$

These equations reduce to a = -b. Hence, setting a = 1 once more,

$$\vec{\Psi}_2 = \begin{pmatrix} 1 \\ -1 \end{pmatrix} .$$

Having obtained the eigenvalues and eigenvectors of K, the general solution, equation (9) becomes

$$\vec{A}(t) = \begin{pmatrix} A_1(t) \\ A_2(t) \end{pmatrix} = a_1\, e^{0\cdot t} \begin{pmatrix} 1 \\ \frac{k_{12}}{k_{21}} \end{pmatrix} + a_2\, e^{-(k_{12}+k_{21})t} \begin{pmatrix} 1 \\ -1 \end{pmatrix} \qquad (12)$$

To obtain the coefficients a_1 and a_2 we need the initial condition of our cycle, i.e., A_1^o and A_2^o. From equation (12) it follows that, setting t = 0,

$$A_1^o = a_1 + a_2 \;; \qquad A_2^o = \frac{k_{12}}{k_{21}} a_1 - a_2 . \qquad (13a)$$

Using the notation of Lasaga (1980), equation (13) can also be written as:

$$\vec{A}^o = \underline{\Psi}\, \vec{a} . \qquad (13b)$$

Where the $\underline{\Psi}$ matrix is comprised of the eigenvectors $\vec{\Psi}_1$ and $\vec{\Psi}_2$:

$$\Psi = \begin{bmatrix} 1 & 1 \\ \frac{k_{12}}{k_{21}} & -1 \end{bmatrix} . \qquad (13c)$$

The solution to equation (13) is

$$a_1 = \frac{k_{21}(A_1^o + A_2^o)}{k_{12} + k_{21}} \;; \qquad a_2 = \frac{k_{12}A_1^o - k_{21}A_2^o}{k_{12} + k_{21}} \qquad (14a)$$

or, once more using the notation in Lasaga (1980):

$$\vec{a} = \underline{\Psi}^{-1}\, \vec{A}^o \qquad (14b)$$

where the inverse of the matrix in (13c) is

$$\underline{\Psi}^{-1} = \frac{1}{k_{12} + k_{21}} \begin{bmatrix} k_{21} & k_{21} \\ k_{12} & -k_{21} \end{bmatrix} \qquad (14c)$$

as can be checked directly in this simple case (see Appendix). Combining equations (12) and (14a) yields the final equation:

$$A_1(t) = \frac{k_{21}\,(A_1^o + A_2^o)}{k_{12} + k_{21}} + \frac{k_{12}\,A_1^o - k_{21}\,A_2^o}{k_{12} + k_{21}}\, e^{-(k_{12}+k_{21})t} \qquad (15a)$$

$$A_2(t) = \frac{k_{12}\,(A_1^o + A_2^o)}{k_{12} + k_{21}} - \frac{k_{12}\,A_1^o - k_{21}\,A_2^o}{k_{12} + k_{21}}\, e^{-(k_{12}+k_{21})t} \; . \qquad (15b)$$

We should now note some important results from even this simple cycle. First, there is only *one* eigenvector with eigenvector E = 0. This means that the cycle will always return to a *unique* steady state. Lasaga (1980) has shown that this property is a general property of *all* linear cycles.

The non-zero eigenvalue of the two-reservoir cycle is $-(k_{12} + k_{21})$. That this is a *negative* eigenvalue is required if the cycle is to be *stable*. Thus the transient term in equation (15) will decay given sufficient time. In fact, the time needed to essentially remove a perturbation and return the system to steady state can be obtained from inspection of the exponential in the second term in equation (15). This time, the *response time* of the cycle is

$$\tau_{\text{response}} = \frac{1}{k_{12} + k_{21}} \; . \qquad (16)$$

To be precise, this is the time needed to reduce the perturbation to 1/e of the original size, a concept widely used in kinetics (see Chapter 1). Equation (16) begins to point out the difference between a simple one-reservoir cycle and multi-reservoir cycles. The *residence times* of our element (at steady state) in reservoirs one and two are given by

$$\tau^{(1)}_{\text{residence}} = \frac{A_1^{\text{steady}}}{(dA_1/dt)_{\text{input}}} = \frac{A_1^{\text{steady}}}{(dA_1/dt)_{\text{output}}} = \frac{A_1^{\text{steady}}}{k_{12}\,A_1^{\text{steady}}}$$

$$\tau^{(1)}_{\text{residence}} = \frac{1}{k_{12}} \qquad (17a)$$

Similarly,

$$\tau^{(2)}_{\text{residence}} = \frac{1}{k_{21}} \; . \qquad (17b)$$

Comparing equations (16) and (17) it is apparent that, unlike the result of the one-reservoir cycle mentioned earlier, the response and residence times of the cycle do *not* coincide. In fact, the response of

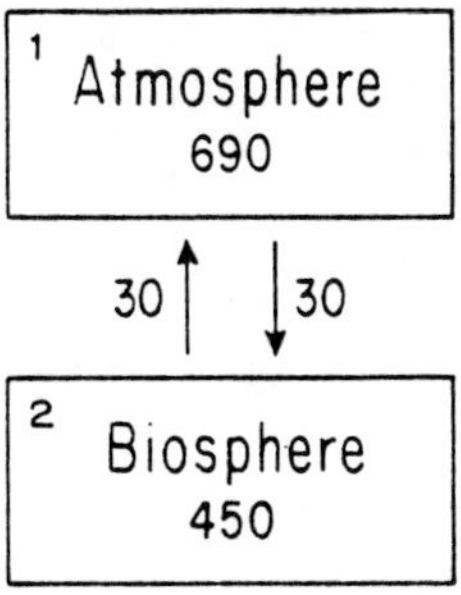

Figure 3. A simplified two-reservoir version of the short-term carbon cycle (see also Fig. 4). The reservoir contents are in units of 10^{15}gC and the fluxes in units of 10^{15}gC/yr. See text for details.

the cycle is *faster* than would be predicted from the individual residence time of the reservoirs. A faster response indicates that the *collective* behavior of the cycle has resulted in a greater stability. Note that if one cycle has a large residence time, the response time is then controlled by the cycle with the small residence time. This is a case of the kinetics being determined by the *fastest* rate (Chapter 1).

Let us apply these equations to a very simplified two-reservoir carbon cycle shown in Figure 3. The reservoir contents are in units of 10^{15} gC and the fluxes in units of 10^{15} gC/yr (Holland, 1978). The cycle in Figure 3 is assumed to be at steady state (i.e., fluxes in equal fluxes out). From the definition of the rate constants

$$k_{12} = \frac{30}{690} = 0.04348 \text{ yr}^{-1} \ ; \quad k_{21} = \frac{30}{450} = 0.06667 \text{ yr}^{-1} \ . \qquad (18)$$

The residence times are the inverse of the k's:

$$\tau^{(1)}_{\text{residence}} = 23 \text{ years} \ ; \quad \tau^{(2)}_{\text{residence}} = 15 \text{ years} \ .$$

However, the collective response time of the cycle is given by

$$\tau_{\text{response}} = \frac{1}{0.04348 + 0.0667} = 9.1 \text{ years} \ .$$

In this case the collective interaction substantially reduces the response time (from the residence times), since the k's are similar.

If, for example, man were to add 50 units (50×10^{15} gC) from the biosphere to the atmosphere (*e.g.*, via deforestation) the initial condition, at the onset of the perturbation, would be

$$A^o_1 = 740 \ ; \quad A^o_2 = 400 \quad (\text{in } 10^{15} \text{ gC units}) \ .$$

Using the rate constants in (18) and equations (15) yields:

$$A_1 = 690 + 50 \ e^{-0.11015 \ t} ; \quad A_2 = 450 - 50 \ e^{-0.11015 \ t} \quad (t \text{ in years}).$$

The atmospheric carbon would have decayed to 707×10^{15} gC in 10 years, during which the biosphere would have increased to 433×10^{15} gC. In 20

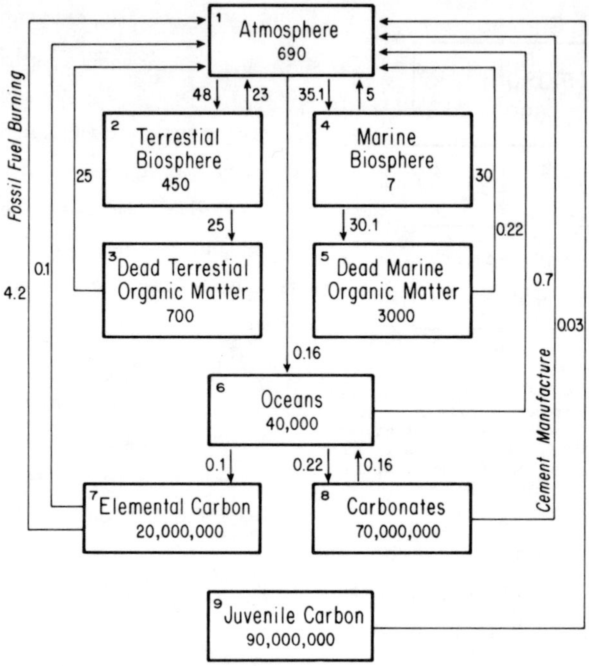

Figure 4. The long-term carbon cycle (modified from Holland, 1978). The reservoir contents are in units of 10^{15}gC and the fluxes in units of 10^{15}gC/yr. See text for details.

years (twice the response time) the system has essentially regained its original unperturbed state (atmosphere content is 696 and biosphere content is 444 in 10^{15} gC units).

On the other hand, if the 50 carbon units were added from fossil fuels, the initial conditions would be $A_1^o = 740$; $A_2^o = 450$. Now the evolution equations (15) yield:

$$A_1(t) = 720.3 + 19.7\, e^{-0.11015\, t} \; ; \quad A_2(t) = 469.7 - 19.7\, e^{-0.11015\, t}$$

(t in years) .

In this case, with the addition of *new* carbon, the cycle reaches a new steady state, i.e., $A_1(t{=}\infty) = 720.3$; $A_2(t{=}\infty) = 469.7$, with both reservoirs ultimately sharing the excess carbon.

The general approach employed by Lasaga (1980) becomes more necessary in the case of a multi-reservoir geochemical cycle, because in this case the feedback processes among the various reservoirs become more intertwined and non-trivial. The treatment can be illustrated with the long-term carbon cycle shown in Figure 4.

Reservoirs 1-5 represent the interactions between the atmosphere and the biosphere; because carbon cycles much faster through these reservoirs, they comprise the short-term carbon cycle. The long-term cycle involves important additional processes. These processes include the burial of a small fraction (<0.01%) of the dead marine organic matter, which escapes oxidation to CO_2 in the water column and is incorporated into sediments. It is this small fraction which is responsible for the net supply of oxygen to the atmosphere. The other major geochemical process in the long-term carbon cycle is the weathering of carbonates and silicates in ancient sediments and the deposition of authigenic carbonates and silicates in new sediments (Garrels and Mackenzie, 1971). CO_2 is the weathering agent in the crust. For example, carbonates weather according to the reaction

$$\underset{\textit{calcite}}{CaCO_3(s)} + CO_2(g) + H_2O \rightleftarrows Ca^{2+} + 2HCO_3^- \, . \tag{19}$$

For each mole of calcite dissolved, one mole of CO_2 is consumed from the atmosphere. The reverse holds during the deposition of carbonates. Likewise, the dissolution and deposition of Mg-, Na-, and K-silicates follow reactions similar to:

$$MgSiO_3 + 3H_2O + 2CO_2 \rightleftarrows Mg^{2+} + H_4SiO_4 + 2HCO_3^- \, . \tag{20}$$

The carbonate reactions are more important than the silicate reactions in the long-term carbon cycle. Therefore, although the silicate reactions should definitely not be ignored in a fuller treatment, our analysis will focus on the carbonate reactions.

The flux of 0.16×10^{15} gC/yr from the atmosphere to the oceans (Fig. 4) and the same flux from the carbonate reservoir to the oceans reflect the dissolution of carbonates, which yields HCO_3^- to the water in rivers and ultimately to the oceans, according to reaction (19). The 0.22×10^{15} gC/yr flux from the oceans to both the atmosphere and sedimentary carbonates reflect, in turn, the deposition of carbonates in new sediments, i.e., the reverse of reaction (19). (See Holland, 1978, for details.) Note that the present carbonate cycle is not balanced.

The situation is altered if the silicate reactions are included. Nonetheless, a CO_2 imbalance is still obtained, which necessitates input from juvenile carbon in the mantle (see Holland, 1978). Figure 4 also

Table 1

Long-Term Carbon Cycle

Non-Zero Rate Constants (in yr^{-1})

$k_{12} = 6.9565 \times 10^{-2}$	$k_{14} = 5.0870 \times 10^{-2}$	$k_{16} = 2.3188 \times 10^{-4}$
$k_{21} = 5.1111 \times 10^{-2}$	$k_{23} = 5.5556 \times 10^{-2}$	$k_{31} = 3.5714 \times 10^{-2}$
$k_{41} = 7.1429 \times 10^{-1}$	$k_{45} = 4.300$	
$k_{51} = 1.0 \times 10^{-2}$	$k_{56} = 3.3333 \times 10^{-5}$	
$k_{61} = 5.50 \times 10^{-6}$	$k_{67} = 2.500 \times 10^{-6}$	$k_{68} = 5.50 \times 10^{-6}$
$k_{71} = 5.000 \times 10^{-9}$	$k_{86} = 2.2857 \times 10^{-9}$	$k_{91} = 3.333 \times 10^{-10}$

includes the oxidation of elemental carbon (7 → 1 flux) in ancient sediments. In addition, Figure 4 shows estimates of the anthropogenic contributions from fossil fuel burning and cement manufacture.

Treating this cycle within the linear approximation and ignoring presently the anthropogenic sources, the non-zero rate constants shown in Table 1 are obtained. Diagonalizing the resulting K matrix for only the first eight reservoirs (i.e., the 9 → 1 flux is also ignored) yields the eigenvalues of Table 2. Note once more that the lowest eigenvalue (E_7) is greater than the lowest individual rate constant (k_{86}) by about a factor of two. This is a consequence of the increased efficiency of collective behavior. For completeness, the eigenvectors of the K matrix are also given in Table 2.

With this information it is now possible to analyze the dynamic behavior of the long-term carbon cycle. The initial reservoir contents are the contents shown in Figure 4. Forming the $\vec{A}^o$ vector from these initial contents, the expansion coefficients are obtained by analogy to equation (13) from

$$\vec{a} = \underline{\Psi}^{-1} \vec{A}^o \tag{21}$$

(see Lasaga, 1980) where $\underline{\Psi}$ is the matrix formed from the eigenvectors of K. The result is

$$\vec{A}(t) = -1.5780 \times 10^{-4} \vec{\Psi}_1 e^{E_1 t} - 0.3848 \vec{\Psi}_2 e^{E_2 t} + 0.6637 \vec{\Psi}_3 e^{E_3 t}$$

$$+ 1.2437 \vec{\Psi}_4 e^{E_4 t} + 1452.04 \vec{\Psi}_5 e^{E_5 t} - 8181.80 \vec{\Psi}_6 e^{E_6 t}$$

Table 2

Long-Term Carbon Cycle

Eigenvalues (yr^{-1})

$E_1 = -5.022$ $\quad E_2 = -1.612 \times 10^{-1}$ $\quad E_3 = -8.488 \times 10^{-2}$

$E_4 = -1.963 \times 10^{-2}$ $\quad E_5 = -5.989 \times 10^{-5}$ $\quad E_6 = -7.157 \times 10^{-6}$

$E_7 = -4.152 \times 10^{-9}$ $\quad E_o = 0$

Eigenvectors

$$\vec{\Psi}_1 = \begin{pmatrix} 0.10865 \\ -0.15378 \times 10^{-2} \\ 0.17135 \times 10^{-4} \\ -0.75444 \\ 0.64732 \\ -0.93140 \times 10^{-5} \\ 0.46370 \times 10^{-11} \\ 0.10201 \times 10^{-10} \end{pmatrix} \quad \vec{\Psi}_2 = \begin{pmatrix} 0.57395 \\ -0.73223 \\ 0.32419 \\ 0.60161 \times 10^{-2} \\ -0.17114 \\ -0.79032 \times 10^{-3} \\ 0.12257 \times 10^{-7} \\ 0.26966 \times 10^{-7} \end{pmatrix} \quad \vec{\Psi}_3 = \begin{pmatrix} -0.20176 \\ -0.64424 \\ 0.72796 \\ -0.20821 \times 10^{-2} \\ 0.11962 \\ 0.50429 \times 10^{-3} \\ -0.14853 \times 10^{-7} \\ -0.32676 \times 10^{-7} \end{pmatrix}$$

$$\vec{\Psi}_4 = \begin{pmatrix} -0.18228 \\ -0.14568 \\ -0.50307 \\ -0.18565 \times 10^{-2} \\ 0.83215 \\ 0.74080 \times 10^{-3} \\ -0.94364 \times 10^{-7} \\ -0.20760 \times 10^{-6} \end{pmatrix} \quad \vec{\Psi}_5 = \begin{pmatrix} -0.10627 \\ -0.69349 \times 10^{-1} \\ -0.10806 \\ -0.10782 \times 10^{-2} \\ -0.46485 \\ 0.86517 \\ -0.36116 \times 10^{-1} \\ -0.79451 \times 10^{-1} \end{pmatrix} \quad \vec{\Psi}_6 = \begin{pmatrix} -0.12839 \times 10^{-1} \\ -0.83736 \times 10^{-2} \\ -0.13028 \times 10^{-1} \\ -0.13025 \times 10^{-3} \\ -0.55860 \times 10^{-1} \\ -0.76266 \\ 0.26660 \\ 0.58630 \end{pmatrix}$$

$$\vec{\Psi}_7 = \begin{pmatrix} 0.12884 \times 10^{-4} \\ 0.84024 \times 10^{-5} \\ 0.13070 \times 10^{-4} \\ 0.13070 \times 10^{-6} \\ 0.56016 \times 10^{-4} \\ 0.23993 \times 10^{-3} \\ 0.70694 \\ -0.70727 \end{pmatrix} \quad \vec{\Psi}_0 = \begin{pmatrix} -0.86386 \times 10^{-5} \\ -0.56339 \times 10^{-5} \\ -0.87638 \times 10^{-5} \\ -0.87638 \times 10^{-7} \\ -0.37559 \times 10^{-4} \\ -0.40689 \times 10^{-3} \\ -0.20344 \\ -0.97909 \end{pmatrix}$$

$$+ 6.3893 \times 10^6 \vec{\Psi}_7 e^{E_7 t} - 7.61157 \times 10^7 \vec{\Psi}_0 e^{E_0 t} . \qquad (22)$$

The fluxes in Figure 4 are not all balanced and hence the initial contents would not represent a steady state situation. To find the steady state of the cycle we need merely allow t to approach infinity in equation (22):

$$\vec{A}_{steady} = -7.61157 \times 10^7 \vec{\Psi}_0 . \qquad (23)$$

Multiplying the entries of the $\vec{\Psi}_0$ eigenvector from Table 2 by -7.61157 x 10^7, the results are

$$A_1^{steady} = 657.5 \quad A_2^{steady} = 428.8 \quad A_3^{steady} = 667.1$$

$$A_4^{steady} = 6.67 \quad A_5^{steady} = 2858.9 \quad A_6^{steady} = 30971$$

$$A_7^{steady} = 1.5485 \times 10^7 \quad A_8^{steady} = 7.4524 \times 10^7$$

all in units of 10^{15} gC. The carbonate imbalance of Figure 4 leads to a significantly reduced carbon content of the oceans as well as a slight depletion in the atmosphere, terrestial biosphere, and dead terrestial organic matter at steady state. The carbonate content increases and that of elemental carbon decreases also at steady state. (Note that the terrestial biosphere as well as the elemental carbon in the lithosphere were initially at steady state.) The approach to this steady state is shown in Figure 5. Since the size of the time responses (i.e., the eigenvalues of Table 2) span a large range, the time scale used is logarithmic. For example, from equation (22) and Table 2, the evolution of atmospheric carbon is given as (in units of 10^{15} gC):

$$\begin{aligned} A_1(t) = & -1.71 \times 10^{-5} e^{-5.02 t} - 0.22 e^{-0.1612 t} - 0.13 e^{-0.0849 t} \\ & - 0.23 e^{-0.01963 t} - 154.32 e^{-5.9893 \times 10^{-5} t} \\ & + 105.04 e^{-7.1568 \times 10^{-6} t} + 82.32 e^{-4.1515 \times 10^{-9} t} \\ & + 657.54 \quad (t \text{ in years}) . \end{aligned} \qquad (24)$$

Because the fast responses (large exponent) all have a negative coefficient, the carbon content initially rises and then decays after 200 million years to its final value of 657 x 10^{15} gC. In the interim it reaches a maximum value of 805 x 10^{15} gC. However, the time scale is rather long; in 100 years the atmospheric carbon content increases by only 1.4 x 10^{15} gC from 690 to 691.4 x 10^{15} gC.

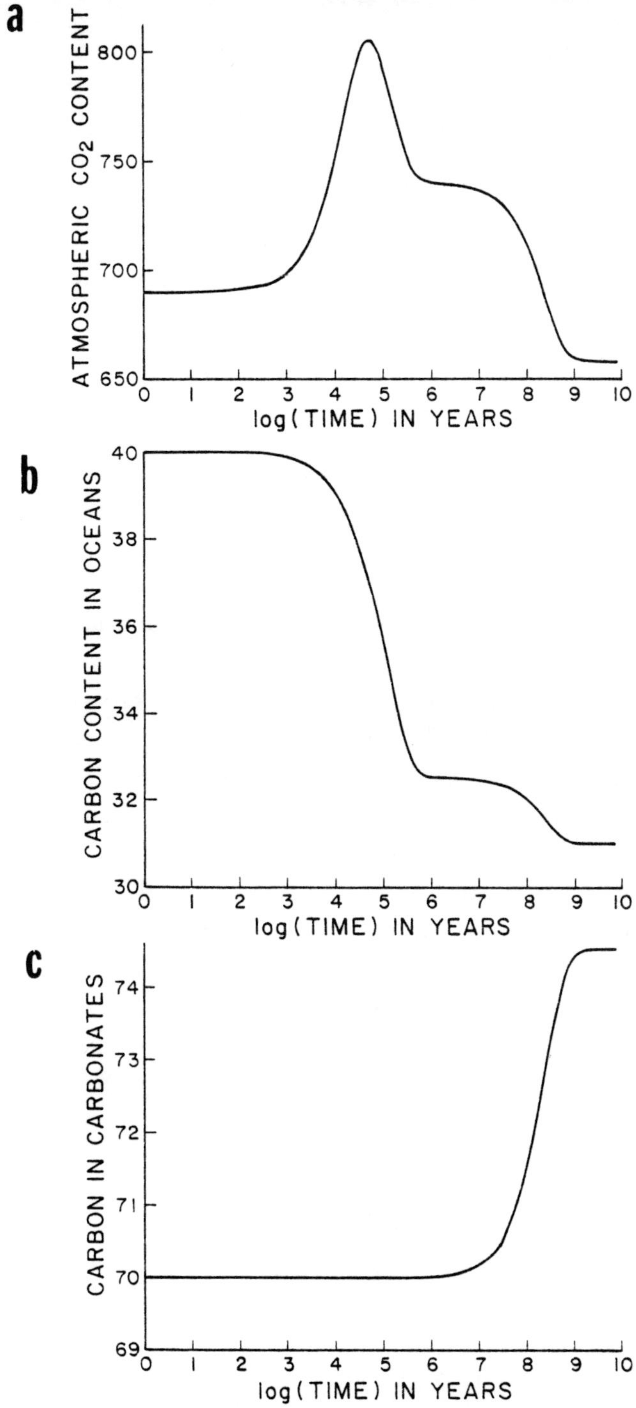

Figure 5. The approach to steady state of several reservoirs in the carbon cycle of Figure 4, as predicted from equation (22). Linear kinetics are assumed. Note logarithmic scale.

(a) the evolution of atmospheric CO_2 content (in 10^{15}gC units)
(b) the evolution of the carbon content in the oceans (in 1000 x 10^{15}gC units)
(c) the evolution of carbon in carbonates (in 1000000 x 10^{15}gC units)

The time evolution of reservoirs 2-5 is very similar to (a). The time evolution of reservoir 7 (elemental carbon in sediments) is nearly a mirror image of curve C. The various plateaus are due to the different time scales of the exponential terms in equation (22).

It is important to realize that the atmospheric imbalance in the cycle of Figure 4 is 0.06×10^{15} gC/yr (i.e., more is coming in than leaving). A simple calculation would lead to an increase in the CO_2 content of the atmosphere of $(0.06 \times 10^{15}$ gC/yr) (10,000 yr) = 600×10^{15} gC in 10,000 years. Such an increase (or a similar decrease) is not possible if life is to be maintained on earth. However, the kinetic treatment just outlined shows that, even without a balancing term from juvenile carbon, the interactions (feedbacks) among reservoirs 1-6 are such that the carbon content of the atmosphere only increases by a much more minor amount. In fact, the kinetic treatment concludes that the new steady state value of the atmosphere will be *lower* than the present value. This underscores the importance of treating the collective dynamics of the entire cycle.

The anthropogenic influences on the carbon cycle are more pertinent to the cycle's evolution over the next 100 years. To include the effects of fossil fuel burning and cement manufacture the appropriate rate constants from Figure 4 are changed. Now (see Fig. 4), $k_{71} = 2.15 \times 10^{-7}$ yr^{-1}; $k_{81} = 1.0 \times 10^{-8}$ yr^{-1}. The new K matrix, which includes these anthropogenic rate constants, has the same large eigenvalues (to six decimal places) as in Table 2 (*e.g.*, E_1, E_2, E_3, and E_4 are the same). The fifth eigenvalue is barely changed to $E_5 = -5.988 \times 10^{-5}$ yr^{-1}. The sixth eigenvalue is changed somewhat, $E_6 = -7.241 \times 10^{-6}$ yr^{-1} and the seventh eigenvalue is drastically changed to $E_7 = -1.503 \times 10^{-7}$ yr^{-1}. Of course, because our fossil fuel burning is not going to last longer than 500 years, the anthropogenic input into the atmosphere becomes basically a constant input over this time period. The result of the anthropogenic input leads to the following evolution of the atmosphere:

$$\begin{aligned} A_1(t) = &-0.0015\, e^{-5.02\, t} - 18.253\, e^{-0.1612\, t} \\ &- 11.074\, e^{-9.08488\, t} - 18.748\, e^{-0.01963\, t} \\ &- 10377.96\, e^{-5.988 \times 10^{-5}\, t} - 11508.78\, e^{-7.241 \times 10^{-6}\, t} \\ &+ 16150.79\, e^{-1.503 \times 10^{-7}\, t} + 6474.029\,. \end{aligned} \qquad (25)$$

In particular, if t = 0, equation (25) yields $A_1 = 690 \times 10^{15}$ gC; this is a useful check on the solution. Figure 6 shows the atmospheric carbon

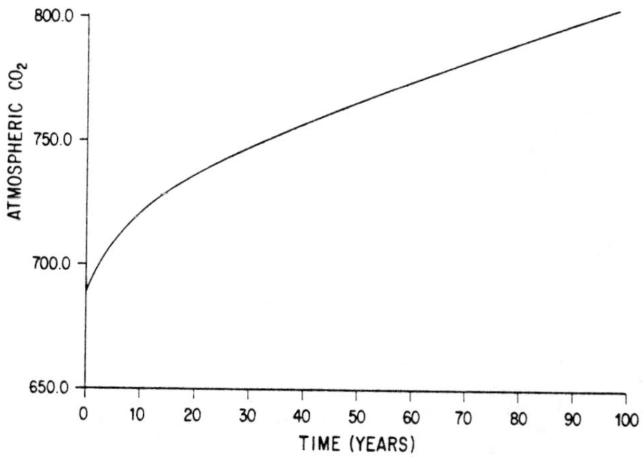

Figure 6. The change in atmospheric CO_2 content (10^{15}gC units) due to anthropogenic influences, as predicted by equation (25).

evolution predicted by equation (25). Note that the "steady state" predicted by (25), 6474, is meaningless in this case, since, as mentioned, the time range is confined to several hundred years. On the other hand, it is important to note that even the small (long time) eigenvalues contribute significantly to the evolution of atmospheric carbon over 100 years. For example, according to (25), atmospheric carbon increases from 690 x 10^{15} gC to 805.5 x 10^{15} gC in 100 years, a net increase of 115.5 x 10^{15} gC. This increase is apportioned among the various terms in equation (25) as follows:

$$\Delta CO_2 = +0.0015 + 18.253 + 11.072 + 16.115 + 61.958 + 8.330 - 0.24 \,. \quad (26)$$

Therefore, even the sixth term in equation (25) is contributing significantly to the net addition of CO_2. In fact, due to the small eigenvalues in the fifth and sixth terms, the exponentials can be expanded:

$$e^{Et} \simeq 1 + Et \,. \quad (27)$$

Consequently, these terms contribute a *linear* and significant time component to the time evolution. This is not the case with the first four terms of equation (25). By comparison, the type of treatment carried out for an isolated reservoir [equation (2)] does not contain any such linear terms, if $t \gtrsim 1/k$, as is the case here ($k \sim 83/690 = 0.12\ yr^{-1}$). The treatment of the collective dynamics, once more, can change the predicted qualitative behavior of a cycle.

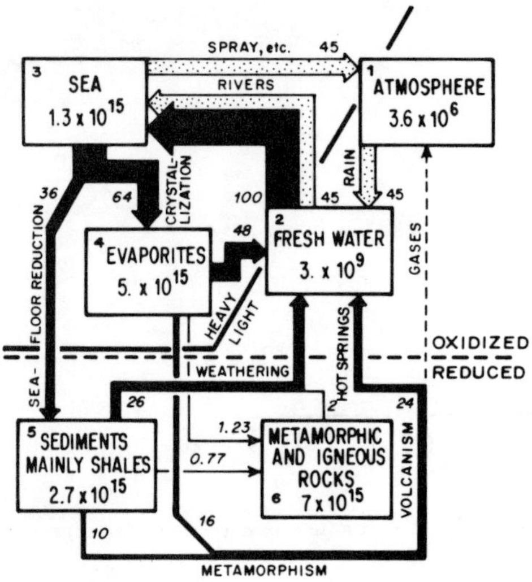

Figure 7. Geochemical cycle of sulfur, modified from Holser and Kaplan, (1966). Contents are in metric tons of sulfur. Most material above the dashed line is oxidized to sulfate, most below this line is reduced to sulfide; above the solid line heavy sulfur predominates $\delta^{34}S >$ + 10 o/oo, below this line light sulfur predominates ($\delta <$ + 5 o/oo). Long-term (dark) and short-term fluxes of sulfur between reservoirs are indicated on the basis of 100 for the long-term component of fresh-water sulfate flowing to the sea.

SULFUR CYCLE

The sulfur cycle is given in Figure 7, patterned after Holser and Kaplan (1966). The cycle, as shown, is at steady state. Holser and Kaplan (1966) give fluxes in relative units, such that 100 units equals the sulfate flux from rivers into oceans. Using the data in Holland (1978) for mean river sulfate content and total water carried to oceans by rivers, the total sulfate flux from rivers can be estimated at 1.7 x 10^{14} gS/yr or 17.3 x 10^{7} tons S/yr. However, the flux in Holser and Kaplan (1966) is taken to be the "non-anthropogenic" sulfate flux. Man has perturbed the sulfate flux in the hydrologic cycle to such an extent by the burning of fossil fuel that only 60 percent of the total river flux is non-anthropogenic (Holland, 1978). Therefore, the "long-term" river flux of sulfate is only 10.4 x 10^{7} tons/yr. Consequently, we will choose one flux unit as 10^{6} tons/yr in the cycle of Holser and Kaplan (1966) and in Figure 7. We have further modified Holser and Kaplan's cycle by apportioning their two flux units from evaporites and shales to metamorphic and igneous rocks, in accordance to the total output from each reservoir (i.e., evaporites or shales). Note that the sulfur cycle in Figure 7 has 36 percent of the total sulfur in *new* sediments as reduced sulfur (pyrite) and 64 percent as sulfate (gypsum). However, the steady state values of the pyrite and gypsum reservoir contents, which also depend on the weathering process, are 35 percent (S^{2-})

Table 3

Sulfur Cycle--Dynamics

Rate Constants (yr^{-1})

$k_{12} = 12.5$	$k_{23} = 0.048333$	
$k_{31} = 3.4615 \times 10^{-8}$	$k_{34} = 4.923 \times 10^{-8}$	$k_{35} = 2.769 \times 10^{-8}$
$k_{42} = 1.255 \times 10^{-8}$	$k_{46} = 2.46 \times 10^{-10}$	
$k_{52} = 1.305 \times 10^{-8}$	$k_{56} = 2.852 \times 10^{-10}$	
	$k_{62} = 2.857 \times 10^{-10}$	

$$K = 10^{-8} \times \begin{bmatrix} -12.5 \times 10^{8} & 0 & 3.4615 & 0 & 0 & 0 \\ 12.5 \times 10^{8} & -4.833 \times 10^{6} & 0 & 1.255 & 1.305 & 0.02857 \\ 0 & 4.833 \times 10^{6} & -11.1535 & 0 & 0 & 0 \\ 0 & 0 & 4.923 & -1.2796 & 0 & 0 \\ 0 & 0 & 2.769 & 0 & -1.33352 & 0 \\ 0 & 0 & 0 & 0.0246 & 0.02852 & -0.02857 \end{bmatrix}$$

Eigenvalues (yr^{-1})

$E_1 = -12.5$	$E_2 = -0.04833$	$E_3 = -8.969 \times 10^{-8}$
$E_4 = -1.314 \times 10^{-8}$	$E_5 = -5.092 \times 10^{-10}$	$E_6 = 0.0$

and 65 percent (SO_4^{2-}) of the total sulfur in sediments. Applying the linear rate law [just as in (18)] to the sulfur cycle yields the rate constants and the K-matrix given in Table 3. The eigenvalues of the system are also given in Table 3. In this case, the very fast response times of the atmosphere and fresh water are not affected by the collective dynamics of the cycle (i.e., they equal the residence times). On the other hand, the longer response times for the other reservoirs are reduced from the corresponding residence times.

The sulfur isotope record in sedimentary rocks suggests that the relative ratio of sulfides and sulfates being precipitated in the oceans has varied with time. Suppose that changing geographic conditions, conducive to evaporite formation, increase the rate constant, k_{34}, to

$8 \times 10^{-8}\ yr^{-1}$. In this case, the initial conditions (i.e., the reservoir contents in Fig. 7) are not at steady state. Furthermore, the new eigenvalues are the same as in Table 3, except for the following:

$$E_3 = -12.040 \times 10^{-8} \qquad E_4 = -1.320 \times 10^{-8}$$

$$E_5 = -5.150 \times 10^{-10} \quad (\text{in } yr^{-1}) \ . \tag{28}$$

Using the contents in Figure 7 as initial values and equation (20), the time evolution given by equation (10) of Lasaga (1980) is

$$A_1(t) = +0.919\ e^{E_3 t} + 0.011\ e^{E_4 t} + 0.030\ e^{E_5 t} + 2.640\ e^{E_6 t} \ (\times 10^6 \text{ tons}) \tag{29a}$$

$$A_2(t) = 0.150\ e^{E_3 t} + 0.010\ e^{E_4 t} + 0.032\ e^{E_5 t} + 2.807\ e^{E_6 t} \ (\times 10^9 \text{ tons}) \tag{29b}$$

$$A_3(t) = 0.332\ e^{E_3 t} + 0.004\ e^{E_4 t} + 0.011\ e^{E_5 t} + 0.953\ e^{E_6 t} \ (\times 10^{15} \text{ tons}) \tag{29c}$$

$$A_4(t) = -0.247\ e^{E_3 t} - 0.784\ e^{E_4 t} + 0.071 e^{E_5 t} + 5.959\ e^{E_6 t} \ (\times 10^{15} \text{ tons}) \tag{29d}$$

$$A_5(t) = -0.086\ e^{E_3 t} + 0.782\ e^{E_4 t} + 0.024\ e^{E_5 t} + 1.980\ e^{E_6 t} \ (\times 10^{15} \text{ tons}) \tag{29e}$$

$$A_6(t) = 0.0007\ e^{E_3 t} - 0.0023\ e^{E_4 t} - 0.1057\ e^{E_5 t} + 7.107\ e^{E_6 t} \ (\times 10^{15} \text{ tons}) \ . \tag{29f}$$

Figure 8. Evolution of the reservoirs in the sulfur cycle after the increase in evaporite deposition rate (see text).

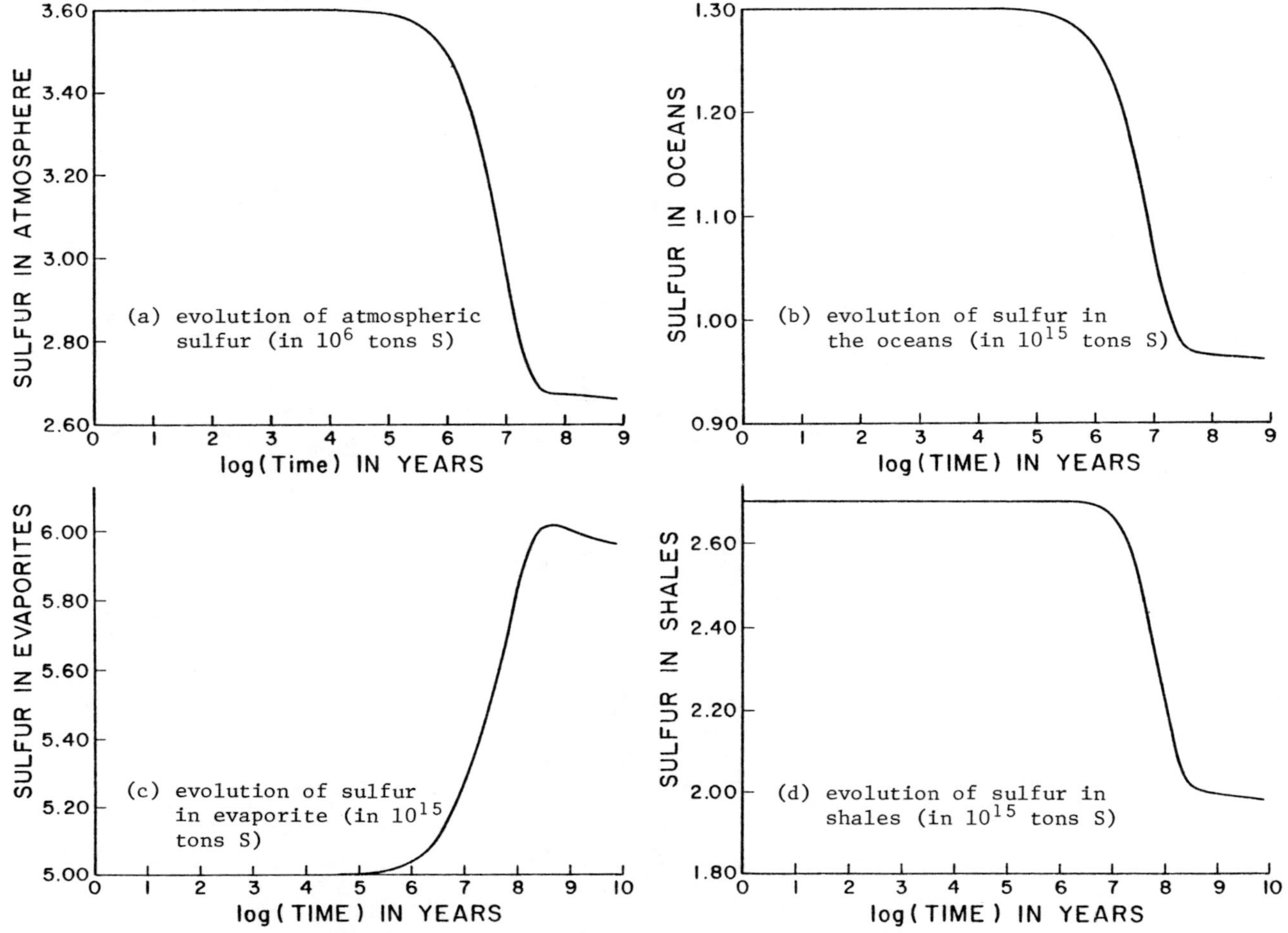

In all cases, the coefficients of the E_1 and E_2 terms are nearly zero ($<10^{-5}$). Note that the eigenvalues in equation (27) refer to the new eigenvalues, i.e., those of equation (26). Also note that the *units* in equation (27) vary from reservoir to reservoir and are shown in parentheses. The coefficients of the E_6 term ($E_6 = 0$) are the new steady state values of the reservoirs. The final steady state values of the reservoirs (1-6) are 2.64×10^6, 2.807×10^9, 0.953×10^{15}, 5.959×10^{15}, 1.980×10^{15}, and 7.107×10^{15} (all in tons S), respectively. The actual evolution is plotted in Figure 8. Obviously, the increased gypsum deposition rate has increased the evaporite sulfur steady state content from 5×10^{15} to 5.96×10^{15} tons S. Note that this increase is proportionately less than expected from the increase in the rate, k_{34}. Concomitantly, the sulfur content of shales is decreased from 2.7×10^{15} to 1.98×10^{15} tons S. Ultimately, the fluxes of sulfur from the oceans to evaporites and shales are 76×10^6 tons/yr and 26×10^6 tons/yr so that only 26 percent of the *new* sulfur deposited in sediments is in the form of pyrite. This is the ratio expected from the new ratio k_{35}/k_{34} ($k_{35} = 2.769 \times 10^{-8}$, $k_{34} = 8 \times 10^{-8}$). Note, once more, that the fraction of reduced sulfur in sediments at steady state, 24.9 percent (using $A_4^{st} = 5.959 \times 10^{15}$ tons, $A_5^{st} = 1.980 \times 10^{15}$ tons), is slightly different from that predicted by the formation of *new* sediments. This difference can be much greater under different circumstances.

These examples indicate that to fully understand the kinetic behavior of geochemical cycles, one must treat all reservoirs and the fluxes between the reservoirs *collectively*. For linear cycles, the tools are straightforward to use and already available. Non-linear cycles, on the other hand, present new scenarios, which will be the topic of the next section.

TREATMENT AND CONSEQUENCES OF NON-LINEAR CYCLES

It is well known that the geochemical cycles of many elements are interdependent. The future development of geochemical cycles must seriously pursue the mode of coupling among various cycles and the effect on the stability of the overall system. In particular, geochemists must quantify the qualitative ideas of positive and negative feedbacks that have pervaded throughout many studies of element distributions.

For example, the cycles of carbon, nitrogen, phosphorus, sulfur, and oxygen, among others, are all intertwined by the formation and destruction of organic compounds. Variation in the C/N ratio of decaying and buried organisms couples the carbon and nitrogen cycles. Variation in the oxygen content of natural waters or the atmosphere affects the fluxes of organic carbon to deep-sea sediments (*cf.* Broecker, 1971). A particularly good illustration of the interrelation among the elemental cycles comes from sulfur deposition in the oceans (see Fig. 7). Isotope data (Holser and Kaplan, 1966) have shown that the sulfide/sulfate ratio of sedimentary rocks has varied through time. Presently, the sulfur in sedimentary rocks is roughly 60 percent SO_4^{2-} and 40 percent reduced sulfur. The actual distribution of sulfur according to Garrels and Perry (1974) is as follows:

$$0.42 \times 10^{20} \text{ moles S as ocean } SO_4^{2-} ,$$

$$1.98 \times 10^{20} \text{ moles S as sedimentary } CaSO_4 ,$$

$$1.47 \times 10^{20} \text{ moles S as pyrite } FeS_2 .$$

The unusually large global gypsum deposits in the rocks of the Permo-Triassic Period, suggest that some tectonic events led to the appearance of suitable climates and basins for the widespread deposition of evaporites. The large amounts of gypsum, deposited during these periods, affect the sulfate output of the oceans in a *linear* way. This linear relation has been used in the previous section. However, many other processes are also affected by the precipitation of gypsum as suggested by Garrels and Perry (1974). For example, since gypsum ($CaSO_4 \cdot 2H_2O$) utilizes Ca^{2+}, the increased output of gypsum transfers Ca^{2+} from the carbonate to the sulfate reservoir. This transfer releases CO_2 to the atmosphere [see equation (19)], which is then removed by additional photosynthesis. The reactions can be written:

$$8Ca^{2+} + 16HCO_3^- + 8SO_4^{2-} + 16H^+ \rightarrow 8CaSO_4 \cdot 2H_2O + 16CO_2 \qquad (30)$$

$$16CO_2 + 16H_2O \rightarrow 16CH_2O + 16O_2 . \qquad (31)$$

These coupled reactions produce O_2. The excess O_2 is then used to oxidize pyrite during weathering and thereby balance the sulfate loss from the oceans in (30):

$$4FeS_2 + 15O_2 + 8H_2O \rightarrow 16H^+ + 8SO_4^{2-} + 2Fe_2O_3 . \qquad (32)$$

Reaction (30) also consumes Ca^{2+} ions; to restore balance in Ca^{2+}, corresponding amounts of carbonates are dissolved:

$$8CaCO_3 + 8CO_2 + 8H_2O \rightarrow 8Ca^{2+} + 16HCO_3^- . \qquad (33)$$

Reactions (31) and (32) still leave an excess of O_2; to remove this O_2, organic matter is then oxidized:

$$O_2 + CH_2O \rightarrow CO_2 + H_2O . \qquad (34)$$

Finally, the combined reactions (33) and (34) require a net supply of CO_2. This supply can be delivered by the transfer of magnesium from the carbonate to the silicate reservoirs, i.e.:

$$7MgCO_3 + 7SiO_2 \rightarrow 7MgSiO_3 + 7CO_2 . \qquad (35)$$

The *net* result of reactions (30) - (35) becomes:

$$4FeS_2 + 8CaCO_3 + 7MgCO_3 + 7SiO_2 + 31H_2O \rightarrow 8CaSO_4 \cdot 2H_2O + 2Fe_2O_3 + 15CH_2O + 7MgSiO_3 . \qquad (36)$$

Reaction (36) involves no net demands on the CO_2 and O_2 contents of the atmosphere (i.e., they remain at steady state). However, the response of the system to the imbalance caused by the increased sulfate output in gypsum evaporites has caused coupled changes in the magnesium, calcium, iron, silica, and organic matter cycles! These various coupling mechanisms lead to a variety of feedback processes that can profoundly affect the dynamical behavior of a geochemical cycle.

In particular, inclusion of the coupling between elemental cycles will lead to non-linear time evolution equations. A number of interesting possibilities arise once we venture into the domain of non-linear feedback cycles. These possibilities include the loss or gain of stability relative to the linear cycle, the existence of multiple steady states, and the onset of oscillatory behavior. These problems have been studied from the chemical reaction point of view by Prigogine and coworkers using the thermodynamics of irreversible processes (i.e., from the point of view of entropy production).

The recent work on the non-linear aspects of irreversible thermodynamics (*cf.* Prigogine, 1980) is of great relevance to geochemical cycles. One of the concepts that has arisen from this broad development is that irreversible processes can play a fundamental constructive role in the coherence and organization of nature. An essential distinction arises at the outset between "adiabatic" and "non-adiabatic" systems. Adiabatic systems are bound by the second law of thermodynamics to achieve the maximum entropy and as a corollary to achieve equilibrium. On the other hand, systems open to energy and/or mass fluxes can under certain circumstances exist with a high degree of structure when far from equilibrium. The contrast between kinetics and thermodynamics is most vividly displayed by the evolution of these nonlinear structures (Procaccia and Ross, 1977a,b), which, although at steady state, are far from equilibrium.

One of the fundamental results states that an open chemical reaction scheme may show (unstable) oscillatory behavior near a steady state or the existence of multiple steady states only in the case that at least one of the reaction steps is *autocatalytic** (Prigogine, 1967; Prigogine and Glansdorff, 1971). A simple example of an autocatalytic step is the reaction $A + X \rightarrow X + X$. In this reaction, the product X is involved as a reactant also and, therefore, catalyzes its own production. Autocatalytic steps represent a strong positive feedback mechanism. The necessity of catalytic behavior for the evolution of complex structures in chemical systems is then akin to requiring strong feedback mechanisms within the overall kinetic processes of geochemical cycles. The existence of autocatalytic steps or strong feedback mechanisms in nature is to be expected. The abundance of these autocatalytic steps in biological systems is well documented. An obvious example would be photosynthetic activity. In this case, the organic biota is reacting to produce more of itself.

An important new concept, which I would like to stress in this section, is the possible role of multiple steady states in modifying drastically the behavior of geochemical cycles. Suppose we focus our attention on the short-term carbon cycle; in fact, only on the marine segment. We will now label the atmosphere as Reservoir 1, the marine biosphere as Reservoir 2, and the dead marine organic matter as Reservoir 3 (see Fig. 9). Furthermore, the photosynthetic carbon flux from the atmosphere to the biosphere will be related not only to the CO_2 content of the atmosphere, as in the linear cycle, but also to the content of carbon in the biosphere. We can postulate a reasonable

* See Chapter 5 for derivation.

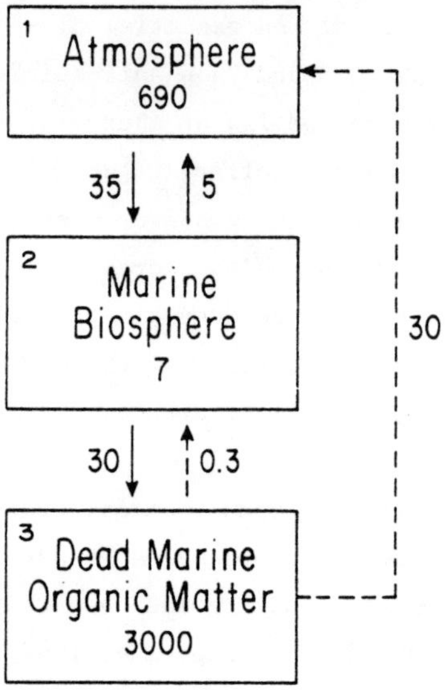

Figure 9. Portion of the short-term carbon cycle used in obtaining equations (37) and (38).

autocatalytic mechanism, which includes the biosphere carbon, labeled as $C^{(2)}$, as shown on the following equation,

$$C^{(1)} + 2C^{(2)} \underset{k_2}{\overset{k_1}{\rightleftharpoons}} 3C^{(2)} \ ;$$

$$C^{(2)} \underset{k_4}{\overset{k_3}{\rightleftharpoons}} C^{(3)} \ . \qquad (37)$$

The top reaction, which is autocatalytic, accounts for the strong feedback role of the biosphere in the photosynthetic flux from Reservoir 1 to Reservoir 2 (to the right) and the respiration flux from Reservoir 2 to Reservoir 1 (to the left). $C^{(1)}$ refers to the CO_2 in the atmosphere (and photosynthetic rates are still simply proportional to it). $C^{(2)}$ refers to the marine biosphere and $C^{(3)}$ to dead organic matter. The particular choice of stoichiometric coefficients in the top reaction is certainly arbitrary at this stage. But the actual numbers are not significant to the general scheme that follows.

In this modified geochemical cycle, which includes autocatalysis, we can solve for the steady state value of the biosphere, i.e., $C^{(2)}$. Since Reservoirs 1 and 3 are much bigger, their values will be set to some constant amount labeled A and B. Obviously, the time scale in the following discussion will be such that A and B change very slowly relative to $C^{(2)}$. Setting $C^{(1)} = A$, $C^{(3)} = B$, and $C^{(2)} = X$, and treating the reaction scheme in (37) as elementary reactions (Chapter 1) yields an expression for dX/dt:

$$\frac{dX}{dt} = -k_2 X^3 + k_1 AX^2 - k_3 X + k_4 B \ . \qquad (38)$$

It should be noted that only the *form* of the rate law in reactions (37) was assumed to be that of an elementary reaction; clearly, each reaction in (37) is *not* an elementary reaction, as defined in Chapter 1. Therefore, for the geochemical cyclist, the autocatalytic requirement is translated into the requirement of a non-linear rate law, such as that in (38). To proceed further we need to obtain values for the rate constants k_i in equations (37) and (38). These values can be obtained from the observed fluxes between the reservoirs as well as from the actual reservoir contents in Figure 9. For example, from the 1 → 2 flux, $k_1 AX^2 = 35 \times 10^{15}$ gC/yr. Using the contents in Figure 9:

$$A = C^{(1)} = 690 \times 10^{15} \text{ gC} \; ; \quad X = C^{(2)} = 7 \times 10^{15} \text{ gC} .$$

Combining these values we arrive at

$$k_1 = 1.0352 \times 10^{-33} \text{ (gC)}^{-2} \text{ yr}^{-1} . \tag{39a}$$

Note the units of k_1, which is now a third-order rate constant (Chapter 1). In a similar fashion, we obtain

$$k_2 = 14.5773 \times 10^{-33} \text{ (gC)}^{-2} \text{ yr}^{-1} \tag{39b}$$

$$k_3 = 4.2857 \text{ yr}^{-1} . \tag{39c}$$

Finally, the rate constant for the incorporation of dead organic matter back into the biosphere, k_4, is arbitrarily set at some low value, $k_4 = 0.0001 \text{ yr}^{-1}$.

Given these rate constants and the values of A and B, equation (38) can be used to solve for the steady state value, X_0, by setting dX/dt to zero:

$$-k_2 X_0^3 + k_1 A X_0^2 - k_3 X_0 + k_4 B = 0 . \tag{40}$$

When this cubic polynomial is solved for X_0, the most important result is that, rather than finding only one real positive steady state as in the linear case, there are now *three* positive real steady states:

$$X_0 = 7.08 \times 10^{13} \text{ gC} \; ; \quad 6.91 \times 10^{15} \text{ gC} \; ; \quad 42.01 \times 10^{15} \text{ gC} . \tag{41}$$

Multiple steady states are possible in highly non-linear systems. However, also unlike the linear case, not all steady states are stable. To test for the stability of any steady state, the system is perturbed slightly from steady state and the time response is analyzed. Suppose the initial amount of $C^{(2)}$ is perturbed to $X_0 + x$, where x is some small number. The amount of X as a function of time will obey an equation of the form

$$X(t) = X_0 + x\, e^{wt} \tag{42}$$

where w represents the time response of the system and is analogous to the E_i in equations (9) or (22). w is the parameter to be determined. Inserting equation (42) into equation (38) yields

$$\frac{d(X_0 + x\,e^{wt})}{dt} = -k_2(X_0 + x\,e^{wt})^3 + k_1\,A(X_0 + x\,e^{wt})^2$$
$$- k_3\,(X_0 + x\,e^{wt}) + k_4\,B\ .$$

Expanding these terms and keeping only the terms which are constant or which are linear in x, the result is

$$x\,w\,e^{wt} = -k_2X_0^3 - 3k_2\,X_0^2\,x\,e^{wt} + k_1\,A\,X_0^2 + 2k_1\,A\,X_0\,x\,e^{wt}$$
$$- k_3X_0 - k_3\,x\,e^{wt} + k_4\,B\ . \qquad (43)$$

The reason that terms involving x^2 or x^3 are ignored is based on the assumed small size of x. Using equation (40) cancels all the terms not containing x on the right of (43); if the remaining terms are divided by $x\,e^{wt}$, the result is

$$w = -3k_2\,X_0^2 + 2k_1\,A\,X_0 - k_3\ . \qquad (44)$$

Equation (44) gives the response of the system to perturbations from steady state. If w is *negative*, the second term in equation (42) will decay in time and the steady state will be *stable* to perturbations. If w is *positive* the deviation in equation (42) will increase with time and thus the steady state is *unstable* (*e.g.*, like the top of a hill).

Inserting the steady state values from equation (41) into equation (44), the results are

$$w = -4.18\ yr^{-1}\ ;\quad +3.50\ yr^{-1}\ ;\quad -21.46\ yr^{-1}\ , \qquad (45)$$

where the order corresponds to that in equation(41). The consequence of equation(45) is that only two of the steady states are stable ($w < 0$), while the third is unstable ($w > 0$). Therefore, in this "slightly" altered carbon cycle the *stable* steady state would correspond to $X = 42 \times 10^{15}$ gC rather than the original 7×10^{15} gC. Let us assume that the system was at its stable steady state at 42×10^{15} gC. It is important to realize that, as far as any small perturbations are concerned, the system would not be "aware" of the existence of other steady states. Any deviation from the steady state at 42×10^{15} gC will decay and the system will return to the original steady state. This isolation of the steady state, which can be viewed as a linear-like behavior, is possible

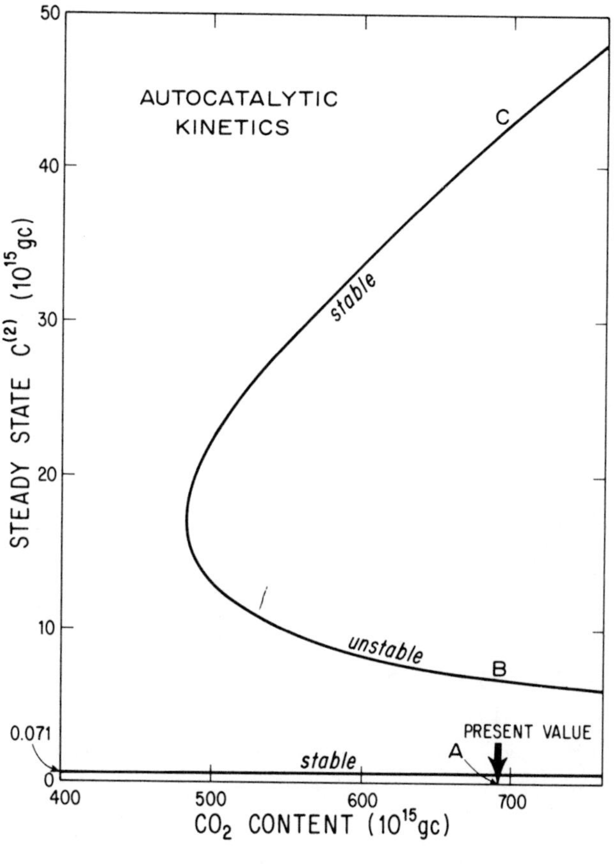

Figure 10. Plot of the possible steady state values of the biosphere as a function of the atmospheric CO_2 content (in 10^{15}gC). The stability of the various steady states is also labeled. Present day value of CO_2 is marked by an arrow in the figure. The corresponding values of the steady states are labeled by A, B and C. Note the disappearance of two steady states at low values of atmospheric CO_2.

because the other steady states are removed by a large carbon content difference from $X_0 = 42 \times 10^{15}$ gC. Conversely, linear-like behavior of a system is no guarantee that no other steady states arising from strong feedback processes are present. These other steady states may be "uncovered" under some circumstances, as shown below.

The isolation of the steady state at 42×10^{15} gC from the other steady states mentioned above breaks down seriously and with drastic consequences under certain circumstances. To see this, the carbon content of the atmosphere, i.e., A, can be externally varied and the previous calculations repeated for each value of A (keeping k_i and B the same). The interesting results are shown in Figure 10. Figure 10 is a plot of the values of the possible steady state contents of $C^{(2)}$ as a function of the atmosphere CO_2 content. The present value of the CO_2 content is shown by an arrow. Note that in some regions there are three steady states, while in others only one steady state is possible. Point C shows the stable steady state corresponding to the present CO_2 value. If the CO_2 content of the atmosphere is increased slightly, the stable steady state in the upper curve of Figure 10 increases slightly. Likewise, a slight decrease in CO_2 leads to a slight decrease in X_0 for the stable upper curve. However, as the CO_2 content decreases even further from the present value, the value of the unstable steady state (middle curve) begins to *increase*. As the stable steady state (upper curve) decreases and the unstable steady state increases with lower CO_2 in the atmosphere, there is a value of the CO_2 content where the two curves meet, i.e., around 490×10^{15} gC (atmosphere CO_2 content) in Figure 10. If the CO_2 content drops below this value, the upper stable and the unstable steady states cease to exist and only one steady state (lower curve) remains. Therefore, near this critical CO_2 content, a slight decrease in CO_2 can drastically and abruptly reduce the steady state value of the biosphere. This scenario raises the specter of *sudden* onsets to new steady states. Even before the CO_2 content of the atmosphere reaches the critical value, where the two curves meet in Figure 10, small fluctuations from the steady state value may take the system slightly beyond the unstable steady state (which is now close to the stable steady state) and thus plunge the $C^{(2)}$ value from the upper curve to the lower stable steady state. The situation may be sketched as in Figure 11. At high CO_2 content of the atmosphere (Fig. 11a), the stable steady state (point C) is isolated substantially from the unstable steady state at B and the low stable steady state at A. Therefore, reasonable fluctuations in $C^{(2)}$ return to point C. However, as the atmospheric CO_2 content drops, points B and C move closer together and the situation is closer to Figure 11b. Now, a small fluctuation from point C may indeed take the system to point A.

The geologic record is full of instances where "sudden" changes have occurred. The onset of glaciation in the Pleistocene may be related to processes similar to that discussed here. The weather patterns of the atmosphere, in fact, are hard to predict at times because of

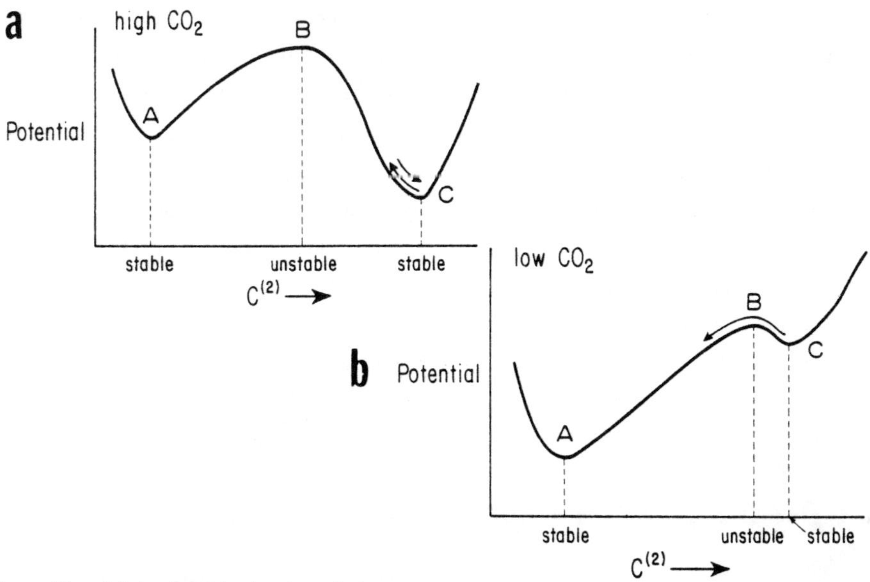

Figure 11. (a) A schematic drawing of the situation in Figure 10 for high values of atmospheric CO_2. The X-axis shows the variation in the $C^{(2)}$ (or X) content. Points A, B, C are the same as in Figure 10. The use of "potential" is meant only in the general sense, i.e. as a driving force. (b) A schematic drawing of the corresponding situation for low values of atmospheric CO_2 (but above the critical CO_2 content). The steady states at B and C are now very close to each other and small fluctuations at C can take the system to point A.

widespread occurrences of *bifurcations*, i.e., the choosing of a new steady state curve (such as in Fig. 10). Natural eco-systems may also have these bifurcations. For example, drastic variations in the amount of buried organic matter in sediments may reflect changes such as discussed here. Even organisms have biochemical bifurcations such that, if the CO_2 content of the ambient drops below a critical value, all photosynthetic activity ceases.

One important consequence of the *coupling* of elemental cycles, such as that illustrated by the gypsum example previously, is the necessity to deal with non-linear kinetic terms. Most transfers of elements among reservoirs can be represented by overall reactions such as those in equations (30) - (36). Following the treatment discussed in Chapter 1, these overall reactions, albeit global ones, have rates, which depend on the products of several of the reactant concentrations. An appraisal of the dynamics of geochemical cycles must quantify these coupled reactions. In a recent paper, Lasaga (1980) initiated the analysis with the carbon-oxygen cycles. The coupling was then generalized to a two-component/two-reservoir geochemical cycle shown in Figure 12. The important conclusion reached is that the response of the coupled cycle to perturbations is faster (more stable) than the response of the corresponding uncoupled linear cycle. For example, let us analyze a perturbation of the linear cycle of element A from steady state, which adds one unit (arbitrary) to A_1 and correspondingly decreases the same unit from A_2. If the steady state is written as A_1^{st} and A_2^{st}, then the initial condition is $A_1^\circ = A_1^{st} + 1$ and $A_2^\circ = A_2^{st} - 1$. If the A-cycle is treated as in the linear section, i.e., ignoring coupling to the B-cycle, then using equations (15) and the steady state condition, $k_{12}\,A_1^{st} = k_{21}\,A_2^{st}$, yields

$$A_1 = A_1^{st} + e^{-(k_{12}+k_{21})t} \quad . \tag{46}$$

As a consequence, the perturbation decays with the previously discussed $1/(k_{12}+k_{21})$ response time, returning the system to the original steady state, A_1^{st}. The exponential in (46) can be expanded for times less than $1/(k_{12}+k_{21})$ to yield

$$A_1 \simeq A_1^{st} + 1 - (k_{12}+k_{21})t \quad . \tag{47}$$

The influence of coupling to the B-cycle can be evaluated within the scheme outlined in Lasaga (1980). In this case, the coupling occurs by making the flux from A_2 to A_1 in Figure 12, also dependent on the amount of B in Reservoir 1, B_1. (The analogy to the C-O system is discussed in Lasaga, 1980.) Now the time evolution of cycles A and B obeys the following equations:

$$\begin{aligned}
\frac{dA_1}{dt} &= -k_{12}\,A_1 + k_{21}^c\,A_2\,B_1 \\
\frac{dA_2}{dt} &= k_{12}\,A_1 - k_{21}^c\,A_2\,B_1 \\
\frac{dB_1}{dt} &= -k'_{12}\,B_1 + k'_{21}\,B_2 + k_{12}\,A_1 - k_{21}^c\,A_2\,B_1 \\
\frac{dB_2}{dt} &= k'_{12}\,B_1 - k'_{21}\,B_2 \quad .
\end{aligned} \tag{48}$$

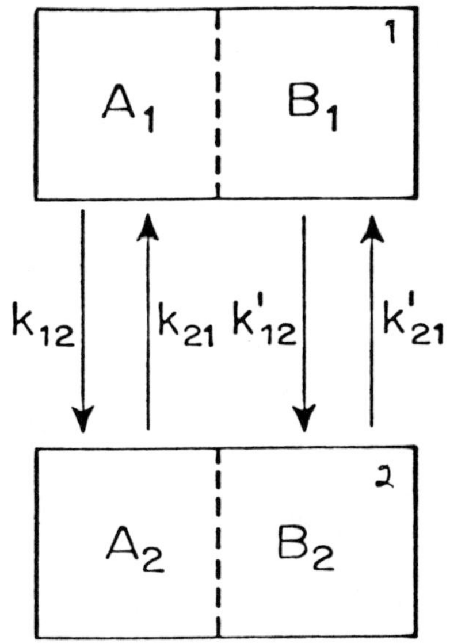

Figure 12. Generalized two-component/two reservoir geochemical cycle used to describe the effects of coupling. A_1 and A_2 are the contents of element A in reservoirs 1 and 2. B_1 and B_2 are the corresponding contents of element B. k_{12}, k_{21} are the first-order rate constants of the A-cycle and k_{12}', k_{21}' are the first-order rate contstants of the B-cycle.

k_{21}^c is now a second-order rate constant. Lasaga (1980) analyzed the response of this coupled A-cycle to the same perturbation posed above ($A_1^\circ = A_1^{st} + 1$ and $A_2^\circ = A_2^{st} - 1$) and obtained that

$$A_1 = A_1^{st} + \frac{b_2}{b_2 - b_1} e^{w_3 t} - \frac{b_1}{b_2 - b_1} e^{w_4 t} \quad (49)$$

where

$$w_3 = 1/2 \{w_1^L + w_2^L - k_{21} R + [(w_2^L - w_1^L + k_{21} R)^2 + 4 k_{21} k_{12}' R]^{1/2}\} \quad (50)$$

$$w_4 = 1/2 \{w_1^L + w_2^L - k_{21} R - [(w_2^L - w_1^L + k_{21} R)^2 + 4 k_{21} k_{12}' R]^{1/2}\} \quad (51)$$

$$1 + b_i = \frac{-(k_{12} + k_{21}) - w_{i+2}}{k_{21} R} \quad i = 1, 2 \quad (52)$$

and:

$$w_1^L = \text{linear eigenvalue of A-cycle} = -(k_{12}+k_{21})$$
$$w_2^L = \text{linear eigenvalue of B-cycle} = -(k_{12}'+k_{21}')$$

and where the ratio R is given by

$$R = A_2^{st}/B_1^{st} . \quad (53)$$

In deriving these equations, the non-linear rate constant k_{21}^c was related to the linear k_{21} by the relation

$$k_{21}^{c} = \frac{k_{21}}{B_1^{st}}$$

since the same flux from A_2 to A_1 is given by $k_{21}\ A_2^{st}$ in one case and by $k_{21}^{c}\ A_2^{st}\ B_1^{st}$ in the other. Equation (49) may be simplified for small times to yield

$$A_1 \simeq A_1^{st} + 1 - (k_{12} + k_{21} + k_{21}R)\ t\ . \tag{54}$$

Comparing equations (49) and (54), the response of the coupled system is seen to be faster, i.e., the perturbation is removed at a faster rate. The increase in the response of the coupled cycle depends on the ratio R; if the A-cycle couples to a small B_1 reservoir, the R ratio is large and the stability is greatly increased. Coupling to a much larger B-cycle (very small R), leaves the dynamics of the A-cycle essentially unchanged, however. The comparison of the response of the linear and the coupled A-cycles to the previous perturbation is shown for all times in Figures 13 and 14.

Figures 13 and 14 compare the linear and coupled A-cycles for different sets of k_i and R. These figures are obtained by using the full equations (46) and (49) and only the decay of the perturbation is plotted (i.e., A^{st} is omitted). It is important to note that the increased response of the coupled A-cycle holds over the *entire* time range. Figure 13 shows the effects of coupling the A-cycle to a faster B-cycle ($k'_{12} + k'_{21} > k_{12} + k_{21}$) and Figure 14 shows the effects, if the coupling is to a slower B-cycle ($k'_{12} + k'_{21} < k_{12} + k_{21}$). The data in Figure 14 indicate clearly that even coupling to a slower B-cycle leads to a *faster* response in the A-cycle, for *all* values of R. The similarities of Figures 13 and 14 suggest that the value of R is more important than the actual rate constants of the B-cycle. In either figure, the stability of the A-cycle is increased drastically if the value of R is high. Therefore, coupling an elemental cycle to a small reservoir of another element can increase the overall stability of the cycle substantially. If R is very small, on the other hand, the coupled A-cycle behaves just like the simple linear A-cycle.

These calculations indicate that coupling of cycles, in general, can change some of our qualitative notions about the stability of the geochemical cycle of a particular element. In the past few years,

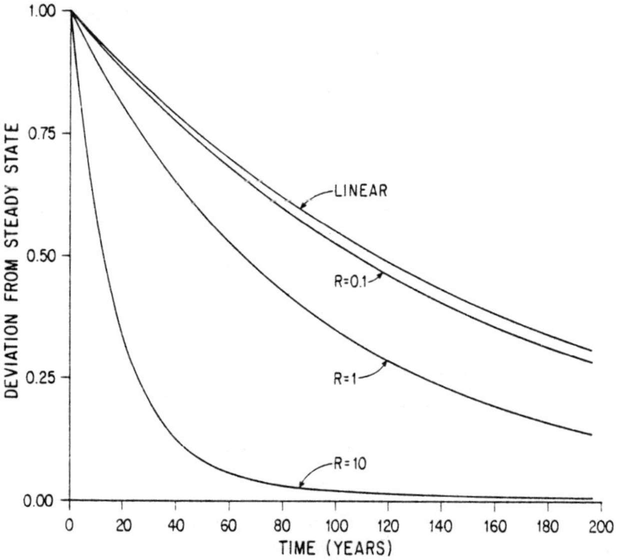

Figure 13. A plot of the decay of the unit perturbation of the A_1 content as a function of time. The linear curve is computed using the second term in equation (46). The other curves show the effect of coupling to a faster B-cycle (see text) of varying size (R ratios) and use the second and third terms of equation (49). The rate constants assumed are:

k_{12} = 0.001 yr^{-1}; k_{21} = 0.005 yr^{-1}; k_{12}' = 0.002 yr^{-1}; k_{21}' = 0.01 yr^{-1}.

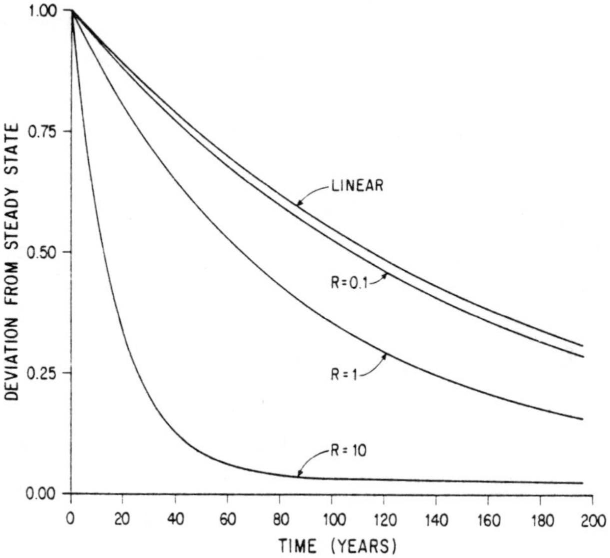

Figure 14. Same as Figure 13 but coupling is now to a slower B-cycle. The rate constants assumed are:

k_{12} = 0.001 yr^{-1}; k_{21} = 0.005 yr^{-1}; k_{12}' = 0.002 yr^{-1}; k_{21}' = 0.001 yr^{-1}.

evidence has been compiled which supports the contention that the contents of the reservoirs of several geochemical cycles have not varied greatly through the last 600 million years (*cf.* Holland, 1972, 1978; Mackenzie, 1975). If we are to understand the reason behind this observation, it seems clear that the effects of coupling on the feedback mechanisms and the kinetics of the elemental cycles must be well understood.

CONCLUSION

This chapter has presented some of the key concepts needed to understand the global kinetics of geochemical cycles. Residence times are related in a simple way to the kinetics of a "cycle" with one reservoir. However, when many reservoirs are interconnected by elemental fluxes, the description of the dynamics of the entire cycle must go beyond merely calculating the residence times. *Collectively*, the reservoirs in a cycle can attain more efficient kinetics. The treatment of collective behavior in geochemical cycles can be handled succintly with the use of matrices, if the cycles are treated as linear systems. In this fashion, many useful results can be obtained on the dynamics of the cycles -- results, which are often not obvious from superficial inspection of the cycle's reservoir contents, reservoir fluxes, and residence times.

Geochemical cycles may sometimes require non-linear global kinetics. Non-linear systems can behave quite differently from their linear counterparts. Two particularly important aspects have been discussed in this chapter. First, non-linearity may lead to the existence of *multiple* steady state. As detailed in the chapter, under some circumstances systems may incur sudden changes from one steady state to another steady state, which is quite different. These *bifurcations* from one steady state to another most likely have played a non-trivial role in geologic history. Prigogine (1980, p. 106) goes even further to state: "It is interesting that, in a sense, the bifurcation introduces *history* into physics and chemistry, an element that formerly seemed to be reserved for sciences dealing with biological, social and cultural phenomena." The second aspect of non-linearity deals with the effects of coupling (feedback) between the cycles of different elements.

Coupling between elements is well known to geochemists. Whether the coupling occurs within a local environment or in a true global sense, the kinetics analyzed here showed that the coupling can have dramatic effects on the stability of the cycles. Hence, the relative constancy of the amounts of some elements in various global reservoirs can be understood from the point of view of coupling of the global kinetics.

CHAPTER 2 APPENDIX -- MATRIX FUNDAMENTALS

Since this chapter (and also Chapter 6) employs matrices, some of the basic definitions and notions are discussed here for the reader. It is interesting to note that, historically, ignorance of matrices delayed the development of quantum mechanics in physics by at least 15 years, until Heisenberg remedied the situation. The author hopes the same will not be applicable to geochemistry. A *matrix* is most often a *square* array of numbers (some are rectangular, but these are not relevant here). The elements of a matrix, *K*, are therefore labeled as K_{ij}, where i refers to the row of the element and j refers to the column of the element in the matrix array. If the *K* matrix is an n x n matrix, then we have:

$$K = \begin{bmatrix} K_{11} & K_{12} & K_{13} & \cdots & K_{1n} \\ K_{21} & K_{22} & K_{23} & \cdots & K_{2n} \\ K_{31} & K_{32} & K_{33} & \cdots & K_{3n} \\ \vdots & \vdots & \vdots & & \vdots \\ K_{n1} & K_{n2} & K_{n3} & \cdots & K_{nn} \end{bmatrix} . \quad \text{(A1)}$$

Of course, assigning numbers to various positions in an array is of no great use until the rules of matrix operation are defined. A vector is a column (or a row!) of numbers. Hence, a vector $\vec{A}$ with n elements (an *n-vector*) is given by

$$\vec{A} = \begin{pmatrix} A_1 \\ A_2 \\ A_3 \\ \vdots \\ A_n \end{pmatrix} \quad \text{(A2)}$$

A vector or a matrix may be multiplied by a number in the obvious way, i.e., b $\vec{\Psi}$, multiplies *each* of the numbers in $\vec{\Psi}$ by b and b *K* multiplies each of the elements of K by the number b. A square n x n matrix transforms an n-vector (*e.g.*, $\vec{A}$) into another n-vector (say $\vec{B}$). The multiplication of a matrix times a vector obeys the following rule:

$$\vec{B} = K\,\vec{A} \tag{A3}$$

where

$$B_i = \sum_{j=1}^{n} K_{ij}\,A_j \ . \tag{A4}$$

Note that with the definition in (A4), it follows that

$$K \cdot (\vec{A} + \vec{C}) = K\,\vec{A} + K\,\vec{C} \tag{A5}$$

i.e., the multiplication of K times a sum of vectors is the same as the sum of K multiplied times each vector. Hence, matrices are really a short-hand notation for functions that take vectors into other vectors; or, in simple words, a matrix takes a set of numbers into a related different set of numbers.

The eigenvectors and eigenvalues are the most fundamental properties of a given matrix, K. An *eigenvector* of a matrix, K, is a vector, $\vec{\Psi}$, which satisfies the following equation:

$$K\,\vec{\Psi} = E\,\vec{\Psi} \ . \tag{A6}$$

The number, E, multiplying the eigenvector, $\vec{\Psi}$, in equation (A6) is then termed the *eigenvalue*. Eigenvectors are very special vectors. For example, take the matrix

$$K = \begin{bmatrix} -1 & 2 \\ 1 & -2 \end{bmatrix} . \tag{A7}$$

If one were to choose a vector at random, *e.g.*,

$$\begin{pmatrix} 1 \\ 3 \end{pmatrix}$$

and multiply K times $\begin{pmatrix} 1 \\ 3 \end{pmatrix}$, using the rule in (A4), the answer is

$$\underline{K}\begin{pmatrix} 1 \\ 3 \end{pmatrix} = \begin{bmatrix} -1 & 2 \\ 1 & -2 \end{bmatrix}\begin{pmatrix} 1 \\ 3 \end{pmatrix} = \begin{pmatrix} -1.1 + 2.3 \\ 1.1 - 2.3 \end{pmatrix} = \begin{pmatrix} 5 \\ -5 \end{pmatrix} .$$

But is there is *no* number, E, such that

$$\begin{pmatrix} 5 \\ -5 \end{pmatrix} = E \begin{pmatrix} 1 \\ 3 \end{pmatrix}$$

The same problem will be found with most other vectors picked at random. On the other hand, the special vector $\begin{pmatrix} 1 \\ -1 \end{pmatrix}$ is an eigenvector of the matrix in (A7) since

$$\begin{bmatrix} -1 & 2 \\ 1 & -2 \end{bmatrix} \begin{pmatrix} 1 \\ -1 \end{pmatrix} = \begin{pmatrix} -1.1 + 2 \cdot (-1) \\ 1.1 + (-2)(-1) \end{pmatrix} = \begin{pmatrix} -3 \\ 3 \end{pmatrix} = -3 \begin{pmatrix} 1 \\ -1 \end{pmatrix}$$

The eigenvalue, in this case, is -3. An important point is that, if $\vec{\Psi}$ is an eigenvector, then so is $a\vec{\Psi}$ for *any* number a. Using geometrical language, an eigenvector defines a *direction* and the length of the eigenvector is not important. For example, the vector $\begin{pmatrix} 2 \\ -2 \end{pmatrix}$ can easily be checked to be also an eigenvector of the matrix in (A7) with an eigenvalue of -3. An important result of matrix theory is that most n x n matrices will have exactly n eigenvectors (treating $\vec{\Psi}$ and $a\vec{\Psi}$ as the same). The n x n matrices which do not have n eigenvectors are termed "pathological" matrices and are unusual. In this latter case, the matrix has *fewer* than n eigenvectors. Under no circumstances can a matrix have *more* than n eigenvectors.

Knowledge of the eigenvectors and eigenvalues of a matrix is very useful in most matrix applications. The quantum energy levels in quantum mechanics are obtained from the eigenvalues of a special matrix, called the Hamiltonian matrix. In fact, the set of eigenvalues of a matrix is often called the *spectrum* of the matrix, in analogy to the lines in spectroscopy. In the usual case that an n x n matrix has n independent eigenvectors (such a matrix is said to possess *completeness*), any n-vector, $\vec{A}$, can be decomposed in terms of these eigenvectors:

$$\vec{A} = \sum_{j=1}^{n} a_j \vec{\Psi}_j \ . \qquad \text{(A8)}$$

If the numbers a_j in (A8) are known, the action of K on $\vec{A}$ can be readily computed using (A5):

$$K \vec{A} = \sum_{j=1}^{n} K (a_j \vec{\Psi}_j) \ .$$

Using (A6), we then have

$$K \vec{A} = \sum_{j=1}^{n} a_j E_j \vec{\Psi}_j \qquad \text{(A9)}$$

where E_j is the eigenvalue corresponding to $\vec{\Psi}_j$. In fact, we can generalize (A9) to

$$K^m \vec{A} = \sum_{j=1}^{n} a_j E_j^m \vec{\Psi}_j \qquad \text{(A10)}$$

where m is any real number! Equation (A10) can be further extended to functions of matrices, *e.g.*,

$$e^K \vec{A} = \sum_{j=1}^{n} a_j e^{E_j} \vec{\Psi}_j \ . \qquad \text{(A11)}$$

The point is that introduction of the eigenvectors and eigenvalues of K and the decomposition of a general vector, $\vec{A}$, in terms of the $\vec{\Psi}_j$, enable a complete description of the actions of K or any function of K on any vectors. It is this powerful property of eigenvectors which is used in this chapter and also in Chapter 6.

The author is quite aware that the introduction to matrices presented here is simplified. For a more detailed exposition of matrices, the reader is referred to the book by Wilkinson (1965).

CHAPTER 2 REFERENCES

Broecker, W. S. (1971) A kinetic model for the chemical composition of sea water. Quarternary Research, 1, 188-207.

Garrels, R. M. and Mackenzie, F. T. (1971) *Evolution of Sedimentary Rocks.* Norton and Co., New York, 397 p.

——— and Perry, E. A., Jr. (1974) Cycling of carbon, sulfur, and oxygen through geologic time. In *The Sea,* Vol. 5, Goldberg (Ed.), p. 303-336.

Glansdorff, P. and Prigogine, I. (1971) *Thermodynamic Theory of Structure, Stability and Fluctuations.* Wiley-Interscience, New York, 306 p.

Holland, H. D. (1972) The geologic history of sea water -- an attempt to solve the problem. Geochim. Cosmochim. Acta, 35, 637-653.

——— (1978) *The Chemistry of the Atmosphere and Oceans.* John Wiley & Sons, New York, 350 p.

Holser, W. T. and Kaplan, I. R. (1966) Isotope geochemistry of sedimentary sulfates. Chemical Geology, 1, 93-135.

Lasaga, A. C. (1980) The kinetic treatment of geochemical cycles. Geochim. Cosmochim. Acta, 44, 815-828.

——— and Holland, H. D. (1976) Mathematical aspects of non-steady-state diagenesis. Geochim. Cosmochim. Acta, 40, 257-266.

Lerman, A. (1979) *Geochemical Processes: Water and Sediment Environments.* John Wiley & Sons, New York, 481 p.

———, Mackenzie, F. T., and Garrels, R. M. (1975) Modeling of geochemical cycles: phosphorus as an example. Geol. Soc. Amer. Mem., 142, 205-218.

Mackenzie, F. T. (1975) Sedimentary cycling and the evolution of sea water. In *Chemical Oceanography,* Vol. 1, Riley (Ed.), 309-364.

Prigogine, I. (1967) *Introduction to Thermodynamics of Irreversible Processes.* Wiley-Interscience, New York, 147 p.

——— (1980) *From Being to Becoming: Time and Complexity in the Physical Sciences.* W. H. Freeman Co., San Francisco, 272 p.

Procaccia, I. and Ross, J. (1977a) Stability and relative stability in reactive systems far from equilibrium I. Thermodynamic analysis. J. Chem. Phys., 67, 5558-5564.

——— (1977b) Stability and relative stability in reactive systems far from equilibrium II. Kinetic analysis of relative stability of multiple stationary states. J. Chem. Phys., 67, 5565-5571.

Pytkowicz, Rocardo M. (1975) Some trends in marine chemistry and geochemistry. Earth-Science Reviews, 11, 1-46.

Southam, J. R. and Hay, W. W. (1976) Dynamical formulation of Broecker's model for marine cycles of biologically incorporated elements. Math.Geol., 8, 511-527.

Wilkinson, J. H. (1965) *The Algebraic Eigenvalue Problem.* Clarendon Press, London, 657 p.

Chapter 3

KINETICS of WEATHERING and DIAGENESIS

Robert A. Berner

INTRODUCTION

Most solid materials found near the surface of the earth are not thermodynamically stable there. Some outstanding examples are: unweathered igneous feldspars, semi-amorphous "clay" (instead of well-crystallized phyllosilicates), and fine-grained goethite (instead of hematite) in soils, and undecomposed organic matter, aragonite (instead of calcite), and opaline silica (instead of quartz) in marine sediments. This metastability arises from fundamental causes which are physical, biological, and chemical in origin. First, igneous and metamorphic minerals, stable only at elevated temperatures, are carried into the zone of weathering by such physical processes as uplift and volcanism, and the energy necessary for this to occur is derived from within the earth. Second, because of an input of solar energy and its trapping by photosynthesis, organisms are able to biologically synthesize organic matter and other unstable substances such as aragonite and opaline silica. Finally, during weathering and diagenesis less stable minerals can form because of chemical factors which bring about the inhibition of nucleation and growth of more stable phases.

The preponderance of metastability near the earth's surface is a direct consequence of the slowness of chemical reaction rates at low temperatures. Thus, kinetics plays more of a leading role in earth surface geochemistry than it does at higher temperatures. Furthermore, we can study the rates and mechanisms of reactions by observing various stages during reaction both in the field and in the laboratory. The purpose of this chapter is to show how several different aspects of chemical kinetics can be brought to bear on the study of two major earth surface processes: weathering and diagenesis. Before proceeding directly to discussions of weathering and diagenesis, it is necessary to present some fundamental kinetic concepts which will enable the reader to better understand the discussions.

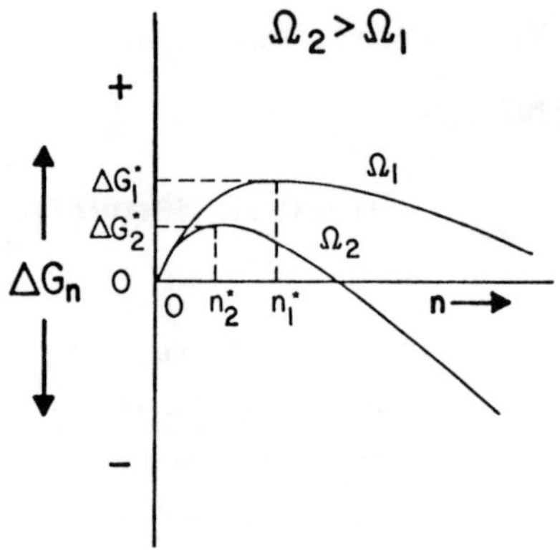

Figure 1. Plot of free energy of precipitation: ΔG_n vs crystal size illustrating nucleation and growth. The free energy change increases up to a maximum due to nucleation and then decreases due to crystal growth. The diagram is for a single crystal containing n atoms, ions or molecules. Note the effect of the degree of supersaturation Ω on the position of each curve. (Modified after Nielsen, 1964 and Berner, 1980)

SOME FUNDAMENTAL CONCEPTS

In this paper we will be mainly concerned with the crystallization and dissolution of minerals from aqueous solution and the bacterial decomposition of organic matter. Basic principles controlling these three processes are presented in this section. Discussion is based on the following references which may be consulted for further details: Nielsen (1964) and Ohara and Reid (1973) for crystallization, Berner (1978a) for dissolution, and Berner (1980) for all three processes.

Crystallization

The crystallization, or precipitation, of a solid substance from aqueous solution can be divided into two processes: nucleation and crystal growth. Nucleation occurs prior to growth and distinction between the two processes can be made in terms of Figure 1. As a body precipitating from solution begins to increase in size (or in the number of atoms it includes) it encounters a free energy barrier to further growth. This barrier is a consequence of the fact that an interface between the growing body and the solution forms, and this results in an increase in free energy due to the creation of the interface. At small sizes the interfacial free energy dominates over the drop in free energy accompanying the relief of supersaturation, and as a result there is a net free energy increase. The free energy increases up to a maximum where the interfacial and bulk terms balance one another. At this point the body is referred to as the *critical nucleus* and the process leading up the free energy "hill" is called nucleation. During nucleation the body is referred to as a crystal embryo.

Once the critical nucleus has formed, further increase in its size can take place spontaneously with a net decrease in free energy.

This process is referred to as crystal growth and the growing body is considered a true *crystal*. Growth (and nucleation of other crystals) continues until enough material is removed from solution that supersaturation is relieved and equilibrium is attained.

As the degree of supersaturation is increased, the ease of nucleation is increased. This is shown in Figure 1 by a decrease in the size of the critical nucleus and a decrease in the free energy of nucleation, ΔG^* as the degree of supersaturation Ω is increased. (The parameter Ω is defined as the actual ion activity product divided by the equilibrium or solubility product.) The rate of nucleation is a strong exponential function of the free energy of nucleation; consequently, at high Ω values nucleation is very fast. Since nucleation and growth compete for dissolved material, at high degrees of supersaturation the rate of nucleation may be so fast that most of the excess solute is precipitated in the form of critical nuclei with little left over for growth. Since critical nuclei are commonly in the size range 10-100 Å, such rapid crystallization can result in the formation of a very fine grained precipitate. In this way the formation of poorly crystallized minerals in nature and in the laboratory can be explained. A common example is the reddish gelatinous precipitate of $Fe(OH)_3$ obtained by the mixing of concentrated solutions of ferric chloride and sodium hydroxide in the laboratory.

Low levels of supersaturation, by contrast, can result in good crystallinity. This comes about because the nucleation rate is so slow that most excess dissolved material is consumed by crystal growth on a limited number of critical nuclei. Most minerals which exhibit some degree of crystallinity from their x-ray diffraction patterns fall in this category. Thus, the rest of the discussion of crystallization will be devoted to crystal growth.

Crystal growth involves the transport of dissolved species to and from the surface of a crystal and various chemical reactions occurring at the surface. The latter includes adsorption, ion exchange, dehydration of ions, formation of two-dimensional nuclei on the surface, diffusion along the surface, ion pair formation, etc. The rate of growth is limited by the slowest step within a whole chain of processes and the nature of the rate-limiting step is not usually known. However, as a first approximation, the rate of crystal growth can be

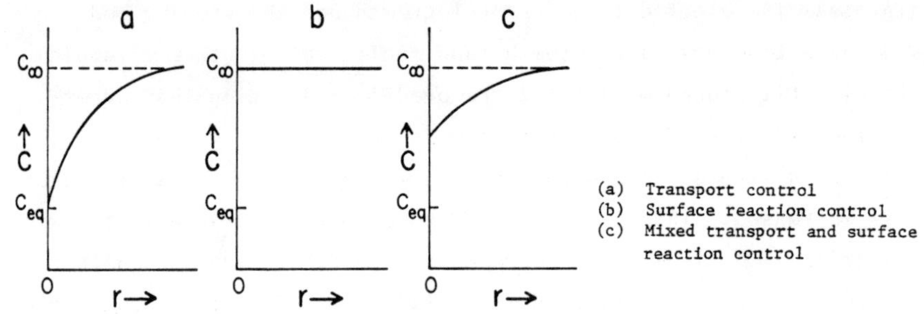

Figure 2. Schematic representation of concentration in solution C as a function of radial distance r, from the surface of a growing crystal. C_{eq} = saturation concentration, C_∞ = concentration out in solution. (After Berner, 1980)

characterized as being controlled either by transport of ions to the surface (transport-controlled), by reactions at the surface (surface-reaction controlled), or by a combination of both processes. A comparison of the three mechanisms is shown in Figure 2.

In pure transport-controlled growth (Fig. 2-a) ions (atoms, molecules) are added to the surface of the crystal so rapidly that migration of ions in solution to take their place cannot keep pace. As a result concentrations in solution adjacent to the crystal surface fall until they almost reach the equilibrium or saturation level. Further growth is limited by the rate at which additional ions can be transported to the crystal surface and the slowest process is that of molecular (ionic) diffusion. Transport to the surface can be accelerated by bulk flow of solution past the growing crystal or by stirring. Thus, transport-controlled growth is a strong function of the hydrodynamic state of the solution.

In pure surface-reaction controlled growth (Fig. 2-b) ion attachment to the surface is so slow that replenishment of ions in solution near the surface is easily accomplished by molecular diffusion and other transport processes. Concentrations in the near-surface zone are little different than those in the bulk solution, and growth rate is not affected by the hydrodynamic state of the solution.

Intermediate situations (Fig. 2-c) arise where ion attachment is sufficiently rapid that concentrations in the near-surface region are lower than they are in the bulk solution but not rapid enough to bring about a lowering to the saturation value. In this case the rate of growth is controlled by both transport and surface reactions.

Discernment of whether the rate of crystal growth of a given mineral is controlled by transport or surface-reaction can be accomplished by comparing measured rates of growth (from laboratory experiments or, preferably, from field methods) with those calculated for growth via molecular diffusion. Molecular diffusion is the slowest process by which crystals can grow and still have their rate controlled by transport. Faster measured rates indicate (advective) transport control whereas slower measured rates point to control by surface chemical reactions. Calculation of the approximate rate of growth via molecular diffusion is normally done using the expression (Nielsen, 1964):

$$\frac{dr_c}{dt} = \frac{\upsilon D_s (C_\infty - C_{eq})}{r_c} \tag{1}$$

where: r_c = average radius of the crystals,
υ = molar volume of the crystalline substance,
D_s = coefficient of molecular diffusion in aqueous solution,
C_∞ = concentration in solution away from the crystal surface,
C_{eq} = equilibrium concentration adjacent to the crystal surface,
t = time.

For constant C_∞ (appropriate to an open system), equation (1) can be integrated to:

$$r_c = [2\upsilon D_s (C_\infty - C_{eq}) t]^{\frac{1}{2}} \tag{2}$$

(Note that equations (1) and (2) are most correct when applied to equi-sized, equidimensional (*i.e.*, cubes, spheres, etc.) crystals separated by at least five diameters.)

Other ways of deducing overall growth mechanisms are through the use of different temperatures and different stirring rates in laboratory crystallization experiments. A stronger temperature dependence than that predicted for aqueous diffusion-controlled growth indicates a surface-reaction mechanism whereas a dependence on stirring rate, as discussed above, indicates transport control.

Actual mechanisms offered to explain surface-reaction controlled growth are varied and numerous. The two most commonly cited ones are surface nucleation control and dislocation control. In the former, rates of ion attachment are limited by the rate at which a new two-

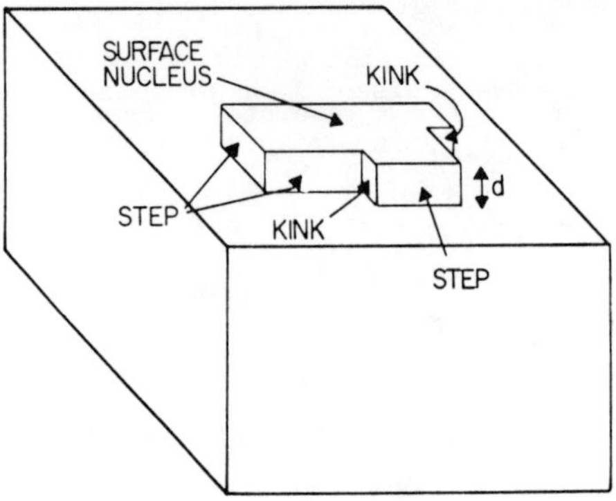

Figure 3. Idealized representation of the surface of a crystal. Dimension d represents one atom, molecule, unit cell, etc. On the flat crystal surface a "two-dimensional" surface nucleus is present which exhibits mon-atomic steps and kinks. (After Berner, 1980)

dimensional nucleus is formed on an otherwise atomically flat crystal face. The new nucleus is needed to provide unit-cell-sized steps and kinks on the surface which are energetically favored points of attachment. This is shown in Figure 3. In dislocation-controlled growth, built-in steps and kinks are provided by the intersection of screw dislocations with the crystal surface. By this mechanism the energy of surface nucleation is already provided by dislocation outcrops and growth spirals emanating from them, and, as a result, growth may occur at very low degrees of supersaturation. Although dislocation-controlled growth has been studied considerably, further discussion is beyond the scope of this paper, and the reader is referred to the book by Ohara and Reid (1973) or the classic work of Burton, Cabrera and Frank (1951) for additional details.

An important process, so far unmentioned, which can appreciably affect the rate of growth (and nucleation) of crystals from aqueous solution, is the adsorption of ions, molecules, etc., from solution onto the crystal surface (see Chapter 1). Natural waters contain many dissolved constituents which readily adsorb onto growing crystals, and some of these can serve as growth inhibitors. If normally available growth sites, such as kinks, are blocked by strongly adsorbed ions, then overall rates of growth can be greatly diminished. Also, preferential adsorption of such "poisons" onto certain crystal faces can result in the deceleration of their growth and a consequent alteration in crystal habit. Octahedral halite grown in the presence of urea is a classical example. The adsorption of inhibitors can so alter growth rates that normal rate dependences on supersaturation, as given for example by the usual models for dislocation-controlled growth (Burton *et al.*, 1951), are not obeyed (*e.g.*, see Ohara and Reid, 1973).

Dissolution

Like crystallization the overall process of dissolution of solids by an aqueous solution is controlled either by transport or by surface chemical reactions. The classification scheme shown in Figure 2 can also be applied to dissolution with the only difference being that concentrations adjacent to the crystal surface, for transport-controlled dissolution, are higher than they are in the bulk solution. Obvisouly, dissolution occurs only where undersaturation of the bulk solution is present, in other words, where $\Omega < 1$.

Methods for the discernment of the rate-controlling mechanism of dissolution are similar to those for crystallization. The rate of diffusion-controlled dissolution can also be obtained from equation (1). Note in this case that dr_c/dt is negative because C_{eq} is greater than C_∞. Comparison of measured rates of dissolution with that calculated for diffusion control, via equation (1), enables elucidation of the overall rate-controlling mechanism. Determination of the effects of temperature variation and stirring in the laboratory can also be used for the same purpose.

Dissolution is dissimilar to crystallization in one important respect. It is that three-dimensional nucleation is not necessary in dissolution since the dissolving crystals are already present. In fact, observations of partially-dissolved crystals can be used as an additional method for deducing the dissolution rate-controlling mechanism. In the case of rate control by surface reaction, slow partial dissolution of the surface results in the formation of crystallographically controlled features, such as large, well-developed etch pits. This arises from the fact that major dissolution originates only at points of excess energy on the surface such as dislocation outcrops. In the case of transport-controlled dissolution, attack of the surface is so rapid and non-specific that etching occurs virtually everywhere, and, as a consequence, only general rounding results. A comparison of morphological features produced by each mechanism for the same mineral, calcite, is shown in Figure 4.

As in the case of crystallization, rates of dissolution are considerably affected by the presence of adsorbed inhibitors. Such inhibition leads to surface-reaction controlled dissolution whose rate

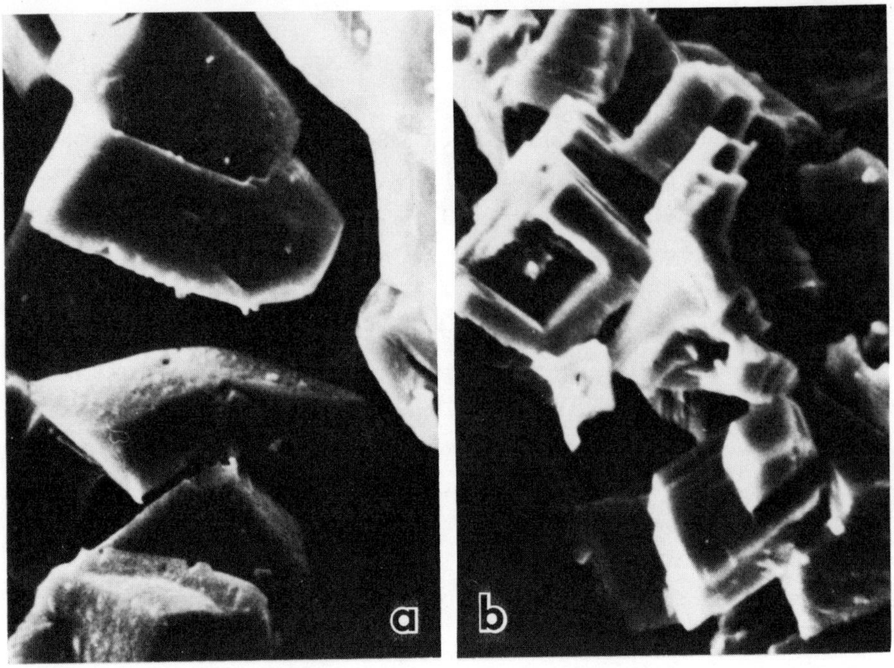

Figure 4. Electron photomicrographs of calcite which has undergone partial dissolution in seawater (X4000)

(a) Transport controlled dissolution; pH = 3.9. Note general rounding.
(b) Surface reaction controlled dissolution; pH = 6.0. Note angular, crystallographically controlled etch features.

(After Berner, 1978a, 1980 and Berner and Morse, 1974)

dependence on the degree of undersaturation is much greater than that predicted by commonly adopted theories. For example, the work of Morse (1978) and Keir (1980) has shown that calcite, under earth-surface conditions dissolves in seawater at a rate that is proportional to the fourth or fifth power of the concentration difference ($C_{eq}-C_{\infty}$). (Compare with equation (1).) This high-order kinetics is believed to be caused by the adsorption of dissolution-inhibiting phosphate ions onto the calcite surface (Berner and Morse, 1974; Morse and Berner, 1979).

Organic Matter Decomposition

Organic matter decomposition is discussed in the present paper because it exerts a major control on the chemistry and authigenic mineralogy of shallowly buried sediments. Rate laws for the decomposition of organic matter in sediments have not been well established. The model used here is that which I have used before (Berner, 1974; 1980). In it, it is assumed that the overall process of organic decomposition (which includes a number of individual microbial steps) to CO_2 and other simple inorganic molecules, is first order with respect to the initial polymeric material undergoing decomposition. In other words:

$$\frac{dG}{dt} = -kG \tag{3}$$

where: G = concentration of organic carbon undergoing decomposition (not total organic carbon),

k = first-order rate constant,

t = time.

The parameter G represents that fraction of the total organic matter actually undergoing decomposition at any given time. The reactivity, or k value, of organic compounds (*i.e.*, metabolizability) varies greatly and this is reflected by large variations in k from sediment to sediment and with depth in a single sediment (Toth and Lerman, 1977; Berner, 1978b; 1980; Jorgensen, 1978). Total organic carbon at the time of deposition is divided into various fractions, G_α, G_β, etc., according to their reactivity or k value. After deposition the most reactive compounds are destroyed first (G_α), followed by the next most reactive compounds (G_β), and so forth. As a result, sediments buried at slow rates contain only the less reactive organic compounds since the more reactive ones are destroyed at the sediment-water interface.

For each organic fraction designated as G the value of k varies with the process of decomposition. Organic matter in sediments is destroyed (microbially) by a variety of oxidizing agents in a definite succession (*e.g.*, Claypool and Kaplan, 1974). Dissolved oxygen is first used until it is entirely consumed, then dissolved nitrate until it is all gone, and then dissolved sulfate. Once all sulfate is used up there are no more inorganic oxidizing agents, and some of the organic carbon then appears as methane. Because of differences in free energy yields and metabolic pathways, decomposition of the same organic compounds by each of these processes can result in a different value of k in each case.

The application of chemical kinetics to weathering is a relatively recent phenomenon. Most theoretical treatments of weathering have been in terms of equilibrium or irreversible, quasi-equilibrium models (Helgeson *et al.*, 1969; Fritz, 1975; Gac *et al.*, 1978; Tardy *et al.*, 1974; Fouillac *et al.*, 1977; Lindsay, 1979). One of the first workers to apply true chemical kinetics to weathering problems was Wollast (1967), who described the results of experiments in which ground K-feldspar was placed in buffered aqueous solutions and the release of silica with time was measured. Wollast found that the rate of addition of silica to solution decreased with time and he explained this decrease in terms of the formation of a protective layer of aluminous precipitate on the surface of the feldspar. The rate of dissolution was assumed to be controlled by the rate at which silica and cations could diffuse through the surface layer. Calculations based on a diffusion model, which is similar to that often used to describe the protective corrosion of metals, resulted in calculated diffusion coefficients D which are of the same order of magnitude as expected for solid diffusion (10^{-15} to 10^{-20} $cm^2 sec^{-1}$).

Subsequent workers (Helgeson, 1971; Pačes, 1973; Busenberg and Clemency, 1976) have expanded on the Wollast model and have suggested that the release of silica and cations from alkali feldspars can be described in terms of "parabolic kinetics" where the concentration in solution builds up in direct proportion to the square root of time. The protective surface layer is presumed to be either a precipitate of aluminum hydroxide and/or clay minerals, or "feldspar" in which the cations have been replaced by hydrogen (or hydronium) ions.

The concepts of parabolic kinetics and a protective surface layer formed on alkali feldspar during weathering have been tested by subsequent work (Petrović *et al.*, 1976; Petrović, 1976; Holdren and Berner, 1979; Berner and Holdren, 1979). In the studies of Petrović *et al.* and Holdren and Berner, ground sanidine and albite were subjected to leaching by aqueous solutions in the laboratory under conditions essentially identical to those employed by Wollast. Parabolic release of silica was found, but unexpectedly, the surfaces of the reacted feldspar grains were not found to be altered in any way. The Wollast model predicts an altered surface of several hundred Ångströms thickness, but, using x-ray photoelectron spectroscopy (XPS), we found that no alteration product any thicker than about 5-15 Å could be present on the feldspar surface. In addition, we found that the parabolic kinetics could not be reproduced if the ground feldspar grains used for experimentation were first treated with HF to remove ultrafine and strained particles produced by grinding. This is shown in Figure 5. We obtained only linear kinetics, or a constant rate of release of silica to solution with time, with HF-treated material. We believe that the "parabolic" kinetics obtained by other workers is the result of varying rates of dissolution of particles produced during grinding. Very fine, submicron-sized particles, as well as strained regions on larger particles, because of excess surface free energy, should dissolve more rapidly than the larger grains. Initial silica release, therefore, should be faster due to preferential dissolution of the fine particles, and deceleration with time should occur as the particles are consumed. We actually observed disappearance of these particles during dissolution. From our results we conclude that feldspar dissolves in the laboratory according to linear kinetics and does not produce an altered surface layer. Parabolic kinetics is most likely an experimental artifact due to grinding.

Our laboratory results are esssentially verified from studies of feldspar grains taken from soils (Berner and Holdren, 1979). Under the scanning electron microscope (SEM) we found clay adhering to the surfaces of feldspar grains, but this clay could not have been protective in the sense envisioned by Wollast (1967). Shrinkage, upon drying (Fig. 6) shows that the clay is highly hydrous and not likely to be protective. (Typical diffusion coefficients in hydrous clay average about 10^{-6} cm^2sec^{-1}, not 10^{-15} to 10^{-20} cm^2sec^{-1} as required by the protective layer model.) Also, in most soils studied the clay does not adhere strongly (as would be expected if it were protective) since it can be removed by ultrasonic cleaning. Finally, examination of the surfaces of ultrasonically cleaned feldspar grains by XPS indicated a surface composition in the outermost few tens of Ångströms essentially identical to grain interiors. No cation depletion or aluminum enrichment, as expected for an altered surface layer, was found.

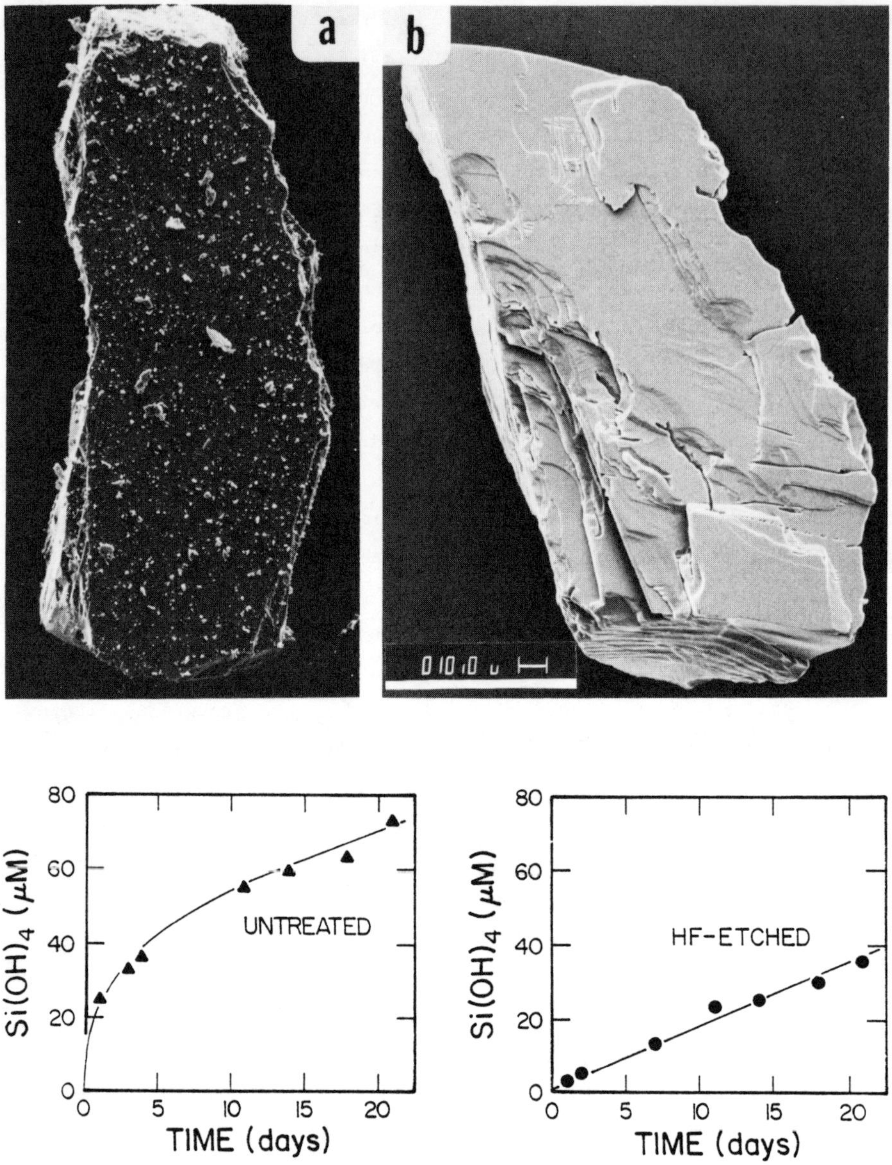

Figure 5. Effect of sample preparation on the dissolution behavior of albite feldspar. Albite grains, shown in SEM photomicrographs were (a) untreated and (b) HF-etched, after grinding. Each silica plot is for material pictured above it. (Adapted from Berner and Holdren, 1979)

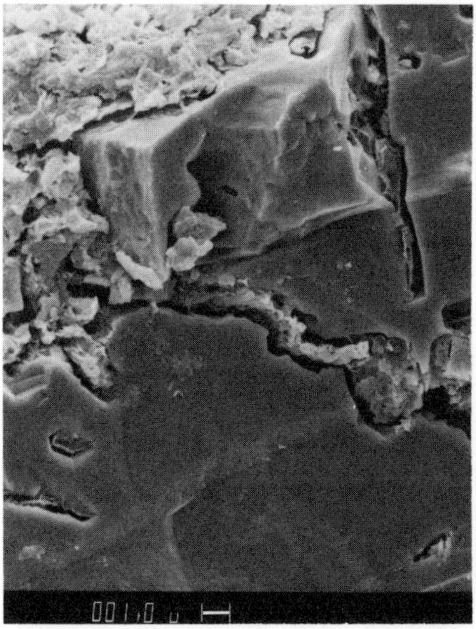

Figure 6. SEM photomicrograph of a soil feldspar grain showing that clay adhering to the feldspar surface after ultrasonic cleaning has contracted and separated from the surface upon drying (see crack in center of photomicrograph). Sample from a gray-brown podzolic soil, Sangre de Cristo Mountains, New Mexico. (After Berner, 1978a and Berner and Holdren, 1979)

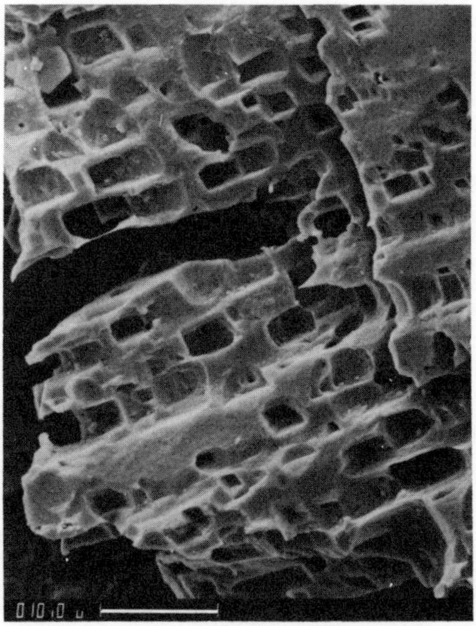

Figure 7. SEM photomicrograph of a feldspar grain (ultrasonically cleaned to remove clay) taken from a lateritic soil near Piedmont, N.C., U.S.A. Note prominent prismatic etch pits. (After Berner, 1978a; and Berner and Holdren, 1979)

If feldspar dissolution during weathering is not controlled by diffusion through a protective surface layer, then how does it dissolve? We feel, along with Aagaard and Helgeson (1981) and Dibble and Tiller (1981), that the limiting step in dissolution is not diffusion but chemical reaction at the feldspar-water interface. Proof of this is provided by SEM observations of ultrasonically cleaned feldspar grains taken from soils. The feldspar grains show the growth and development of characteristic, crystallographically-controlled etch pits on the surface which indicates dissolution rate control by surface chemical reaction (Wilson, 1975; Berner and Holdren, 1979). Some examples are shown in Figure 7. The etch pits indicate selective attack of the feldspar surface by soil acids, at points of excess energy such as dissolution outcrops. General attack of the surface with consequent rounding, as predicted for diffusion-controlled dissolution, is not found.

We discovered that we could duplicate the etch features found on soil grains by treating fresh feldspars with HF for extended periods in the laboratory (Berner and Holdren, 1979). Also, the same type of pitting (prismatic-shaped pits) was found on a variety of K and Na feldspars taken from several different soils where different soil acids should have been present. From this we conclude that pit shape is not sensitive to the type of acid attacking the feldspar as is the case for calcite dissolution (Keith and Gilman, 1960).

Our overall conclusions with regard to the dissolution of feldspar have been corroborated more recently from studies of pyroxene and amphibole dissolution, both in the laboratory and in soils (Berner *et al.*, 1980; Schott *et al.*, 1981). Pyroxenes and amphiboles also dissolve via linear and not parabolic kinetics in the laboratory when HF pretreatment is employed to remove ultrareactive fine particles produced by grinding. Dissolution in soils also produces characteristic etch pits showing that dissolution rate of pyroxenes and amphiboles is controlled by surface chemical reactions (example in Figure 8). One difference is that pyroxenes show some cation depletion, relative to silicon, on their surfaces, but this depletion extends to a depth of only a few Ångströms. This is too thin to be described as a diffusion-inhibiting protective surface layer. Diffusion loses all meaning when the total distance of diffusion is of the same order of magnitude as the size of the diffusing entities.

The mechanism of dissolution of minerals other than feldspars, pyroxenes and amphiboles during weathering has not been determined. Nevertheless, some prediction can be made. This is because there appears to be a reasonably good correlation between the solubility of a mineral and the rate-controlling mechanism by which it dissolves. This is shown in Table 1. The less soluble minerals all dissolve by surface reaction control. AgCl is an apparent exception but this may be caused by photochemical changes during kinetic studies, *e.g.*, see Nielsen and Söhnel (1971). Since most minerals involved in weathering have solubilities falling in the lower range shown in the table, it is likely that the rate of dissolution of many other minerals during weathering is also controlled by surface chemical processes and not by diffusion, either in aqueous solution or through protective surface layers.

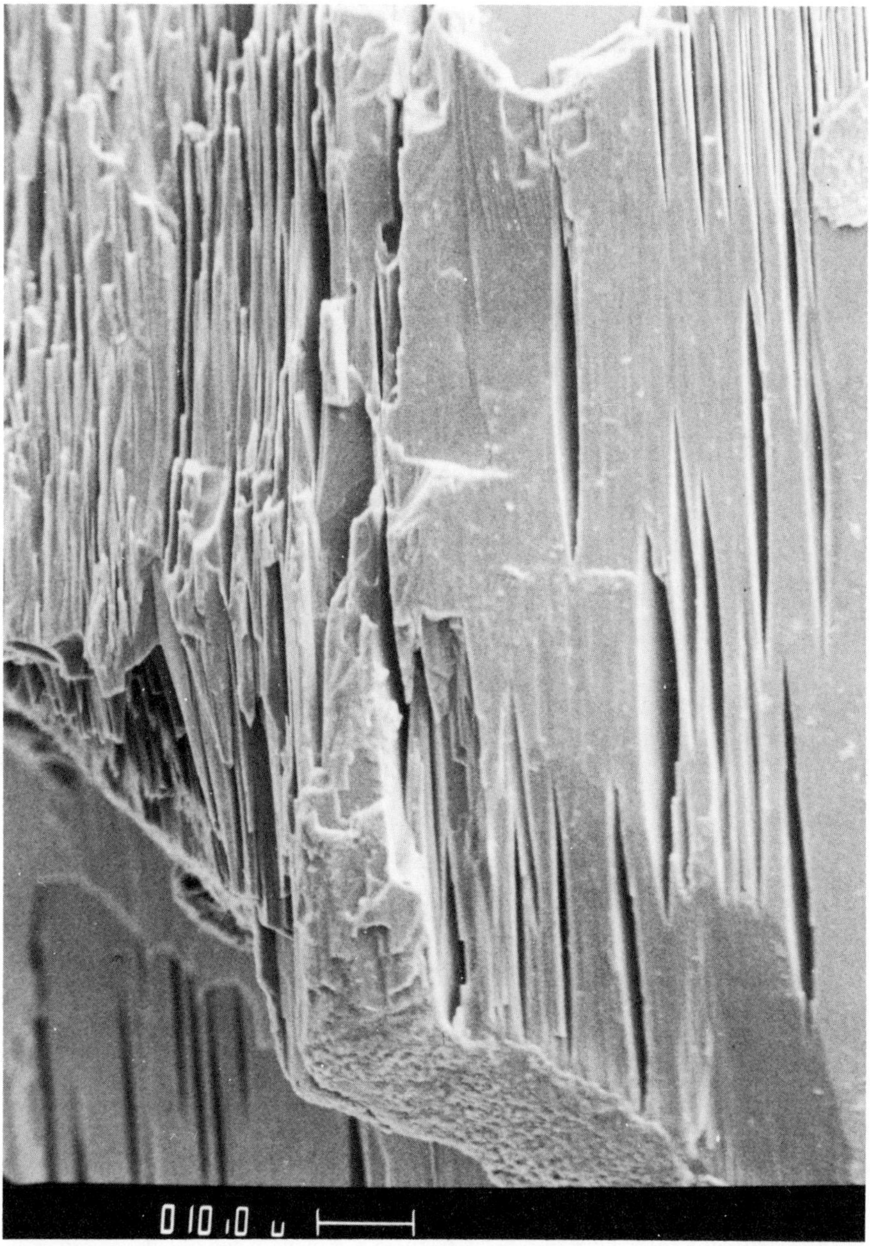

Figure 8. SEM photomicrograph of a hornblende grain (ultrasonically cleaned to remove clay) taken from a soil in Ashe County, North Carolina, U.S.A. Note prominent lens-shaped etch pits.

Table 1. Dissolution rate-controlling mechanism for various substances arranged in order of solubilities in pure water (mass of mineral which will dissolve to equilibrium.) (After Berner, 1978a; 1980)

Substance	Solubility mole per liter	Dissolution rate control
$Ca_5(PO_4)_3OH$	2×10^{-8}	Surface-reaction
$KAlSi_3O_8$	3×10^{-7}	Surface-reaction
$NaAlSi_3O_8$	6×10^{-7}	Surface-reaction
$BaSO_4$	1×10^{-5}	Surface-reaction
$AgCl$	1×10^{-5}	Transport
$SrCO_3$	3×10^{-5}	Surface-reaction
$CaCO_3$	6×10^{-5}	Surface-reaction
Ag_2CrO_4	1×10^{-4}	Surface-reaction
$PbSO_4$	1×10^{-4}	Mixed
$Ba(IO_3)_2$	8×10^{-4}	Transport
$SrSO_4$	9×10^{-4}	Surface-reaction
Opaline SiO_2	2×10^{-3}	Surface-reaction
$CaSO_4 \cdot 2H_2O$	5×10^{-3}	Transport
$Na_2SO_4 \cdot 10H_2O$	2×10^{-1}	Transport
$MgSO_4 \cdot 7H_2O$	3×10^{0}	Transport
$Na_2CO_3 \cdot 10H_2O$	3×10^{0}	Transport
KCl	4×10^{0}	Transport
$NaCl$	5×10^{0}	Transport
$MgCl_2 \cdot 6H_2O$	5×10^{0}	Transport

A simple calculation shows how incorrect it is to assume that a typical, relatively insoluble, mineral dissolves during weathering at a rate predicted for diffusion or transport control in aqueous solution. Using the solubility for K-feldspar at a soil pH of 6 (rather than the value given in Table 1 which is for a hydrolysis pH of 8), typical values of $D_s = 10^{-6}$ $cm^2 sec^{-1}$ and $r_c = 200$ μm, and the reasonable assumption for dilute soil solutions that $C_\infty << C_{eq}$, we obtain from equation (2) the time necessary to completely dissolve 200 μm sized feldspar grains by molecular diffusion in solution. The result is t = 14 months. (This is a maximum value since flow of the soil or ground water has not been considered.) Obviously, this result is incorrect since K-feldspars persist in soils for thousands to millions of years; thus, actual dissolution rate is far slower than that predicted for aqueous diffusion control.

DIAGENESIS

Diagenesis, especially early diagenesis, lends itself readily to a kinetic approach. The principal reason for this is that a time frame is associated with sedimentary layers. In modern unlithified sediments overlain by water, the most convenient frame of reference is the sediment-water interface. Depth below this interface is proportional to time, and burial can be viewed as the flow of sediment downward away from the interface. If one, then, fixes on a given depth in a sediment, processes affecting chemical properties at that depth must also include burial advection. Also, since vertical concentration gradients are generally much greater than horizontal ones, one can treat the properties as being only a function of depth and time. Mathematically, these ideas can be expressed in the form of a diagenetic equation:

$$\left(\frac{\partial C}{\partial t}\right)_x = \frac{dC}{dt} - \omega\left(\frac{\partial C}{\partial x}\right)_t \qquad (4)$$

where: C = concentration, x = depth in the sediment, t = time, and ω = rate of burial (dx/dt). Here the partial derivative refers to changes at any fixed depth x and the total derivative refers to changes occurring in a fixed sediment layer undergoing burial.

A very useful concept is that of steady state diagenesis. This occurs when all diagenetic processes balance one another so that there is no change with time at a given depth. Mathematically, this is expressed as:

$$\left(\frac{\partial C}{\partial t}\right)_x = 0 \qquad (5)$$

so that from (4):

$$\frac{dC}{dt} = \omega\left(\frac{\partial C}{\partial x}\right)_t \qquad (6)$$

Steady state diagenesis can be viewed as a balance set up such that material supplied or depleted from above by burial is consumed or produced by chemical reaction (and diffusion) and, as a result, plots of concentration versus depth always look the same at successive times. This is all illustrated in Figure 9. Note that there is diagenetic

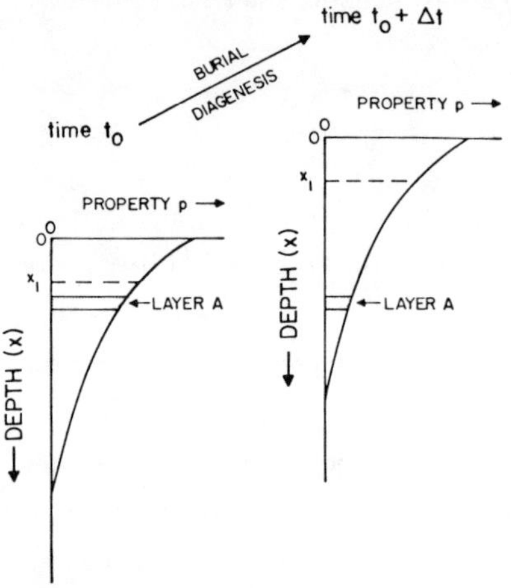

Figure 9. Diagramatic representation of steady state diagenesis. Upon burial, the value of property p does not change for a fixed depth x_1 (or x = 0), but does change for a fixed layer A. (After Berner, 1980)

change within a given layer undergoing burial but no change at a fixed depth. Since the sediment-water interface is also a fixed depth (x = 0), steady state diagenesis means that the composition of material added to the sediment at the time of deposition does not vary with time. Obviously, steady state is most likely where depositional conditions are constant and continuous. This is best exhibited by fine-grained muds deposited in lakes or the oceans.

The treatment of diagenesis in this paper will be restricted to a brief discussion of organic matter decomposition, via sulfate reduction, in fine-grained marine sediments. This is done merely to demonstrate a chemical kinetic approach to diagenesis and is not intended as a thorough treatment of either bacterial sulfate reduction or diagenetic processes. For a more complete discussion of quantitative approaches to diagenesis, the reader is referred to the book by Berner (1980). Here we will ignore the processes of compaction, bioturbation, adsorption, and externally forced water flow, and will concentrate only on chemical reaction, burial advection, and interstitial molecular diffusion. Also, only steady state diagenesis will be considered. In the case of sulfate reduction these assumptions are reasonably valid if we restrict our discussion to sediment depths below the top 10-20 cm where bioturbation (the disturbance of bottom sediment by macro-organisms) and consequent water movement can be appreciable. Also, sulfate does not strongly adsorb on solid particles in most sediments.

Bacterial sulfate reduction involves the reduction of dissolved sulfate in the interstitial water of a sediment to H_2S with a concomitant oxidation of organic matter to CO_2 and HCO_3^-. The overall process can be summarized by the reaction:

$$2CH_2O + SO_4^{2-} \rightarrow H_2S + 2HCO_3^- \qquad (7)$$

This process can take place only under strictly anoxic conditions and as a result it is normally restricted to buried sediments since most all surface waters contain dissolved oxygen. Sulfate reduction is an important process both for the decomposition of organic matter and for the formation of pyrite (which forms by the reaction of H_2S with detrital iron minerals in the sediment).

The rate of organic matter decomposition via sulfate reduction can be assumed to follow first-order kinetics as discussed earlier in this paper (see also Chapter 1). In other words, in terms of equation (3):

$$\frac{dG}{dt} = -kG \tag{8}$$

where G now refers to the concentration of CH_2O in reaction (7). (For reasons that will soon become apparent, G is expressed in units of mass of solid organic carbon per unit volume of enclosing pore water.) If we assume that the organic matter is present in a solid form so that the only processes affecting it are burial advection and decomposition, and that it consists of only one set of organic compounds with one k value (which is reasonable for sediments below the zone of bioturbation), then from equations (6) and (8) at steady state:

$$-\omega \frac{\partial G}{\partial x} - kG = 0 \tag{9}$$

Solution of this equation for the boundary conditions;

$$x = 0 \qquad x \to \infty$$
$$G = G_o \qquad G \to 0$$

yields:

$$G = G_o \exp[-\frac{k}{\omega} x] \tag{10}$$

Now from the stoichiometry of reaction (7), the rate of bacterial sulfate reduction is given by:

$$\frac{dC}{dt}_{bact} = \frac{1}{2}\frac{dG}{dt} \tag{11}$$

where C = concentration of dissolved sulfate in mass per unit volume of interstitial water. (Equation (11) is simple since we have expressed G in terms of mass per unit volume of pore water.) From (8), (10), and (11), finally:

$$\frac{dC}{dt}_{bact} = -\frac{kG_o}{2} \exp[-\frac{k}{\omega} x] \tag{12}$$

The diagenetic equation for sulfate, since it is a dissolved species, involves diffusion as well as bacterial reduction and burial. At steady state the diagenetic equation is:

$$D_s \frac{\partial^2 C}{\partial x^2} - \omega \frac{\partial C}{\partial x} - \frac{kG_o}{2} \exp \left[- \frac{k}{\omega} x\right] = 0 \qquad (13)$$

where D_s = molecular diffusion coefficient of sulfate in the sediment. The first term on the left assumes that Fick's Second Law of diffusion describes the effect of diffusion on sulfate concentration. Solution of equation (13) for the boundary conditions:

$$x = 0 \qquad x \to \infty$$
$$C = C_o \qquad C \to C_\infty$$

yields:

$$C = (C_o - C_\infty) \exp \left(- \frac{k}{\omega} x\right) + C_\infty \qquad (14)$$

$$C_o - C_\infty = \frac{\omega^2 G_o}{2(\omega^2 + kD_s)} \qquad (15)$$

(The asymptotic concentration C_∞ strictly represents the concentration of sulfate when organic matter is exhausted, in other words, when G goes to zero. However, in most sediments sulfate is exhausted before organic matter and in this case C_∞ becomes a negative number and is used only as a curve fit parameter.)

From plots of measured dissolved sulfate concentrations versus depth, such as that shown in Figure 10, the values of k and G_o can be obtained by a combination of curve fitting to obtain C_o, C_∞, and k/ω and the use of independently determined values of ω and D_s. This has been done for a wide variety of sediments (Toth and Lerman, 1977; Berner, 1978b; 1980; Murray *et al.*, 1978; Filipek and Owen, 1980) resulting in values of k ranging over six orders of magnitude. Values of G_o are much more uniform. In this way, use of diagenetic equations along with sediment chemical data can be used to obtain fundamental rate data for organic matter decomposition.

The model presented above assumes that the rate of sulfate reduction is a function only of the concentration of metabolizable organic carbon and not also of sulfate. From a number of laboratory studies using natural sediments (*e.g.*, Martens and Berner, 1977) this is justified but only at sulfate concentrations greater than about 25% of that

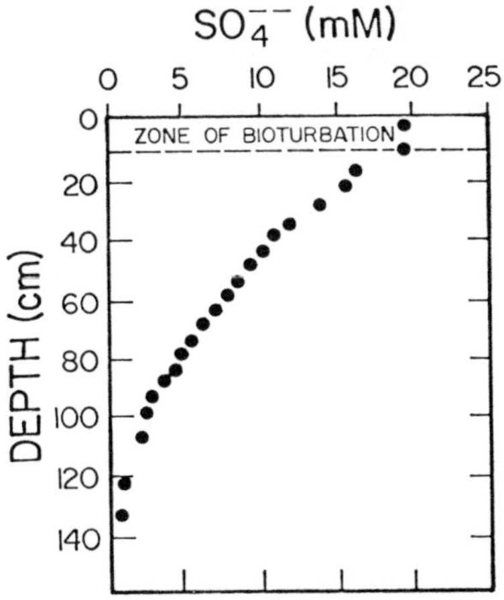

Figure 10. Representative sulfate concentration profile from an organic-rich near-shore sediment (FOAM site) from Long Island Sound, N.Y., U.S.A (Modified after Goldhaber, 1977)

found in the overlying seawater. As sulfate is depleted with depth to very low values, the rate of sulfate reduction in most sediments must become limited by sulfate itself. Recent unpublished laboratory experiments by Joseph Westrich at Yale University shows this to be the case. Westrich has found that in order to describe bacterial sulfate reduction at all depths and all sulfate concentrations, equation (12) must be modified to:

$$\frac{dC}{dt}_{bact} = -\frac{kG_o}{2}\exp[-\frac{k}{\omega}x]\left(\frac{C}{K_M+C}\right) \quad (16)$$

where K_M = curve fit parameter with the units of concentration. This dependence upon sulfate concentration shows that sulfate reduction obeys the Michaelis-Menten equation (Cornish-Bowden, 1976) which is a common rate law for bacterial processes. In this case K_M is referred to as the Michaelis constant. For sediments buried well below the zone of bioturbation Westrich found the value of K_M to average around 1.5 mM. Thus, at most sulfate concentrations, such as shown in Figure 10, equation (16) is sufficiently accurately represented by the much more simple equation (12) and the type of diagenetic modeling discussed above is justified. Use of equation (16), instead of (12), in the diagenetic equation would lead to a more complicated solution since in this case the diagenetic equation becomes non-linear.

CONCLUSION

In this paper I have tried to show how the subjects of weathering and diagenesis can be approached from a kinetic standpoint. This type of work depends upon an intimate interplay between field measurements, laboratory experiments, and mathematical theory. Because of the accessibility of soils and sediments, models based on laboratory measurements and theoretical calculations can be used along with field data to calculate rate parameters which are otherwise difficult to measure in the laboratory. Also, theoretical models can be tested in the field. Much more work needs to be done in the application of chemical kinetics to geological problems, and it is hoped that this chapter has shown some of the ways this can be done for the surficial environment.

ACKNOWLEDGMENTS

Research was supported by National Science Foundation Grants EAR 80-07815 and OCE 79-06919.

Aagaard, P. and Helgeson, H. C. (1981) Thermodynamic and kinetic constraints on reaction rates among minerals and aqueous solutions. 1. Theoretical considerations. Amer. J. Sci. (in press).

Berner, R. A. (1974) Kinetic models for the early diagenesis of nitrogen, sulfur, phosphorus, and silicon in anoxic marine sediments. In E. D. Goldberg (ed.), *The Sea*, vol. 5, Wiley, New York, p. 427-450.

________ (1978a) Rate control of mineral dissolution under earth surface conditions. Amer. J. Sci., 278, 1235-1252.

________ (1978b) Sulfate reduction and the rate of deposition of marine sediments. Earth Planet. Sci. Letters, 37, 492-498.

________ (1980) *Early Diagenesis: A Theoretical Approach.* Princeton Univ. Press, Princeton, New Jersey, 241 p.

________ and Holdren, G. R., Jr. (1979) Mechanism of feldspar weathering -- II. Observations of feldspars from soils. Geochim. Cosmochim. Acta, 43, 1173-1186.

________ and Morse, J. W. (1974) Dissolution kinetics of calcium carbonate in seawater: IV. Theory of calcite dissolution. Amer. J. Sci., 274, 108-134.

________, Sjöberg, E. L., Velbel, M. A., and Krom, M. D. (1980) Dissolution of pyroxenes and amphiboles during weathering. Science, 207, 1205-1206.

Burton, W. K., Cabrera, N., and Frank, F. C. (1951) The growth of crystals and the equilibrium structure of their surfaces. Royal Soc. London Philos. Trans., A-243, 299-358.

Busenberg, E. and Clemency, C. V. (1976) The dissolution kinetics of feldspars at 25°C and 1 atm CO_2 partial pressure. Geochim. Cosmochim. Acta, 40, 41-50.

Claypool, G. and Kaplan, I. R. (1974) The origin and distribution of methane in marine sediments. In I. R. Kaplan (ed.), *Natural Gases in Marine Sediments*, Plenum, New York, p. 99-139.

Cornish-Bowden, A. (1976) *Principles of Enzyme Kinetics*, Butterworths, London, 206 p.

Dibble, W. E., Jr. and Tiller, W. A. (1981) Non-equilibrium water/rock interactions -- I. Model for interface-controlled reactions. Geochim. Cosmochim. Acta, 45, 79-92.

Filipek, L. H. and Owen, R. M. (1980) Early diagenesis of organic carbon and sulfur in outer shelf sediments from the Gulf of Mexico. Amer. J. Sci., 280, 1097-1112.

Fouillac, C., Michard, G., and Bocquier, G. (1977) Une méthode de simulation de l'évolution des profils d'altération. Geochim. Cosmochim. Acta, 41, 207-213.

Fritz, B. (1975) Étude thermodynamique et simulation des réactions entre mineraux et solutions. Application a la géochimie des altérations et des eaux continentales. Univ. Louis Pasteur de Strasbourg, Inst. de Geologie, Mem. 41, 153 p.

Gac, J. Y., Badaut, D., Al-Droubi, A. and Tardy, Y. (1978) Compartement du calcium du magnésium et de la silice en solution. Précipitation de calcite magnésienne, de silice amorphe, et de silicates magnésiens au cors de l'evaporation des eaux du Chari (Tchad). Sci. Géol. Bull. (France), 31, 185-193.

Goldhaber, M. B., Aller, R. C., Cochran, J. K., Rosenfeld, J. K., Martens, C. S. and Berner, R. A. (1977) Sulfate reduction, diffusion, and bioturbation in Long Island Sound sediments: Report of the FOAM group. Amer. J. Sci., 277, 193-237.

Helgeson, H. C. (1971) Kinetics of mass transfer among silicates and aqueous solutions. Geochim. Cosmochim. Acta, 35, 421-469.

________, Garrels, R. M., and MacKenzie, F. T. (1969) Evaluation of irreversible reactions in geochemical processes involving minerals and aqueous solutions -- II. Application. Geochim. Cosmochim. Acta, 33, 455-481.

Holdren, G. R., Jr. and Berner, R. A. (1979) Mechanism of feldspar weathering -- I. Experimental studies. Geochim. Cosmochim. Acta, 43, 1161-1171.

Jørgensen, B. B. (1978) Comparison of methods for the quantification of bacterial sulfate reduction in coastal marine sediments. II. Calculation from mathematical models. Geomicrobiology J., 1, 20-47.

Keir, R. S. (1980) The dissolution kinetics of biogenic calcium carbonates in seawater. Geochim. Cosmochim. Acta, 44, 241-252.

Keith, R. E. and Gilman, J. J. (1960) Dislocation etch pits and plastic deformation in calcite. Acta Metall., 8, 1-10.

Lindsay, W. (1979) *Chemical Equilibria in Soils.* John Wiley, New York, 449 p.

Martens, C. S. and Berner, R. A. (1977) Interstitial water chemistry of anoxic Long Island Sound sediments. I. Dissolved gases. Limnology and Oceanography, 22, 10-25.

Morse, J. W. (1978) Dissolution kinetics of calcium carbonate in sea water: VI. The near equilibrium dissolution kinetics of calcium carbonate-rich deep sea sediments. Amer. J. Sci., 278, 344-353.

________ and Berner, R. A. (1979) The chemistry of calcium carbonate in the deep oceans. In E. A. Jenne (ed.), *Chemical Modeling--Speciation, Sorption, Solubility and Kinetics in Aqueous Systems.* Amer. Chem. Soc. Symposium Series, No. 93, 499-535.

Murray, J. W., Grundmanis, V. and Smethie, W. M. (1978) Interstitial water chemistry in the sediments of Saanich Inlet. Geochim. Cosmochim. Acta, 42, 1011-1026.

Nielsen, A. E. (1964) *Kinetics of Precipitation.* MacMillan, New York, 151 p.

________ and Söhnel, O. (1971) Interfacial tensions electrolyte crystal-aqueous solution from nucleation data. Jr. Crystal Growth, 11, 233-242.

Ohara, M. and Reid, R. C. (1973) *Modeling Crystal Growth Rates from Solution.* Prentice-Hall, Englewood Cliffs, New Jersey, 272 p.

Pačes, T. (1973) Steady state kinetics and equilibrium between ground water and granitic rock. Geochim. Cosmochim. Acta, 37, 2641-2663.

Petrović, R. (1976) Rate control in feldspar dissolution -- II. The protective effect of precipitates. Geochim. Cosmochim. Acta, 40, 1509-1522.

________, Berner, R. A., and Goldhaber, M. B. (1976) Rate control in dissolution of alkali feldspars -- I. Study of residual feldspar grains by X-ray photoelectron spectroscopy. Geochim. Cosmochim. Acta, 40, 537-548.

Schott, J., Berner, R. A., and Sjöberg, E. L. (1981) Mechanism of pyroxene and amphibole weathering. I. Experimental studies of iron-free minerals. (Submitted to Geochim. Cosmochim. Acta.)

Tardy, Y., Cheverry, C., and Fritz, B. (1974) Neoformation d'une argile magnésienne dans les dépressions interdunaires du lac Tchad. Application aux domaines de stabilité des phyllosilicates alumineux, magnésiens, et ferrifères. C. R. Acad. Sci., Paris, Ser. D., 278, 1999-2002.

Toth, D. J. and Lerman, A. (1977) Organic matter reactivity and sedimentation rates in the ocean. Amer. J. Sci., 277, 265-285.

Wilson, M. J.,(1975) Chemical weathering of some primary rock-forming minerals. Soil Sci., 119, 349-355.

Chapter 4

TRANSITION STATE THEORY

Antonio C. Lasaga

INTRODUCTION

Most kinetic phenomena from the microscopic viewpoint consist of a sequence of elementary steps, each of which requires overcoming a potential energy barrier. The microscopic dynamics may be likened to a geologist who travels from valley to valley by climbing over successive ridges and hopefully finds the easiest mountain passes between potential energy peaks. The task of transition state theory (TST) is to characterize precisely the climbing rates of reactants over these potential barriers. This powerful theory dates back to a series of papers in 1935 by Henry Eyring (Eyring, 1935a,b; Glasstone *et al.*, 1941) who developed a statistical mechanical calculation of reaction velocities. We can explain the tenets of the theory as follows. An *elementary* reaction involves the molecular collision of reactant species to produce specified products. Both reactants and products are situated at the bottom of well-defined potential energy wells; i.e., they are stable (or metastable) species. However, in proceeding from reactants to products, the species must usually travel over a potential energy barrier (see Fig. 1a). The theory at this point focuses on the molecular configuration at the top of this barrier, (ABC)*, a configuration roughly midway between that of reactants and products. The potential energy surface has a maximum at this configuration in the direction of the so-called *reaction coordinate*, but the potential energy increases in all other directions. Therefore, the potential energy surface topology in the neighborhood of this configuration, termed the *activated complex*, resembles that of a saddle. The reaction coordinate (dashed line in Fig. 1a) is the net motion which takes reactants to products. For example, for the reaction A + BC → AB + C, motion along the reaction coordinate (see Fig. 1a) alters the configuration of the atoms as follows:

$$\vec{A} - - - \overleftarrow{B} - - - \vec{C} \; . \tag{1}$$

This motion illustrates the decrease in the r_{AB} distance and the concomitant increase in the r_{BC} distance needed to form the products.

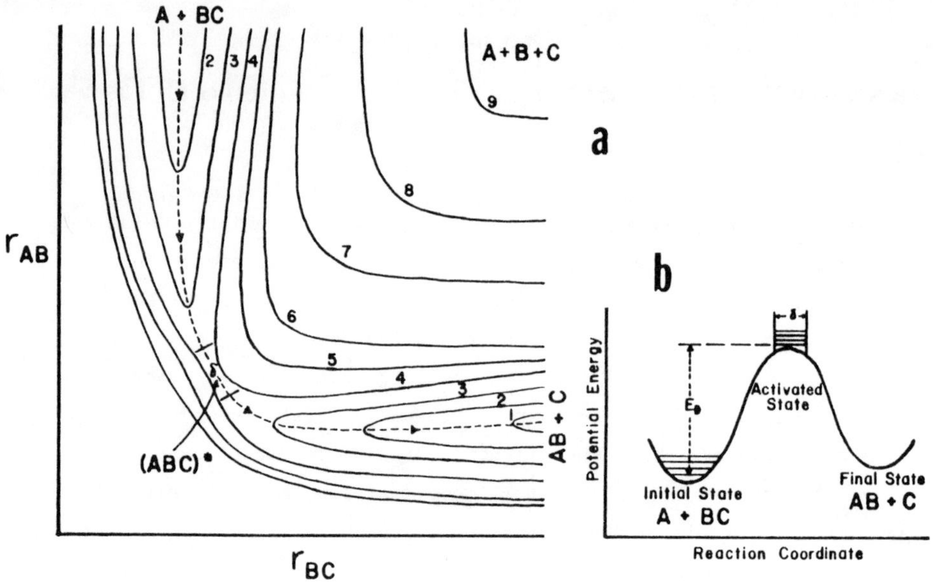

Figure 1. (a) Potential surface in the form of a contour diagram for a hypothetical one-dimensional reaction A + BC → AB + C. The number on the contours are values of the potential in arbitrary units. The activated complex is in a region of length δ.
(b) Potential-energy curve for reaction showing imaginary potential box containing the activated state.

The theory proceeds to treat this activated complex as a true chemical species and furthermore assumes that the initial reactants are always in *equilibrium* with the activated complexes. This is a crucial assumption, which must be checked out carefully for each application. By calculating the number of activated complexes and the rate of decomposition of each activated complex, the theory can then obtain the reaction rate.

Before quantifying the statistical treatment outlined above, we should point out that the application of this theory has been quite successful in determining the reaction rates of numerous processes (*cf.* Connor *et al.*, 1979; Benson, 1976). A good test of the theory has come out of the substantial experimental data on the hydrogen abstraction reaction $H + H_2 \rightarrow H_2 + H$ and its various isotopic variants. Figure 2 compares the results from transition state theory with experimental data on the reactions $H + D_2 \rightarrow HD + D$ and $D + H_2 \rightarrow DH + H$. The agreement over a wide range of temperature and several orders of magnitude

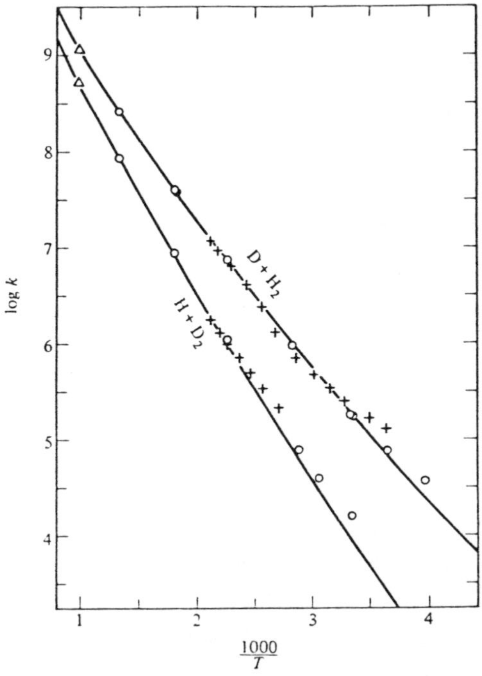

Figure 2. Rate constants for the reactions D + H_2 and H + D_2. Solid line, calculated values using transition state theory; points, experimental values. (From Weston and Schwartz, 1972.)

in the bimolecular rate constant is quite excellent. These reactions provide the best test of the theory, since in these cases we have quite accurate theoretical potential energy surfaces. The lack of adequate potential energy surfaces for general reactions -- including those of interest to the geologist -- is a serious problem with the quantitative application of the theory. Another problem arises with changes in which large numbers of atoms move cooperatively such as in martensitic growth. The elementary reaction nature assumed in transition state theory breaks down in these cases. Nonetheless, with proper caution, some remarkable and useful predictions, relevant to geochemical rates, can still be made with the theory as will be shown below. Furthermore, transition state theory forms the most widely used theoretical structure in kinetics because the molecular description of most kinetic processes can be cast easily in terms of the activated complex.

REVIEW OF PARTITION FUNCTIONS

To present the logic behind transition state theory some statistical mechanics must be used. Therefore, we feel that a brief introduction of partition functions is helpful for the general reader. In a system at temperature T the probability of observing a molecule in a state of energy E_j is proportional to $e^{-E_j/kT}$, the Boltzmann factor (see Nash, 1965, for an easy-to-follow derivation of this result). Therefore, according to the usual way of computing probabilities, the probability of observing such a state with energy E_j is

$$P_j = \frac{e^{-E_j/kT}}{\Sigma_i \; e^{-E_i/kT}} \tag{2}$$

where the sum in equation (2) is carried out over all allowed states of the molecule. The denominator in equation (2) normalizes the probabilities so that $\sum_j P_j = 1$. This important term

$$Q \equiv \Sigma_i \; e^{-E_i/kT} \tag{3}$$

is called the *partition function*. Its importance is readily seen by computing the *mean* energy $\bar{E}$ of a molecular species (reactant, activated complex or product): $\bar{E} = \Sigma_j E_j P_j$.

Inserting equations (2) and (3) for P_j yields $\bar{E} = \frac{1}{Q} \Sigma_j \; E_j \; e^{-E_j/kT}$.

This equation can be rewritten as

$$\bar{E} = -\frac{k}{Q} \frac{dQ}{d(1/T)} \quad \text{or} \quad \bar{E} = kT^2 \frac{\partial \, ln\, Q}{\partial T} . \tag{4}$$

Thermodynamics states that $E = -T^2 \frac{\partial (A/T)}{\partial T}$, where A is the Helmholtz free energy. Combining equation (4) and the last equation yields the key result

$$A = -kT \; ln \; Q . \tag{5}$$

Equations (3) and (5) provide the basic link between the microscopic states of the system, E_i, the temperature, and the macroscopic thermodynamic state, A, via the partition function Q. Once Q is known, E and A can be computed [equations (4) and (5)] and therefore all other thermodynamic functions can be obtained. For example, the entropy is obtained from $S = \frac{E - A}{T}$. As a consequence, all the thermodynamic properties of a system are obtained if the summation in equation (3) can be carried out! The field of statistical mechanics is devoted to the methods available to carry out such a sum.

In a system with N_B distinguishable molecules of type B, the energy terms, E_i, in equation (3) are the sums of the energy contribution from *each* molecule. Therefore, the energy of the system, E_i, can be written as $E_i = \varepsilon_1 + \ldots + \varepsilon_{N_B}$, where ε_i can be any of the allowed energy levels fo. one B molecule. In this case, the partition function for the system, Q, becomes

$$Q = \Sigma_{N_B} \; \Sigma_{\varepsilon_i} \; e^{-(\varepsilon_1 + \ldots + \varepsilon_{N_B})/kT} . \tag{6}$$

Defining $$q_B = \Sigma_j e^{-\varepsilon_j/kT} , \quad (7)$$

where ε_j are the available energies to an individual B molecule, we can rewrite equation (6) as

$$Q = (\Sigma_j e^{-\varepsilon_j/kT})^{N_B} = q_B^{N_B} .$$

If the molecules are indistinguishable, then the previous equation over-counts the energy states by the number of permutations of the N_B molecules. Hence, in this case,

$$Q_B = \frac{1}{N_B!} q_B^{N_B} . \quad (8)$$

The chemical potential μ_B can be obtained from A:

$$\mu_B = \left(\frac{\partial A}{\partial N_B}\right)_{T,V} = -kT\left(\frac{\partial \, ln \, Q_B}{\partial N_B}\right)_{T,V} .$$

Using equation (8) and the Stirling approximation, $ln \, N! = N \, ln \, N - N$, yields an equation for μ_B:

$$\mu_B = -kT \, ln \, (\frac{q_B}{N_B}) . \quad (9)$$

We can use this result to compute equilibrium constants. For a reaction $\nu_A A + \nu_B B \rightleftarrows \nu_C C + \nu_D D$, the condition of equilibrium is that $\nu_A\mu_A + \nu_B\mu_B = \nu_C\mu_C + \nu_D\mu_D$. Inserting equation (9) into the previous equation yields

$$-kT \, ln \, (\frac{q_A}{N_A})^{\nu_A} - kT \, ln \, (\frac{q_B}{N_B})^{\nu_B} = -kT \, ln \, (\frac{q_C}{N_C})^{\nu_C} - kT \, ln \, (\frac{q_D}{N_D})^{\nu_D}$$

or, re-arranging: $$\frac{N_C^{\nu_C} \, N_D^{\nu_D}}{N_A^{\nu_A} \, N_B^{\nu_B}} = \frac{q_C^{\nu_C} \, q_D^{\nu_D}}{q_A^{\nu_A} \, q_B^{\nu_B}} .$$

Dividing both sides by the volume of the system yields an expression for the equilibrium constant:

$$K = \frac{C_C^{\nu_C} \, C_D^{\nu_D}}{C_A^{\nu_A} \, C_B^{\nu_B}} = \frac{(q_C/V)^{\nu_C} \, (q_D/V)^{\nu_D}}{(q_A/V)^{\nu_A} \, (q_B/V)^{\nu_B}} . \quad (10)$$

Equation (10) with the definition in equation (7) is the essential result used in transition state theory.

DERIVATION OF THE RATE CONSTANT

The assumptions of transition state theory simplify the kinetics of a reaction and allow a calculation of the rate from the number of activated complexes, i.e., how many reactants make it to the "top of the hill," and the rate with which these complexes decompose to yield products. To obtain the rate of an elementary reaction, we therefore need to evaluate the product of two terms:

$$\text{Rate} = (\#\text{ of activated complexes}) \left\{\begin{matrix}\text{frequency with which any} \\ \text{complex crosses barrier}\end{matrix}\right\} . \quad (11)$$

Let us work out the second term on the right of equation (11). To do this, we need to evaluate the mean velocity of the activated complexes *along* the reaction coordinate, i.e., along the top of the hill. If the "effective" mass of the complex is m* (m* will cancel in the end), then by definition of *mean* velocity

$$\bar{v} = \frac{\int_0^{\infty} v e^{-\frac{1}{2}m^* v^2/kT} dv}{\int_{-\infty}^{\infty} e^{-\frac{1}{2}m^* v^2/kT} dv} = \left(\frac{kT}{2\pi m^*}\right)^{\frac{1}{2}} . \quad (12)$$

Equation (12) uses the Boltzmann distribution of velocities ($\frac{1}{2}m^*v^2$ is the kinetic energy) and integrates the numerator only over velocities in the positive direction (i.e., towards products in Fig. 1b). The time needed for an activated complex to cross the potential barrier, i.e., to go from one side to the other of the "box" of length δ (Fig. 1b), is

$$\tau = \frac{\delta}{\bar{v}} = \delta \left(\frac{2\pi m^*}{kT}\right)^{\frac{1}{2}} \text{ sec} . \quad (13)$$

The length, δ, along the reaction coordinate in Figures 1a and 1b, which is considered to circumscribe the region of the activated complex, is arbitrary. Nonetheless, this quantity will also cancel at the end. If it takes τ time units per complex to cross the barrier, the frequency of crossing per unit time is $\frac{1}{\tau} = \left(\frac{kT}{2\pi m^*}\right)^{\frac{1}{2}} \frac{1}{\delta} \text{ sec}^{-1}$. Inserting this last result into equation (11) yields

$$\text{Rate} = C^{\dagger} \left(\frac{kT}{2\pi m^*}\right)^{\frac{1}{2}} \frac{1}{\delta} \quad (14)$$

where $C^{\dagger}$ = concentration of activated complexes. The rates of elementary reactions are always related to the concentrations of reactants (see

Chapter 1). Consequently, for a reaction involving A and B as reactants, the rate is written as k C_AC_B, where C_A, C_B, ... are the concentrations of the reactants and k is the rate constant of the reaction. In what follows, A and B -- rather than A and BC, etc. -- will be used in a formal way to stand for any reactants, atoms, or molecules. Equation (14) and the usual form of the rate law (k C_AC_B) allow the rate constant to be given as

$$k = \frac{C^\dagger}{C_A\ C_B\ \cdots}\left(\frac{kT}{2\pi m^*}\right)^{\frac{1}{2}}\frac{1}{\delta}\ . \tag{15}$$

If the activated complex is in equilibrium with the reactants, then

$$\frac{a^\dagger}{a_A\ a_B\ \cdots} = K_{eq}\ . \tag{16}$$

Writing $a_i = c_i\gamma_i$ and inserting (16) into (15) yields

$$k = K_{eq}\frac{\gamma_A\ \gamma_B\ \cdots}{\gamma^\dagger}\left(\frac{kT}{2\pi m^*}\right)^{\frac{1}{2}}\frac{1}{\delta}\ . \tag{17}$$

K_{eq} can be computed from equation (10), i.e.,

$$K_{eq} = \frac{(q^\dagger/V)}{(q_A/V)(q_B/V)\ \cdots}\ . \tag{18}$$

But the partition function for the activated complex, $q^\dagger$, has a vibration frequency which is imaginary, since the potential energy of the complex is at a *maximum* with respect to the reaction coordinate. For example, if the atoms in the A-B-C activated complex were to vibrate along the reaction coordinate illustrated earlier [equation (1)], there would not be a restoring force. On the other hand, motion perpendicular to the dashed line in Figure 1a will increase the potential energy and hence will incur a restoring force. The latter motion constitutes a normal vibrational degree of freedom. This "vibration" mode (or degree of freedom) along the reaction coordinate must be treated in a special way in computing $q^\dagger$. Since the motion is almost free along the reaction coordinate at the top of the barrier, it can be treated as a simple translation. To compute the partition function for a translation, where $E_i = m^*v_i^2/2 = \frac{p_i^2}{2m^*}$ (p = momentum = m*v), the Boltzmann term must be integrated over all possible p_i, i.e., $-\infty$ to $+\infty$:

$$q^\dagger_{translation} = \frac{1}{h}\int_{-\infty}^{\infty} e^{-\frac{p^2}{2m^*kT}}\,dp\int dx \tag{19}$$

(see p. 74, Hill, 1960, or pp. 29-31, Nash, 1965 for further discussion). Doing the p integration in equation (19)

$$q_{translation} = \frac{\sqrt{2m^*\pi kT}}{h} \int dx = \frac{\sqrt{2m^*\pi kT}}{h} \delta$$

The x-integration is over the length of the "box" at the top of the barrier; hence, δ is obtained. Note that the usual sum in the definition of a particular function, equation (7), is replaced by an integral in (19). Treating the motion along the reaction coordinate separately, then, the equilibrium constant in equation (18) can be rewritten

$$K_{eq} = \frac{(q^{\dagger'}/V)}{(q_A/V)(q_B/V)\ \ldots} q^{\dagger}_{translation} \ . \qquad (20)$$

$q^{\dagger'}$ is the partition function for the activated complex *without* including the imaginary vibrational frequency. Using equations (19) and (20) in equation (17) yields

$$k = \text{rate constant} = \frac{(q^{\dagger'}/V)}{(q_A/V)(q_B/V)\ \ldots} \frac{\gamma_A\gamma_B\ \cdots}{\gamma^{\dagger}} \frac{\sqrt{2m^*kT\pi}}{h} \delta \left(\frac{kT}{2\pi m^*}\right)^{\frac{1}{2}} \frac{1}{\delta}$$

or

$$k = \frac{kT}{h} \frac{(q^{\dagger'}/V)}{(q_A/V)(q_B/V)\ \ldots} \frac{\gamma_A\gamma_B\ \cdots}{\gamma^{\dagger}} e^{-\frac{E_o}{kT}} \ . \qquad (21)$$

In equation (21) each partition function is evaluated from the zero point of energy of each molecule; hence, the difference in energy between reactants and complex has been taken out as E_o, the *activation energy*. The energy terms in equation (7), ε_j, must be evaluated relative to some zero energy level. If the zero of energy is set at the bottom of the potential energy well for the reactants, then the energy levels for the activated complex are given by $\varepsilon_j^{\dagger} = \varepsilon_j^{\dagger'} + E_o$ where $\varepsilon_j^{\dagger'}$ are the energy differences in Figure 1b between the energy levels of the activated complex and the *top* of the potential energy barrier. The E_o in the equation for $\varepsilon_j^{\dagger}$ gives rise to the $\exp(-E_o/RT)$ term in equation (21). In this case, the partition function $q^{\dagger'}$ is evaluated using the terms $\varepsilon_j^{\dagger'}$ in equation (7). One often incorporates the partition functions and the activation energy E_o into an equilibrium constant term

$$\frac{(q^{\dagger'}/V)}{(q_A/V)(q_B/V)\ \ldots} e^{-\frac{E_o}{kT}} \equiv K^{\dagger} \ . \qquad (22)$$

Therefore, equation (21) becomes

$$k = \frac{kT}{h} \frac{\gamma_A \gamma_B}{\gamma^\dagger} K^\dagger . \tag{23}$$

$K^\dagger$ can be written in the usual thermodynamic fashion

$$K^\dagger = e^{-\frac{\Delta G^\dagger}{RT}} = e^{\frac{\Delta S^\dagger}{R}} e^{-\frac{\Delta H^\dagger}{RT}} \tag{24}$$

where $\Delta G^\dagger$, $\Delta H^\dagger$, and $\Delta S^\dagger$ are the *standard* Gibbs free energy, enthalpy, and entropy change from reactants to activated complex (i.e., at unit concentration). (The usual superscript o is left out to avoid complexity of representation.) Equation (23) then becomes

$$k = \frac{kT}{h} \frac{\gamma_A \gamma_B}{\gamma^\dagger} e^{\frac{\Delta S^\dagger}{R}} e^{-\frac{\Delta H^\dagger}{RT}} . \tag{25}$$

It should be stressed that the standard state chosen for $\Delta S^\dagger$, $\Delta H^\dagger$, $\Delta G^\dagger$, depends on the units used for concentration in equation (15). If C is given in moles/cm^3 then the standard state is 1 mole/cm^3, i.e., $\Delta G^\dagger$ is the change in the Gibbs free energy of the activated complex reaction, when all reactants and the complex have concentrations of 1 mole/cm^3. Analogous standard states follow if molalities, molarities, etc. are used for C in both the rate law and equation (15).

ARRHENIUS PARAMETERS AND ENTROPY OF ACTIVATION

The Arrhenius form of the rate constant has been introduced in Chapter 1: $k = A\, e^{-E_a/RT}$. Experimentally, E_a is obtained from the slope of a plot of ln k vs. 1/T, i.e., from the equation

$$\frac{d\, ln\, k_{expt}}{dT} = \frac{E_a}{RT^2} . \tag{26}$$

Using equation (25) and assuming that $\Delta H^\dagger$ and $\Delta S^\dagger$ do not vary with T (i.e., $\Delta C_p^\dagger = 0$) yields

$$\frac{d\, ln\, k}{dT} = \frac{1}{T} + \frac{\Delta H^\dagger}{RT^2} .$$

Therefore,

$$E_a = RT + \Delta H^\dagger \tag{27a}$$

and so

$$A = e \frac{kT}{h} \frac{\gamma_A \gamma_B}{\gamma^\dagger} e^{\frac{\Delta S^\dagger}{R}} . \tag{27b}$$

The term RT in equation (27a) arises from incorporating the kT/h term in the temperature variation of the rate constant. Generally, RT is about 1 Kcal/mole and so the difference between E_a and $\Delta H^\dagger$ is only minor. Furthermore, $\Delta H^\dagger$ is essentially the same as the potential energy difference, E_o, in equation (21). In fact, $E_o = \Delta H^\dagger - \Delta(PV)^\dagger = \Delta H^\dagger - P\,\Delta V^\dagger$, which follows from the definition of enthalpy. If pressure is also varied and $\Delta H^\dagger$ assumes a standard state of 1 bar, then we must rewrite (25) as

$$k = \frac{kT}{h}\,\frac{\gamma_A\gamma_B}{\gamma^\dagger}\,e^{\frac{\Delta S^\dagger}{R}}\,e^{-\frac{\Delta H^\dagger}{RT}}\,e^{-\frac{(P-1)\Delta V^\dagger}{RT}} \tag{28}$$

where $\Delta V^\dagger$ is the change in volume upon forming the activated complex and is assumed independent of pressure in equation (28). In this case,

$$E_a = RT + \Delta H^\dagger + (P-1)\Delta V^\dagger\ . \tag{29}$$

However, in most instances $\Delta V^\dagger$ is also a negligible term.

Comparing equations (22), (24), and (27b), it becomes apparent that the evaluation of the partition functions allows a calculation of the standard entropy of activation, $\Delta S^\dagger$:

$$\frac{q^{\dagger\prime}/V}{(q_A/V)(q_B/V)} = e^{\frac{\Delta S^\dagger}{R}}\ .$$

$\Delta S^\dagger$, in turn, allows an estimate of the pre-exponential factor, A. The evaluation of $\Delta H^\dagger$, or E_a, requires computing the potential energy surface and is a harder task.

The entropy of activation, $\Delta S^\dagger$, is a widely used indicator of the configuration of the activated complex. Almost all experimentally derived $\Delta S^\dagger$ (i.e., from the A factor) are *negative* because $A < kT/h$. This decrease in entropy upon forming the activated complex stems from the restricted motion imposed by the binding of the reactant molecules. In fact, there is a loss of translational and rotational freedom when reactants combine to form the activated complex. If the activated complex is "loose," i.e., if the reactant molecules are separated by long bonds in the complex, then the decrease in entropy will not be large and A, the pre-exponential factor, will be high. If the activated complex is "tight" then $\Delta S^\dagger$ will be a large negative number and A will be low (see Benson, 1976). For some reactions of the type $A + B \rightarrow (AB)^\dagger \rightarrow AB$,

$\Delta S^{\dagger}$ may be very similar to ΔS^{o}, where ΔS^{o} is the standard state entropy change for the A + B → AB reaction. This similarity indicates that the activated complex $(AB)^{\dagger}$ is similar in structure to the product molecule, AB. For some examples see Golden (1979). The estimation of $\Delta S^{\dagger}$ (and also A) from partition functions and hence from vibrational data in *gaseous* reactions is discussed at length in Benson (1976).

IMPORTANT CONSEQUENCES OF TRANSITION STATE THEORY

Fundamental Frequency

The first term in front of equation (23), kT/h, which has units of $(\text{time})^{-1}$ is termed the *fundamental frequency*. This frequency is indeed fundamental to *any* activated process according to theory. At room temperature, T = 300°K, the value of the fundamental frequency is $\frac{kT}{h} = 6 \times 10^{12}\ \text{sec}^{-1}$. Aside from activity coefficient corrections, rate constants have the form $k_r = \frac{kT}{h} K^{\dagger}$. In effect, the dynamical aspects of a reaction have all been incorporated in the simple kT/h frequency term. The difficult step, obviously, is obtaining the activated complex equilibrium constant $K^{\dagger}$.

Relation Between Rates and ΔG

The use of the equilibrium constant $K^{\dagger}$ in equation (23) has important implications for the relation between kinetics and thermodynamics. Note that equation (23) can be employed equally well in studying the forward or the reverse reaction of an elementary step (Fig. 1). Let us take the case of the formal reaction

$$A + B \rightleftarrows C + D \ . \qquad (30)$$

The forward rate is given by

$$R_{+} = k_{+}\, C_A\, C_B \qquad (31)$$

and the reverse rate is given by

$$R_{-} = k_{-}\, C_C\, C_D \ . \qquad (32)$$

Equations (23) and (24) dictate that the two rate constants are

$$k_{+} = \frac{kT}{h}\, \frac{\gamma_A \gamma_B}{\gamma^{\dagger}}\, e^{-\frac{\Delta G^{\dagger}}{kT}} \qquad (33)$$

$$k_- = \frac{kT}{h}\frac{\gamma_C\gamma_D}{\gamma^{\dagger}} e^{-\frac{\Delta G^{\dagger'}}{kT}} \tag{34}$$

where $\Delta G^{\dagger}$ is the *standard* free energy difference between the activated complex and the reactants A, B:

$$\Delta G^{\dagger} = G^{\circ\dagger} - G^{\circ}_A - G^{\circ}_B \ . \tag{35}$$

Likewise,

$$\Delta G^{\dagger'} = G^{\circ\dagger} - G^{\circ}_C - G^{\circ}_D \ . \tag{36}$$

If the activated complex is the *same* in either direction, the ratio of the rate constants can be obtained from (33) and (34):

$$\frac{k_+}{k_-} = \frac{\gamma_A\gamma_B}{\gamma_C\gamma_D} e^{-\frac{(\Delta G^{\dagger}-\Delta G^{\dagger'})}{kT}} = \frac{\gamma_A\gamma_B}{\gamma_C\gamma_D} e^{-\frac{\Delta G^{\circ}}{kT}}$$

where ΔG° is the standard Gibbs free energy difference between reactants and products. The previous equation can also be expressed as

$$\frac{k_+}{k_-} = \frac{\gamma_A\gamma_B}{\gamma_C\gamma_D} K_{eq} = K'_{eq} \tag{37}$$

where K'_{eq} is the concentration equilibrium constant between reactants and products. Equation (37) is expected from the principle of detailed balancing (see Chapter 1). Furthermore, from equations (31), (32), and (37)

$$\frac{R_+}{R_-} = \frac{k_+ \ C_A \ C_B}{k_- \ C_C \ C_D} = \frac{\gamma_A \ \gamma_B}{\gamma_C \ \gamma_D}\frac{C_A \ C_B}{C_C \ C_D} e^{-\frac{\Delta G^{\circ}}{kT}}$$

$$\frac{R_+}{R_-} = \frac{a_A a_B}{a_C a_D} e^{-\frac{\Delta G^{\circ}}{kT}} = e^{-\Delta G/kT} \tag{38}$$

where ΔG is the *actual* free energy difference. Of course, at equilibrium $\Delta G = 0$ and $R_+ = R_-$.

Equation (38) can be put in yet another useful form. The *net* rate of an elementary reaction is the difference between the forward and reverse rates, i.e., $R_{net} = R_+ - R_-$. Using equation (38), R_{net} can be rewritten as

$$R_{net} = R_+(1 - e^{\Delta G/RT}). \tag{39}$$

If the reaction is close enough to equilibrium ($\Delta G = 0$) that $|\Delta G| < RT$, the exponential in equation (39) may be expanded to obtain

$$R_{net} = -\frac{R_{+}\ \Delta G}{RT} \ . \qquad (40)$$

Equation (40) establishes a *linear* relationship between the net rate of an elementary reaction and the free energy difference of the reaction, as long as the system is not far from equilibrium. Kinetics and thermodynamics have been smoothly connected in the near-equilibrium region. The linear relation in equation (40) is but a special case of the much more general linear relations studied in the theory of irreversible thermodynamics (see Chapter 5). $-\Delta G$, sometimes called the affinity A (see Prigogine, 1967), is, in a sense, the driving force of the reaction. If $-\Delta G$ is positive ($\Delta G < 0$), the net rate of the reaction is also positive and the kinetics proceeds from left to right in reaction (30). Likewise, a negative $-\Delta G$ ($\Delta G > 0$) leads to a negative net rate of reaction, i.e., from right to left in equation (30). It is important to note, however, that the linear relation in (40) is only guaranteed *near* equilibrium. On the other hand, the linear relation holds for *both* elementary and complex overall reactions; i.e., it is a fully general result (see Prigogine, 1967, Chapter 3). The reason linearity extends to overall reactions stems from the applicability of the principle of detailed balancing to these reactions as discussed in Chapter 1. For example, the expression for the dissolution of calcite discussed in Chapter 1 was Rate = $k(1-\Omega^{n})$, where the saturation ratio was defined by

$$\Omega = \frac{a_{Ca^{2+}}\ a_{CO_3^{2-}}}{K_{sp}}$$

and K_{sp} is the solubility product of calcite. (The area terms are included in k for simplicity.) Ω can be rewritten as

$$\Omega = a_{Ca^{2+}}\ a_{CO_3^{2-}}\ e^{\frac{\Delta G^{\circ}}{RT}} = e^{\frac{\Delta G}{RT}}$$

where ΔG is the free energy difference of the reaction

$$CaCO_3\ (calcite) \rightarrow Ca^{2+} + CO_3^{2-}$$

and the activity of calcite is set to unity (ignore Mg^{2+}). Near equilibrium the rate law can still be simplified to

$$\text{Rate} = k\,(1 - e^{\frac{n\Delta G}{RT}}) \simeq -k\,\frac{n\Delta G}{RT}$$

and linearity is again obtained, even for high values of n.

The main conflict arises with expressions such as

$$\text{Rate} = k\,(1-\Omega)^n\ . \tag{41}$$

These expressions are typical of crystal growth/dissolution processes involving defects on the surface. For example, screw dislocation controlled processes typically yield (Nielsen, 1964): Rate α $(1-\Omega)^2$. No matter how close C is to C_{eq} or Ω to 1 the latter rate laws will never reduce to simple linear relations. Is the theory of irreversible thermodynamics refuted in these cases? The answer lies in the essential role played by dislocations and similar defects in expressions such as (41). Dislocations are *not* equilibrium defects (see Chapter 7). Therefore, as long as non-equilibrium surface defects play an essential role in the kinetics, the overall reactions cannot be considered to be close to equilibrium even if $\Omega \sim 1$. This is the explanation for the non-linear kinetics.

In experimental petrology, one often uses equation (40) at least implicitly. When experimental P-T brackets are determined for a given reaction, the experimentalist actually measures the direction of the net reaction, which, according to equation (40), indicates the sign of ΔG. This linear relation will be discussed further in Chapter 5.

Implication of Transition State Theory on Ionic Strength Effects

The transition state theory (TST) equation (25) predicts that the rate constant will depend on the activity coefficients. Suppose a reaction occurs in an aqueous medium. If all the terms, which are *independent* of the nature of the solvent in an aqueous reaction, are lumped into some reference k_o, equation (25) can be rewritten as

$$k = k_o\,\frac{\gamma_A\gamma_B}{\gamma^{\dagger}}\ . \tag{42}$$

The influence of ionic strength in the case that A or B are ions is easy to evaluate. Let the charge on A be z_A and that on B be z_B. Then the charge of the activated complex is $z_A + z_B$. For not too concentrated solutions, the extended Debye-Huckel formula can be used to estimate the activity coefficients:

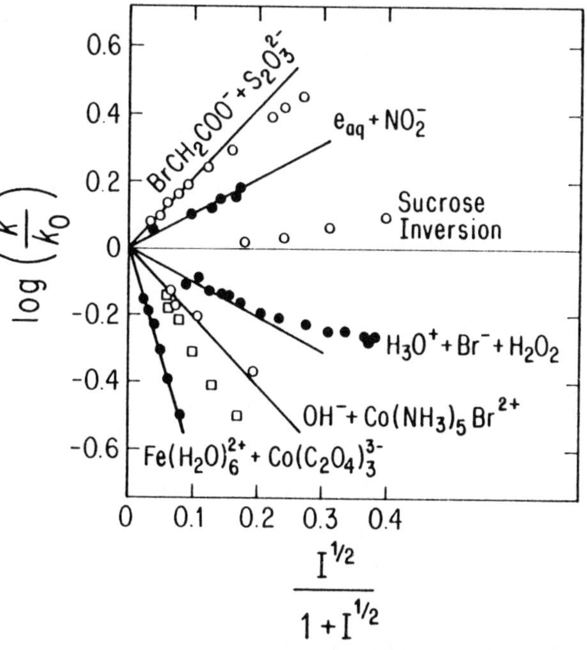

Figure 3. Dependence of the rate constant of several reactions on the ionic strength.

$$\log \gamma_i = -\frac{A z_i^2 \sqrt{I}}{1 + \bar{a} B\sqrt{I}} . \tag{43}$$

Therefore, from equations (42) and (43)

$$\log k = \log k_o + \log \gamma_A + \log \gamma_B - \log \gamma^{\dagger}$$

$$= \log k_o - \frac{A\sqrt{I}}{1 + \bar{a} B\sqrt{I}} [z_A^2 + z_B^2 - (z_A + z_B)^2]$$

or

$$\log k = \log k_o + \frac{A\sqrt{I}}{1 + \bar{a} B\sqrt{I}} 2z_A z_B . \tag{44}$$

Equation (44) predicts that the logarithm of the rate constant of an elementary reaction involving ions will depend on the square root of the ionic strength of the solution, $\sqrt{I}$. For high values of the charge on the ions, the dependence would be more pronounced. If A and B have the same charge, increasing I will increase the rate; while if A and B have opposite charge, increasing I will decrease the rate. Figure 3 shows some examples of the ionic strength effect in various reactions. Experimental data uphold the predictions of transition state theory rather well. Note that since equation (44) is a logarithmic relation, the effects of I may be significant!

In some geochemical applications (*e.g.*, Rimstidt and Barnes, 1980), recent papers have sometimes rewritten the forward rate of a reaction such as that in equation (30) as $R_+ = k_+ a_A a_B$, using activities rather than using concentrations as in equation (31). In this case, the rate constant is given by

$$k_+ = \frac{kT}{h} \frac{1}{\gamma^\dagger} e^{\frac{\Delta S^\dagger}{R}} e^{-\frac{\Delta H^\dagger}{RT}} \tag{45}$$

and will have a different dependence on ionic strength. In fact, it can be easily checked that now k_+ satisfies

$$\log k_+ = \log k_o - \frac{A(z_A + z_B)^2\sqrt{I}}{1 + \bar{a}\, B\sqrt{I}} . \tag{46}$$

Compensation Law

Relations between the pre-exponential factors and the activation energies are termed "compensation laws." An example of a compensation law in solid state diffusion is given in Chapter 7. Equations (27a) and (27b) give a qualitative explanation for compensation. Generally, $\Delta H^\dagger$ and $\Delta S^\dagger$ tend to change in the same way with temperature or as the chemical system changes; i.e., both increase or both decrease. Decreases in enthalpy, $\Delta H^\dagger$, are associated with "tighter" binding of the complex (more exothermic). But this tighter binding is also responsible for a decrease in entropy, $\Delta S^\dagger$. If these changes are *linearly* related, then $\Delta S^\dagger = n\Delta H^\dagger$. Using equations (27a) and (27b), it follows that

$$ln\ A = ln\left(\frac{ekT}{h}\right) - nT + \frac{n}{R}E_a \quad \text{or} \quad ln\ A = \text{constant} + \frac{n}{R}E_a$$

(if T is constant). This last equation, relating the preexponential factor and the activation energy, is similar to that used in Chapter 7.

MISCONCEPTIONS AND LIMITATIONS OF TRANSITION STATE THEORY

Several stringent assumptions have been made in deriving equation (25). Some common misconceptions have arisen from their interpretation. For a good recent discussion see Golden (1979).

Meaning of $K^\dagger$

Although the terms in equation (22) are written as an equilibrium constant, $K^\dagger$, it is important to note that $K^\dagger$ is *not* really the full equilibrium constant K_{eq} between reactants and activated complexes given by equation (16) or (18). The reason is that one degree of freedom was removed from the partition function of the activated complex to obtain the kT/h term. Therefore, $K^\dagger$ lacks this one degree of freedom. However, Glasstone *et al.* (1941) and others have pointed out that the term removed can be made to equal 1 so that *numerically* $K^\dagger$ is equal to the equilibrium constant K_{eq} between reactants and complexes.

An interesting paradox arises in employing TST for both directions of an elementary reaction as done earlier in equations (33) and (34). Returning to equation (30), if A and B are in equilibrium with $(AB)^\dagger$ and C and D are in equilibrium with $(CD)^\dagger$ then A + B and C + D should be in equilibrium with each other as long as $(AB)^\dagger = (CD)^\dagger$. In this case, there should be no *net* reaction between A + B and C + D! This paradox merely asserts the fact that TST employs an equilibrium assumption between reactants and complexes, and this assumption is really fully valid only when reactants and products *are* in equilibrium. Of course, the fact that there is no *net* reaction at equilibrium does not invalidate the calculation of the rate constants k_+ and k_-, since equilibrium merely states that the forward and reverse rates (not k_+ and k_-!) are equal. Hence, the TST formulae are most correct near equilibrium, although in many cases they can be extended to far from equilibrium situations.

The Value of $\gamma^\dagger$

A problem with the use of TST arises when values for the activity coefficient of the activated complex, $\gamma^\dagger$, are needed. It is usually very hard to evaluate $\gamma^\dagger$, and it is certainly even harder to experimentally measure $\gamma^\dagger$. An exception can be made in the case that the complex is an ion in solution (as was detailed above), since in this case the simple electrolyte theories may help in obtaining $\gamma^\dagger$.

Elementary versus Overall Reactions

A possibly misleading use of transition state theory arises from its application to *overall* chemical reactions such as

$$NaAlSi_3O_8 + H^+ + 7H_2O \rightarrow Al(OH)_3(s) + Na^+ + 3H_4SiO_4 \ . \tag{47}$$

The fundamental equations of this chapter [*e.g.*, equations (21), (23), (25), or (28)] are only applicable to *elementary* reactions. An overall reaction such as (47) involves a series of elementary reactions. Examples would be the adsorption of H^+ onto the feldspar surface, H^+ exchange with Na^+ in the feldspar, dissolution of protonated feldspar surface to yield Al^{3+} and Si^{4+} in solution, hydrolysis and hydration of the ions, nucleation of gibbsite or growth of gibbsite onto an existing gibbsite surface. Note some of the above reactions may themselves be further broken down. When faced with such a series of elementary steps, great care must be taken in applying transition state theory to the overall reaction. The most obvious problem arises when considering the form of the rate law as was done in Chapter 1. An overall reaction can have a rather complex rate law. On the other hand, an elementary reaction will always have the rate law *assumed* by the derivation of transition state theory, i.e., the step following equation (14). Therefore, analysis of reaction (47) by TST requires that each elementary step of the overall reaction be studied individually.

For example, the ozone decomposition reaction was analyzed in Chapter 1 and the rate constant found to be given by

$$k_{net} = \frac{2}{3} \frac{k_1 \, k_2}{k_{-1}} \tag{48}$$

where the rate constants are defined by

$$O_3 \underset{k_{-1}}{\overset{k_1}{\rightleftarrows}} O_2 + O \quad \text{and} \quad O + O_3 \xrightarrow{k_2} 2O_2$$

and the overall reaction is $2O_3 \xrightarrow{k_{net}} 3O_2$.

Use of equation (25) to obtain the individual elementary rate constants in (48) yields

$$k = \frac{2}{3}\frac{kT}{h} \, e^{\frac{\Delta S_1^\dagger + \Delta S_2^\dagger - \Delta S_{-1}^\dagger}{R}} \, e^{-\frac{\Delta H_1^\dagger + \Delta H_2^\dagger - \Delta H_{-1}^\dagger}{RT}} \tag{49}$$

where the activity coefficients were set to unity and the activation entropies and enthalpies are labeled just as the k_i. In this case, the experimental activation energy [equation (27)] is given by

$$E_a = RT + \Delta H_1^{\dagger} + \Delta H_2^{\dagger} - \Delta H_{-1}^{\dagger} = RT + \Delta H_r^{\circ} + \Delta H_2^{\dagger} \qquad (50)$$

where ΔH_r° is the standard enthalpy of the reaction $O_3 \rightarrow O_2 + O$. It is obvious that E_a is not just determined from the activation enthalpy of the rate determining step, $\Delta H_2^{\dagger}$. Furthermore, equation (50) clearly shows that ΔH° for the overall reaction is not simply related to the predicted activation enthalpy. On the other hand, once the mechanism is understood, application of transition state theory can be useful to obtain estimates of $\Delta H_2^{\dagger}$ and thereby obtain E_a, via equation (50).

APPLICATION OF TRANSITION STATE THEORY

In this section transition state theory (TST) will be used to elucidate the rates of several processes of geochemical significance. The successful applications of TST to obtain rate constants for gas phase reactions are numerous in the chemical literature (see Weston and Schwarz, 1972; and especially Benson, 1976). It is useful and instructive to illustrate some of the basic computational concepts with the simple molecular reaction discussed earlier:

$$A + BC \rightarrow A\text{---}B\text{---}C \rightarrow AB + C \; . \qquad (51)$$

The $H + D_2 \rightarrow HD + D$ and $D + H_2 \rightarrow HD + H$ reactions, for example, fall in this category (see Fig. 2). Moreover, if A, B, and C are considered as molecules instead of atoms and C, in particular, is part of a solid, reaction (51) can be applied also to geochemical reactions. Presently, we shall treat A, B, and C in (51) as atoms. Using equation (21) and ignoring the activity coefficients, the rate constant for reaction (51) is:

$$k = \frac{kT}{h} \frac{q_{ABC}^{\dagger}}{q_A \, q_{BC}} \, e^{-\frac{E_o}{kT}} \qquad (52)$$

where E_o is the potential energy difference between the activated complex and the reactants (Fig. 1). The molecular partition functions q_A, q_{BC}, and q_{ABC} are evaluated by standard techniques. Each molecule has three different types of degrees of freedom (modes of motion): translational, vibrational, and rotational. Since the energy of a molecule can be written as the *sum* of the energies in each of these categories, the partition function [equation (7)] is just the *product* of the individual partition functions for each degree of freedom. Each translational degree of freedom has a term just like the one computed in equation (19),

$$q_{\text{translation}} = \frac{(2\pi m\ kT)^{\frac{1}{2}}}{h} L \tag{53}$$

where m is the mass of the molecule, L is the length of the "box" containing the molecules ($V^{1/3}$), and h is Planck's constant. The rotational energy of a linear molecule (including both degenerate rotations) is given by

$$\varepsilon_j^{\text{rotation}} = \frac{j(j + 1)h^2}{8\pi^2 I} \tag{54}$$

where h is Planck's constant, I is the moment of inertia, and j can take on any integer values j = 0, 1, 2, ... (see Hill, 1960). Furthermore, for each such energy level there are 2j + 1 rotational states that have energy $\varepsilon_j^{\text{rotation}}$. Therefore, equation (7) for a rotation yields

$$q_{\text{rotation}} = \sum_{j=0}^{\infty} (2j+1)\ e^{-\frac{j(j+1)h^2}{8\pi^2 I\ kT}} \quad \text{or} \quad q_{\text{rotation}} = \frac{8\pi^2 I\ kT}{h^2} .$$

For certain molecules with high degree of symmetry only some rotational states are allowed, and so we add the *integer* σ (usually σ = 1) to obtain

$$q_{\text{rotation}} = \frac{8\pi^2 I\ kT}{\sigma\ h^2} . \tag{55}$$

Generally, the value of q_{rotation} lies between 10 - 100. Finally, to estimate the vibrational partition function we need the formula for the allowed vibrational energy levels:

$$\varepsilon_j^{\text{vibration}} = (j + 1/2h\nu) \quad j = 0, 1, \ldots \tag{56}$$

where ν is the particular frequency of vibration. (Note: We are treating each vibration and rotation as essentially independent.) Equation (56) leads to

$$q_{\text{vibration}} = \sum_{j=0}^{\infty} e^{-\frac{(j+1/2)h\nu}{kT}} = e^{-\frac{h\nu}{2kT}} \sum_{j=0}^{\infty} \left(e^{-\frac{h\nu}{kT}}\right)^j$$

This last sum is nothing but the sum of a geometric series; therefore, the sum yields

$$q_{vibration} = \frac{e^{-\frac{h\nu}{2kT}}}{1 - e^{-\frac{h\nu}{kT}}} . \tag{57}$$

$q_{vibration}$ usually has a value between 1 - 10. These three formulae for the partition functions can, in fact, be used to compute the equilibrium constants and thermodynamic properties of ideal gases to equal or better precision than experiment! An atom A has only three translational (x, y, or z directions) degrees of freedom (it cannot rotate or vibrate). Hence, using (53) and setting $V = L^3$:

$$\frac{q_A}{V} = q^3_{translation} = \frac{(2\pi m_A kT)^{\frac{3}{2}}}{h^3} . \tag{58}$$

Molecule BC has two atoms and hence 2 x 3 or 6 degrees of freedom. Three of these degrees of freedom are translational, two are rotational, and the remaining one is vibrational. Therefore, $q_{BC} = q^3_{translation}\, q_{rot}\, q_{vib}$. Note that q_{rot} takes care of the *two* rotations. Inserting equations (53), (55), and (57)

$$\frac{q_{BC}}{V} = \frac{[2\pi(m_B + m_C)kT]^{\frac{3}{2}}}{h^3} \frac{8\pi^2 I_{BC} kT}{\sigma_{BC}\, h^2} \frac{e^{-\frac{h\nu_{BC}}{2kT}}}{1 - e^{-\frac{h\nu_{BC}}{kT}}} \tag{59}$$

where ν_{BC} and I_{BC} are the fundamental frequency of the diatomic BC and the moment of inertia, respectively.

Finally, the linear activated complex, A---B---C, has three atoms and hence 3 x 3 or 9 degrees of freedom. Removing the three translation and the two rotation degrees of freedom leaves four vibrational degrees of freedom. However, one of the fundamental vibration frequencies is imaginary [corresponding to the vibration in (1)], since the complex lies at the top (not bottom) of a potential surface "hill." (Remember this is the degree of freedom removed, which gave rise to kT/h.) Therefore, there are only three *real* vibrational frequencies $\nu_1^\dagger$, $\nu_2^\dagger$, and $\nu_3^\dagger$ for the activated complex. The partition function for the complex becomes

$$q^\dagger_{ABC} = \frac{(2\pi(m_A+m_B+m_C)kT)^{3/2}}{h^3} \frac{8\pi^2 I_{ABC} kT}{\sigma_{ABC}\, h^2} \prod_{i=1}^{3} \frac{e^{-\frac{h\nu_i^\dagger}{2kT}}}{1 - e^{-\frac{h\nu_i^\dagger}{kT}}} . \tag{60}$$

Inserting equations (58) - (60) into (52) yields the final expression for the rate constant

$$k = \frac{kT}{h} \frac{h^3}{(2\pi kT)^{3/2}} \frac{(m_A + m_B + m_C)^{3/2}}{m_A^{3/2} (m_B + m_C)^{3/2}} \frac{\sigma_{BC} I_{ABC}}{\sigma_{ABC} I_{BC}}$$

$$\times \frac{\prod_{i=1}^{3} e^{-\frac{h\nu_i^\dagger}{2kT}} \left(1 - e^{-\frac{h\nu_i^\dagger}{kT}}\right)^{-1}}{e^{-\frac{h\nu_{BC}}{2kT}} \left(1 - e^{-\frac{h\nu_{BC}}{kT}}\right)^{-1}} e^{-\frac{E_o}{RT}} . \tag{61}$$

From the shape of the potential energy surface in the neighborhood of the activated complex, the three vibrational frequencies $\nu_i^\dagger$ can be obtained and from the geometry of the activated complex (*e.g.*, the A-B and B-C bond lengths in A---B---C) the moment of inertia I_{ABC} can be obtained. Knowledge of $\nu_i^\dagger$, I_{ABC}, and E_o allows immediate calculation of the rate constant from equation (61) since all other quantities are trivially obtained. The accuracy of the results when compared to experiment has been already shown in Figure 2 for hydrogen abstraction reactions.

The usefulness of transition state theory is best exemplified by equation (61) which calculates the rate constant of any elementary tri-atomic reaction from basic atomic parameters. There are no fudge or ambiguous parameters in the equation. The drawback of the theory is that the potential energy surface must be known, at least in the neighborhood of the activated complex. For reactions involving low atomic number atoms, reasonable potential energy surfaces have been computed in recent years (*e.g.*, see Benson, 1977). For these cases, TST has been very useful in predicting rate constants. However, the stress in this section is on geochemical and mineralogical applications and hence we will focus next on the use of TST in condensed media.

Diffusion in Liquids

The concepts of transition state theory can be applied to any elementary kinetic process with an *activation barrier*. Therefore, TST is not restricted to gas phase reactions but can also treat transport processes such as diffusion. The diffusion coefficient of a species

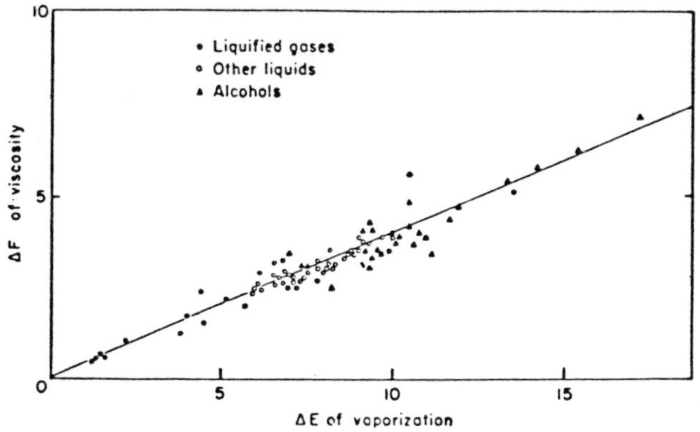

Figure 4. Empirical relation between free energy of activation in liquids, ΔF, and energy of evaporation, ΔE. (From Jost, 1960.)

with a jump distance, λ, in a condensed phase is given by

$$D = \lambda^2 \Gamma_o \tag{62}$$

where Γ_o is the mean jump frequency (in the units of sec^{-1}); i.e., Γ_o is the number of jumps per species per second (*e.g.*, see Chapter 6). If a species is diffusing in a liquid it must move the solvent molecules apart to execute a "jump." This formation of a solvent hole requires energy. Given this diffusion potential barrier, the principles of transition state theory can be applied to obtain Γ_o. Both the activated complex and the reactant, in this case, are the diffusing molecule or atom. Therefore, equation (21) can be used to write Γ_o as

$$\Gamma_o = \frac{kT}{h} \frac{q^\dagger}{q} e^{-E_o/kT}$$

where E_o is the size of the diffusion barrier.
$q^\dagger$ is the partition function of the diffusing species at the top of the potential barrier and q is the partition function of the same species in its usual stable well within the liquid. The simplest approximation to make is to assume that q and $q^\dagger$ have identical factors except for the degree of freedom along the diffusion path (*e.g.*, all other vibrations are the same in the activated state and in the normal state). Remember that $q^\dagger$ is missing a contribution from the degree of freedom along the reaction coordinate, since we removed it explicitly in the derivation of equation (21) and incorporated it into the kT/h term. If all other degrees of freedom are the same for q and $q^\dagger$, their contributions cancel in the ratio $q^\dagger/q$ and all that is left is the contribution of the degree

of freedom along the diffusion path to the normal partition function, q. To proceed further, the diffusing species in the liquid can be visualized as being inside a molecular "cage" with a certain "free" volume, $\bar{V}_f$. If we treat the extra degree of freedom in q as a *translation* within this given volume of liquid, we can, once more, use equation (53) to obtain

$$q_{\text{diffusion path}} = \frac{(2\pi m\ kT)^{\frac{1}{2}}}{h} \bar{V}_f^{\frac{1}{3}} \tag{64}$$

where m is the mass of the diffusing species. In equation (64) we have replaced L by $\bar{V}_f^{\frac{1}{3}}$, where $\bar{V}_f$ is the "free" molar volume in the liquid. Using some straightforward thermodynamic considerations Glasstone *et al.* (1941) estimate the free molar volume from the heat of vaporization of the liquid, ΔE_{vap}, as

$$\bar{V}_f = \frac{8(kT)^3\ \bar{V}}{\Delta E_{vap}^3} \tag{65}$$

where $\bar{V}$ is the molecular volume of the liquid. Using equation (64) we have

$$q^{\dagger}/q = \frac{h}{(2\pi m\ kT)^{\frac{1}{2}}\ \bar{V}_f^{\frac{1}{3}}} \tag{66}$$

and so

$$\Gamma_o = \frac{kT}{h} \frac{h}{(2\pi m\ kT)^{\frac{1}{2}}\ \bar{V}_f^{\frac{1}{3}}}\ e^{-E_o/kT} . \tag{67}$$

Finally, E_o is also estimated from the heat of vaporization. Since evaporating a molecule results in a hole in the liquid, it is reasonable to expect a relation between E_o and ΔE_{vap} as well as between $\bar{V}_f$ and ΔE_{vap}. From data on several liquids (see Fig. 4), it is found that

$$E_o \approx \frac{\Delta E_{vap}}{3} . \tag{68}$$

Altogether, then, the absolute diffusion coefficient in a liquid is given by

$$D = \lambda^2 \frac{kT}{h} \frac{h}{(2\pi m\ kT)^{\frac{1}{2}}\ \bar{V}_f^{\frac{1}{3}}}\ e^{-\frac{\Delta E_{vap}}{3RT}} . \tag{69}$$

Let us apply equation (69) to diffusion of H_2S in water at 300°K. The enthalpy of vaporization of water is ΔH_{vap} = 40 866 Joule/mole = 9767 cal/mole (Robie *et al.*, 1978). If we treat water vapor as an ideal gas at 300°K (a crude approximation) and ignore the molar volume of water, then

$$\Delta E_{vap} = \Delta H_{vap} - \Delta(PV) = \Delta H_{vap} - RT = 9171 \text{ cal/mole at } 300°K \ .$$

The molar volume of water is 18.07 cm^3/mole (Robie *et al.*, 1978); equation (65) then yields

$$\bar{V}_f = \frac{8(1.987 \cdot 300)^3(18.07)}{(9171)^3(6.023 \times 10^{23})} = 6.591 \times 10^{-26} \text{ cm}^3\text{/molecule.}$$

The jump distance, λ, can be approximated by the mean water distance, i.e.,

$$\lambda \approx \bar{V}^{\frac{1}{3}} = \left(\frac{18.07 \text{ cm}^3\text{/mole}}{6.023 \times 10^{23}}\right)^{\frac{1}{3}} \approx 3.107 \times 10^{-8} \text{ cm} \ .$$

Using these values and m_{H_2S} = 34.076 g/mole = 5.658 x 10^{23} g/molecule in equation (69):

$$D = (3.107 \times 10^{-8})^2 \ \frac{(1.38044 \times 10^{-16} \cdot 300) \quad e^{-\frac{9171}{3(1.987)(300)}}}{(2\pi \cdot 5.66 \times 10^{-23})^{1/2}(6.591 \times 10^{-26})^{1/3}} \text{ cm}^2\text{/sec}$$

$$D = 1.53 \times 10^{-5} \text{ cm}^2\text{/sec} \ .$$

The calculated value is indeed in agreement with experimental data on diffusion in water. (D_{H_2S} = 2.01 x 10^{-5} cm^2/sec, extrapolated to pure water solvent from Farouque and Fahidy, 1978.)

Note that the dependence of the diffusion coefficient on the nature of the diffusing species enters only in the $m^{-\frac{1}{2}}$ factor of equation (69). For neutral species such as H_2S the dependence on the square root of the mass is reasonable. However, for charged species the charge density (z/r) will also play an important role. The charge of the species will influence the strength and size of the hydration sphere around the ion. Since the hydration sphere moves along with the ion, the diffusing species, in this case, includes the waters of hydration. Because m in equation (69) refers to the mass of the diffusing species, m includes the mass of the waters of hydration of diffusing ions.

An often used relation to estimate diffusion coefficients is the *Stokes-Einstein* equation. This relation, which stems from a classical hydrodynamical treatment of the motion of the species in the solvent, states that

$$D = \frac{kT}{6\pi\eta r} \tag{70}$$

where η is the viscosity of the solvent and r is the "radius" of the diffusing species (assumed spherical). Equation (70) predicts that the individual diffusion coefficients vary as 1/r for constant temperature in a fixed solvent. Equations (69) and (70) would predict comparable dependences on the nature of the species only if r was proportional to $m^{\frac{1}{2}}$. In the case of hydrated ions, it may be more appropriate to expect that r is proportional to $m^{\frac{1}{3}}$, but even this relation breaks down if the ion is a heavy ion (relative to water). This slight discrepancy points out the approximate nature of both equations (69) and (70).

Solid State Diffusion

Transition state theory can also be applied to diffusion in crystals. To do this let us combine equations (62) and (63):

$$D = \lambda^2 \frac{kT}{h} \frac{q^\dagger}{q} e^{-E_o/kT} X_v \ . \tag{71}$$

Note that equation (71) incorporates the mole fraction of vacancies, X_v, since an atom must have a vacancy available to achieve a "jump" in a vacancy diffusion process (see Chapter 7). The evaluation of E_o, the size of the migration barrier, from basic principles is discussed in Chpater 7. We will discuss the ratio $q^\dagger/q$ which will yield $\Delta S^\dagger$ or the pre-exponential factor. In a crystal with N atoms all the 3N degrees of freedom (except for six which can be ignored) are vibrational degrees of freedom. Each of the 3N vibrations (phonons) with a vibrational frequency ν_i will contribute a term to the partition function given by [see equation (57)]

$$q_{vib} = \frac{e^{-\frac{h\nu_i}{2kT}}}{1 - e^{-h\nu_i/kT}} \ . \tag{72}$$

Both q and $q^\dagger$ consist of *products* of terms such as in (72) for each crystal vibrational frequency. In a simple approach similar to the one discussed in the previous section, the vibrational terms in q and $q^\dagger$ are assumed the same; i.e., the crystal vibrational frequencies are assumed unchanged as the ion moves to the saddle point. However, since $q^\dagger$ is missing one degree of freedom, the ratio $q^\dagger/q$ will cancel all but *one* of the vibration terms in this simple approach. If we label the extra frequency in q by ν^*, then from equations (71) and (72) and the assumption made above

$$D = \lambda^2 \frac{kT}{h} (1 - e^{-h\nu^*/kT})\; e^{-\frac{E_o - \frac{1}{2}h\nu^*}{kT}}\; X_v \;. \qquad (73)$$

The frequency ν^* is often related to the Debye frequency, ν_D, of the crystal (Flynn, 1972), for example, $\nu^* = 3/4\ \nu_D$. The Debye temperature Θ_D, defined by $h\nu_D/k$, has typical values in the range 500-1000°K (see Shaw, 1976) in minerals. Therefore, $h\nu^*/k$ will be in the range 375-750°K. If $h\nu^*/k << T$, then

$$1 - e^{-\frac{h\nu^*}{kT}} \approx \frac{h\nu^*}{kT}$$

and equation (73) simplifies at high temperature to

$$D \simeq \lambda^2\ \nu^*\ e^{-\frac{E_o}{kT}}\ X_v \qquad (74)$$

where we have also assumed that E_o is much greater than $h\nu^*$ (a very good assumption -- see Table 7 in Chapter 7). In this case, ν^* in equation (74) can be given a simple interpretation, i.e., ν^* is the frequency with which the ion or atom attempts to surmount the migration barrier. Let us apply equation (74) to forsterite. The Debye temperature for forsterite, Mg_2SiO_4, is Θ_D = 759°K (Shaw, 1976). Therefore,

$$\nu^* = \frac{3}{4}\nu_D = \frac{3}{4}\frac{k}{h}(759°K) = 1.19 \times 10^{13}\ sec^{-1}\ .$$

The size of ν^* indicates that at atom certainly "tries" to jump quite a number of times in one second! In the case of Mg^{2+} diffusion λ can be taken as the distance between two Mg ions in forsterite, i.e., $\lambda \approx 3.0$ Å. Therefore, equation (74) would predict

$$D = (3 \times 10^{-8}\ cm)^2 (1.19 \times 10^{13}\ sec^{-1})\ e^{-\frac{E_o}{kT}}\ X_v$$

or

$$D = 1.07 \times 10^{-2}\ e^{-E_o/kT}\ X_v\ cm^2/sec\ . \qquad (75)$$

Equation (75) can be compared with the high temperature (T > 1125°C) results from Buening and Buseck for cation diffusion:

$$D_{Mg} = 1.74 \times 10^{-2}\, e^{-\frac{2.52\ ev}{kT}}\ cm^2/sec\ .$$

If X_v contributes essentially to the exponential (see Chapter 7) in equation (75), then the comparison is fairly good. However, D_{Mg} in olivines depends on f_{O_2} (see Chapter 7) and also on the direction of diffusion so the comparison should be done with great caution! Note that at T < 1125°C Buening and Buseck obtain for diffusion along the c-axis

$$D_{Mg} = 3.71 \times 10^{-7}\, e^{-\frac{1.25\ ev}{kT}}\ cm^2/sec$$

with a quite different D_o terms. This reflects a constant impurity-controlled X_v which will now contribute to D_o in (75).

Returning to equation (73), if the temperature is low enough so that $h\nu^* >> kT$, then

$$D = \lambda^2 \frac{kT}{h} e^{-\frac{E_o}{kT}} X_v\ . \tag{76}$$

If equation (76) holds, then the only difference between the D_o of different minerals within the same temperature range [since T is involved in equation (76)] and in the intrinsic region would arise from differences in the size of the diffusion jump, λ! Furthermore, D_o will now depend on the temperature, i.e.,

$$D_o = \lambda^2 \frac{kT}{h}\ .$$

We can generalize the transition state formulation of diffusion in a solid. The high temperature (classical) limit of the vibrational term in equation (72) is

$$q_{vib} = \frac{kT}{h\nu_i}\ . \tag{77}$$

Therefore, the general expression for D in the high T limit (see Flynn, 1972) is

$$D = \lambda^2 \frac{\prod_{i=1}^{3n-9} \nu_i}{\prod_{i=1}^{3n-10} \nu_i^{\dagger}}\ e^{-E_o/kT} \tag{78}$$

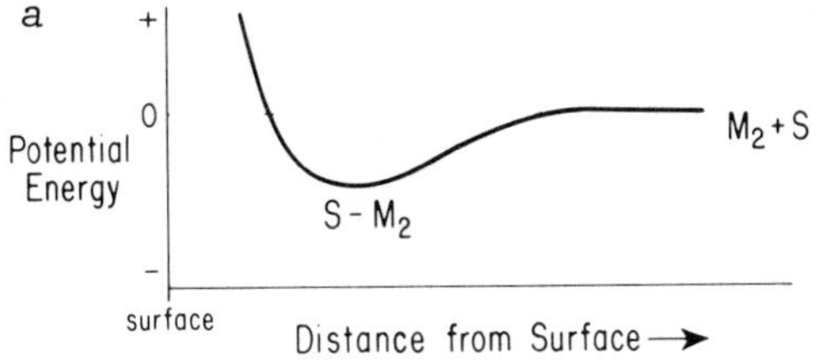

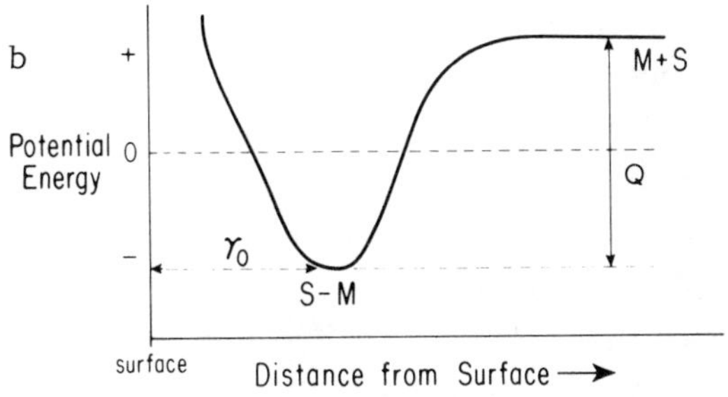

Figure 5. (a) Potential energy curve relating to the physical adsorption of a molecule M_2 on the surface of a solid S.
(b) Schematic representation of the chemisorption of a reactive atom M on the surface of a solid, S. Q is the heat or energy of adsorption and r_0 the equilibrium separation distance.

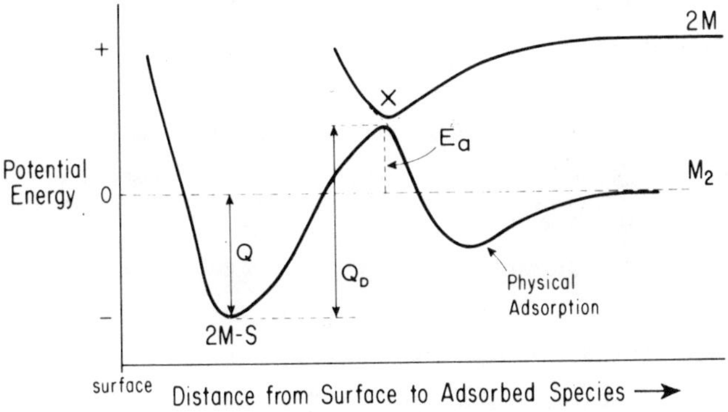

Figure 6. Potential energy curves including both physical adsorption and chemisorption. The intersection of the two curves is slightly rounded, as expected from the non-crossing rule between two states. Q is the energy of chemisorption; Q_D is the energy of activation for adsorption and E_a is the energy of activation for chemisorption.

where the 3n-9 ν_i's are the normal vibration frequencies of a crystal, which is perfect except for a vacancy at one site, and the 3n-10 $\nu_i^{\dagger}$'s (one ν missing) are the vibrational frequencies of the same crystal but with one ion lying between two vacant sites (i.e., the activated complex).

Energetics of Surface Reactions

Heterogeneous reactions are important in many geochemical processes such as mineral dissolution or recrystallization. In these reactions the surface of a mineral becomes one of the "reactants" and it may either act as a catalyst or be changed during the reaction. The use of transition state theory has been successful in homogeneous reactions; if one thinks of the surface of a solid as a large molecule, then the same tools can be transferred to heterogeneous reactions.

Chapter 1 discussed the difference between physical adsorption and chemisorption. A potential energy curve for physical adsorption of a diatomic M_2 molecule is shown in Figure 5a. It follows a van der Waal's potential, i.e., attractive at moderate distances from the surface and repulsive at short distances. Note that there is essentially no activation energy barrier to physical adsorption; this is in accord with the very fast rates of physical adsorption. The size of the well in Figure 5a is the energy of adsorption, which is usually several Kcal/mole. Of course, the energy of adsorption becomes the activation energy for physical desorption. Physical adsorption energies can be calculated theoretically using a Lennard-Jones potential between molecules (see Adamson, 1976).

Figure 5b is a schematic representation of the potential energy curve for the chemisorption of an atom M (not a diatomic M_2). The potential is similar to that of normal chemical bonds. The size of the well, Q, is the chemisorption energy.

Both physical adsorption and chemisorption can occur and so Figure 6 illustrates the *combined* potential curve for the case of M_2 molecules in a fluid. Note that potential energy curves are not allowed to cross and hence the intersection in Figure 6 is rounded to produce the overall topology shown. The size of Q depicts the energy of adsorption of M_2 as two chemisorbed M atoms. An important consequence of the non-crossing rule is that a *maximum* occurs at position X in the figure. Therefore, while physical adsorption of an M_2 molecule requires no activation energy, the chemisorption of M_2, as two chemisorbed M atoms, does require overcoming an activation energy, E_a in Figure 6. The activation energy for desorption is also given by Q_d in Figure 6.

The role of surfaces in providing an important path for chemical reactions in a fluid has been discussed in Chapter 1. For example, molecules A and B may react in the fluid phase to product products:

$$A + B \rightarrow \text{Products} . \tag{79}$$

On the other hand, A and B may react more readily on the solid surface:

$$A + B + \underset{|\ \ |}{-S-S-} \rightleftarrows \overset{A\ \ B}{\underset{|\ \ |}{-S-S-}} \rightarrow \text{Products} \tag{80}$$

(*e.g.*, see the $CO + O_2$ reaction in Chapter 1). Transition state theory gives the rate of (79) as

$$\text{Rate}_{\text{homogeneous}} = C_A C_B \frac{kT}{h} \frac{(q^{\dagger}_{\text{hom}}/V)}{(q_A/V)(q_B/V)} e^{-\frac{E_{\text{hom}}}{RT}} \tag{81}$$

where C_A and C_B are the fluid phase concentrations. Likewise, the heterogeneous rate is given by

$$\text{Rate}_{\text{heterogeneous}} = C_A\, C_B\, C_{S_2} \frac{kT}{h} \frac{(q^{\dagger}_{\text{het}}/V)}{(q_A/V)(q_B/V)(q_{S_2}/V)} e^{-\frac{E_{\text{het}}}{RT}} \tag{82}$$

where C_{S_2} is the concentration of bare but adjacent surface sites (see Chapter 1). If the activated complex in the heterogeneous reaction is considered virtually immobile, its partition function can be taken as *unity* [i.e., set all the $\varepsilon_i >> kT$ in equation (7)]. The same applies to the partition function for S_2, which can only vibrate ($q_{vib} \sim 1$). Therefore,

$$\text{Rate}_{\text{heterogeneous}} = C_A\, C_B\, C_{S_2} \frac{kT}{h} \frac{1}{(q_A/V)(q_B/V)} e^{-\frac{E_{\text{het}}}{RT}} . \tag{83}$$

Taking the ratio of (81) and (83)

$$\frac{\text{Rate}_{\text{het}}}{\text{Rate}_{\text{hom}}} = \frac{C_{S_2}}{(q^{\dagger}_{\text{hom}}/V)} e^{\frac{\Delta E}{RT}} \tag{84}$$

where $\Delta E = E_{\text{hom}} - E_{\text{het}}$. ΔE is the adsorption energy (see Chapter 1). The activated complex partition function can be evaluated using equations (55) and (58) as

$$q_{hom}^{\dagger} = \left(\frac{2\pi m^{\dagger} kT}{h^2}\right)^{\frac{3}{2}} V \left(\frac{8\pi^2 I^{\dagger} kT}{\sigma h^2}\right)^{\frac{3}{2}} q_{vib} \quad (85)$$

To obtain an estimate of the order of magnitude of $q_{hom}^{\dagger}$, the following values can be used

$m^{\dagger} \sim 10$ amu $= 1.67 \times 10^{-25}$ g, $k = 1.38 \times 10^{-16}$ erg/K,

$h = 6.62 \times 10^{-27}$ erg·sec, $I^{\dagger} = \mu r^2 = (5 \text{ amu})(2°\text{A})^2 = 33.4 \times 10^{-40}$ g cm^2,

$\sigma^{\dagger} = 1$, $T = 298°$K.

Inserting these values into (85) yields $q_{hom}^{\dagger} = (3 \times 10^{22} \text{ cm}^{-3})V(4 \times 10^3)q_{vib}$. Therefore,

$$\frac{q_{hom}^{\dagger}}{V} \simeq 10^{26\pm2} \text{ cm}^{-3} \quad . \quad (86)$$

The concentration of paired vacant sites C_{S_2} can also be estimated. For example, spinel ($MgAl_2O_4$) has oxygen atoms in cubic closest packing (face-centered cubic) and the side of each face-centered cube has a length of $a_o/2$, where a_o is the lattice parameter. Therefore, each cube face [(100), (010), or (001)] has an area of $a_o^2/4$ and contains two oxygen atoms. Using $a_o = 8.11$ Å (DeJong, 1959), the number of oxygen atoms on 1 cm^2 of surface area parallel to (100), (010), or (001) is given by

$$\text{oxygen sites} = \frac{10^{16} \text{ Å}^2}{(8.11/2)^2 \text{ Å}^2} \times 2 = 1.22 \times 10^{15} \text{ per cm}^2 \; .$$

Likewise, forsterite (hcp packing of oxygens) has eight oxygens, four oxygens and four oxygens on the (100), (010), and (001) planes of a unit cell, respectively. Using $a = 4.752$ Å, $b = 10.193$ Å, $c = 5.977$ Å (25°C data, Hazen, 1976), the number of oxygen sites per cm^2 on the different surfaces is given by

$$(100): \quad \frac{10^{16}}{(10.193)(5.977)} \times 8 = 1.31 \times 10^{15} \text{ sites/cm}^2$$

$$(010): \quad \frac{10^{16}}{(4.752)(5.977)} \times 4 = 1.41 \times 10^{15} \text{ sites/cm}^2$$

$$(001): \quad \frac{10^{16}}{(4.752(10.193)} \times 4 = 0.83 \times 10^{15} \text{ sites/cm}^2 \; .$$

While there are clearly variations in the number of sites, a reasonable estimate for C_{S_2} is 10^{15} sites/cm^2. Using this estimate for C_{S_2} and equation (86) for $q^{\dagger}_{hom}$ in equation (84) yields

$$\frac{\text{Rate}_{het}}{\text{Rate}_{hom}} = 10^{-11\pm2}\ e^{\frac{\Delta E}{RT}} . \tag{87}$$

It is clear that, if the heterogeneous reaction did not lower the activation energy, the much lower pre-exponential factor in (87) would make the heterogeneous path highly unfavorable. Equation (87) is based on 1 cm^2 of surface area and 1 cm^3 of fluid volume; i.e., it assumes that the specific area is 1 cm^2 surface/cm^3 solution. Using a typical specific surface area of 10 m^2/g or 10 m^2/cm^3 fluid, the ratio now becomes

$$\frac{\text{Rate}_{het}}{\text{Rate}_{hom}} = 10^5 \times 10^{-11\pm2}\ e^{\frac{\Delta E}{RT}} = 10^{-6\pm2}\ e^{\frac{\Delta E}{RT}} . \tag{88}$$

Nonetheless, to make the heterogeneous reaction favorable, the activation energy reduction via the surface reaction (ΔE) must be such that

$$10^{-6}\ e^{\frac{\Delta E}{RT}} > 1$$

or

$$\Delta E > 13.8\ RT . \tag{89}$$

At 25°C, (89) requires that ΔE be greater than 8 Kcal/mole. This requirement is certainly satisfied by most chemisorption energies. Note that the required minimum ΔE increases with temperature; at 300°C (*e.g.*, hydrothermal solutions) the decrease in the activation energy must be at least 15.7 Kcal/mole.

The application of transition state theory to heterogeneous reactions has clarified the mechanism whereby the rates are altered from their homogeneous values. By computing the appropriate partition functions, as dictated by TST, the pre-exponential factors can be evaluated. The interesting result is that a heterogeneous reaction actually has a much lower pre-exponential factor, orders of magnitude smaller, in fact, than the homogeneous pre-exponential factor. Therefore, the heterogeneous path is favored only if the activation energy is lowered enough to counterbalance the decrease in the pre-exponential factor.

CONCLUSION

Transition state theory is one of the cornerstones of the molecular approach to kinetics. The introduction of the activated complex permits a simple model of the crossing of potential barriers. With this model, the rate constant of any elementary reaction can be calculated, if the structure of the activated complex is known. The general results of transition state theory lead to an understanding of some important predictions of geochemical interest. For example, the effect of ionic strength on rates or the compensation law can be analyzed using TST. The theory can also be applied to transport in condensed media as well as to heterogeneous reactions. Although the potential energy surfaces needed to calculate the rates of geochemical reactions can only be estimated in a rough way, transition state theory provides an important framework, which allows a molecular approach to the kinetics. It is this unifying characteristic of the theory that makes it worthwhile for use in geochemical kinetics.

ACKNOWLEDGMENTS

The author is grateful to Gregory E. Muncill and Randall T. Cygan for assistance in the preparation of the manuscript and Mary Frank for help with the initial typing. Support from NSF Grant OCE-7909240 is also gratefully acknowledged.

CHAPTER 4 REFERENCES

Adamson, A. W. (1976) *Physical Chemistry of Surfaces.* John Wiley & Sons, New York, 698 p.

Benson, S. W. (1976) *Thermochemical Kinetics,* 2nd ed. John Wiley & Sons, New York, 320 p.

Boyd, R. K. (1978) Some common oversimplifications in teaching chemical kinetics. J. Chem. Education, 55, 84-89.

Buening, D. K. and Buseck, P. R. (1973) Fe-Mg lattice diffusion in olivine. J. Geophys. Res., 78, 6852-6862.

Connor, J. N. L., Jakubetz, W. and Lagana, A. (1979) Comparison of quasi-classical, transition state theory, and quantum calculations of rate constants and activation energies for the collinear reaction $X + F_2 \rightarrow X F + F$ ($X = \mu u$, H, D, T). J. Phys. Chem., 83, 73-78.

DeJong, W. F. (1959) *General Crystallography.* W. H. Freeman & Co., San Francisco, 281 p.

Eyring, H. (1935a) The activated complex in chemical reactions. J. Chem. Phys., 3, 107-120.

________ (1935b) The activated complex and the absolute rate of chemical reactions. Chem. Rev., 17, 65-82.

Farooque, M. and Fahidy, T. Z. (1978) The electrochemical oxidation of hydrogen sulfide in the Tafel region and under mass transport control. J. Electrochem. Soc., 125, 544-546.

Flynn, C. P. (1972) *Point Defects and Diffusion.* Clarendon Press, Oxford, 826 p.

Glasstone, S., Laidler, K. and Eyring, H. (1951) *The Theory of Rate Processes.* McGraw-Hill, New York, 600 p.

Golden, D. M. (1979) Experimental and theoretical examples of the value and limitations of transition state theory. J. Phys. Chem., 83, 108-113.

Hazen, R. M. (1976) Effects of temperature and pressure on the crystal structure of forsterite. Amer. Mineral., 61, 1280-1293.

Hill, T. L. (1960) *An Introduction to Statistical Thermodynamics.* Addison-Wesley, Reading, MA, 508 p.

Jost, W. (1960) *Diffusion in Solids, Liquids, Gases.* Academic Press, New York, 558 p.

Karplus, M., Porter, R. N. and Sharma, R. D. (1965) Exchange reactions with activation energy. I. Simple barrier potential for (H_1H_2), J. Chem. Phys., 43, 3259-3287.

Nash, L. K. (1965) *Notes on Statistical Mechanics.* McGraw-Hill, New York.

Nielsen, A. E. (1964) *Kinetics of Precipitation.* Pergammon Press, New York, 148 p.

Prigogine, I. (1967) *Introduction to Thermodynamics of Irreversible Processes.* Interscience Pub., New York, 147 p.

Rimstidt, J. D. and Barnes, H. L. (1980) The kinetics of silica-water reactions. Geochim. Cosmochim. Acta, 44, 1683-1699.

Robie, R. A., Hemmingway, B. S. and Fisher, J. R. (1978) *Thermodynamic Properties of Minerals and Related Substances at 298.15K and 1 bar (10^5 pascals) Pressure and at Higher Temperatures.* U. S. Geol. Surv. Bull., U.S. Printing Office, Washington, D. C., 456 p.

Shaw, G. H. (1976) Calculation of entropies of transition and reaction and slopes of transition and reaction lines using Debye theory. J. Geophys. Res., 81, 3031-3035.

Weston, R. E. and Schwartz, H. H. (1972) *Chemical Kinetics.* Prentice-Hall Inc., Englewood Cliffs, N. J., 274 p.

Chapter 5
IRREVERSIBLE THERMODYNAMICS in PETROLOGY
George W. Fisher & Antonio C. Lasaga

INTRODUCTION

Beginning with the pioneering work of Bowen (1913) and Goldschmidt (1911), the concepts of equilibrium thermodynamics have contributed hugely to our understanding of geochemistry and petrology. Today the geological sciences probably make more extensive use of thermodynamic concepts and data than any other field. This alliance has proved successful because thermodynamics has provided a simple but powerful conceptual framework linking a vast array of experimental data and petrologic information. Given the requisite thermochemical data, measurements of mineral compositions can yield quantitative information on conditions of formation. If the experimental data are not yet available, thermodynamics can point the way to those experiments which will be most helpful in understanding a particular petrologic problem. Thermodynamics provides a unique method of organizing our thinking about geologic systems, forcing us to ask new and genetically important questions.

Despite its successes, classical thermodynamics is severely limited by its restriction to systems in equilibrium. Time is one of the most fundamental geologic parameters, and a full understanding of geological processes demands a kinetic description. Ideally, a complete kinetic description should be based on a series of detailed mechanistic models. At the present state of our understanding, however, such a complete description is rarely possible. Consequently, it is often desirable to utilize a more general approach to kinetics, independent of specific kinetic or statistical models. Irreversible thermodynamics provides one such approach. In favorable cases it can provide quantitative estimates of the rates of geochemical processes. Even when the requisite kinetic coefficients are unknown -- as is all too often the case -- irreversible thermodynamics provides a conceptual framework which can be immensely helpful in organizing our thinking about geologic processes. Just as equilibrium thermodynamics can lead us to ask the relevant questions about geologic *systems*, irreversible thermodynamics can help us to formulate important new ways of looking at geologic *processes*.

This chapter will outline some aspects of irreversible thermodynamics which are particularly relevant to petrologic processes, and illustrate some simple applications of the ideas developed. We cannot provide detailed derivations of all of the equations used in the space available. Instead, we will concentrate on attempting to give some physical insight into the relations used. For a more detailed discussion, refer to the treatments of de Groot and Mazur (1962), Katchalsky and Curran (1967) and Prigogine (1967).

THE FUNDAMENTAL KINETIC EQUATIONS

The first step in any kinetic formulation is to express the rate of a process as a function of some appropriate driving force. For many processes, the simplest possible relation -- a linear equation -- provides an adequate description. For example, the rate of diffusion of component i, written J_i^D (measured in mol/cm^2·s), can be written as a linear function of either the concentration gradient (in mol/cm^4) or the chemical potential gradient (in cal/mol·cm), each multiplied by an appropriate kinetic coefficient. For one-dimensional diffusion*

$$J_i^D = L_{ii}^D \frac{d}{dx}(-\mu_i) = D_{ii}\frac{d}{dx}(-c_i) \tag{1}$$

where L_{ii}^D is expressed in mol^2/cal·cm·s, and D_{ii} in cm^2/s. Because chemical potential can be expressed in terms of concentration (for an ideal solution, $\mu_i = \mu_i^o + RT\ln c_i$), the kinetic coefficients are also related ($L_{ii}^D = c_i D_{ii}/RT$ for an ideal solution). Consequently, the two formulations are fundamentally equivalent, and a choice between them is dictated by convenience. In many petrologic systems it is much easier to determine the chemical potential gradient for a component than it is to estimate the concentration gradient along the diffusion path, so it is often easier to write diffusion rates as a function of chemical potential gradients. We shall do so here. Representative values of the L_{ii}^D are listed in Table 1.

----

*For a complete description the fluxes and forces should be written as vectors in three-dimensional space. The physical concepts involved can, however, be presented more simply if we restrict ourselves to systems which can be described in terms of a single dimension (diffusion between two plane parallel rock layers, for example).

TABLE 1

Representative Values for Kinetic Coefficients in Metamorphism

Coefficient of grain boundary diffusion.

$$L_{ii}^{D} = \frac{2.54 \times 10^{-2}}{X_{gr} RT} \exp \frac{-2.0 \times 10^{-4}}{RT} \ \mathrm{mol^2/cal \cdot cm \cdot s}$$

Coefficient of thermal conductivity.

$$K = 7 \times 10^{-3} \ \mathrm{cal/cm \cdot s \cdot {}^{\circ}C}$$

Coefficient of reaction between silicate and aqueous fluid.

$$L_{rr}^{R} = \frac{0.46}{X_{gb} RT} \exp \frac{-1.50 \times 10^{4}}{RT} \ \mathrm{mol^2/cal \cdot cm^3 \cdot s}$$

Data from Fisher (1978), and Handbook of Physics and Chemistry.

Similarly, the heat flux (J_q, in cal/cm^2·s) can be written as a linear function of the temperature gradient:

$$J_q = k \frac{d}{dx} (-T) \tag{2}$$

where k is the thermal conductivity, measured in cal/cm·s·deg (Table 1).

Finally, in systems not too far from equilibrium chemical reaction rates, denoted J_r^R (measured in mol/cm^3·s), can be written as linear functions of reaction affinity, A_r (measured in cal/mol), defined as the negative sum of the chemical potentials of all components or species participating in the reaction, each multiplied by its stoichiometric coefficient, ν_i^r (a dimensionless number). Formally, $A_r = -\sum_i \nu_i^r \mu_i$, and

$$J_r^R = L_{rr}^R A_r \tag{3}$$

where L_{rr}^R has the units mol^2/cal·s·cm^3 (Table 1). Individual components or species are then produced at the rate

$$J_i^R = \nu_i^r J_r^R = \nu_i^r L_{rr}^R A_r \ . \tag{4}$$

Equations (1) and (2) are generally valid for a wide range of systems, even those quite far from equilibrium. The range of validity of equation (3) is much more restricted, and many geologic processes operate sufficiently far from equilibrium that reaction rates must be expressed in more complex functional form. We will examine some of the consequences of this fact in a later section, but for now we will adopt the linear formulation of equations (3) and (4).

In systems involving both diffusion and chemical reaction, the net change in concentration of a component is given by the mass balance equation,

$$\frac{\partial c_i}{\partial t} = \nu_i^r \, J_r^R - \frac{dJ_i^D}{dx} , \qquad (5)$$

sometimes called the continuity equation. This relation simply expresses the fact that at any point in a system a difference between the local production of a component by reaction ($\nu_i^r \, J_r^R$) and the net loss of that component by diffusion (dJ_i^D/dx, the local *change* in the diffusive flux along the diffusion path) must produce a change in concentration.

Many petrologic problems can be analyzed in terms of these equations alone. Given estimates of the relative values of the L_{ii}^D for different components, the diffusion equation (1) and the mass balance equation (5) may be combined to predict the sequence, modal composition and relative thickness of reaction zones formed by diffusive exchange between two incompatible assemblages (Fisher, 1977; Joesten, 1977). Or, conversely, measurements of zone compositions can be used to estimate relative L_{ii}^D values for different components (Joesten, 1977). Given estimates of the absolute value of the kinetic coefficients, the mass balance equation can be combined with the rate equations for diffusion, reaction and heat flow to estimate absolute growth times for a variety of metamorphic structures (Fisher, 1978). For example, consider the growth of a layer of a mineral, C, by reaction between layers of minerals A and B (Fig. 1). The rate of growth of a layer of C is given by

$$\frac{dX_C}{dt} = \bar{v}_C \, J_C^R \, X_{gb}$$

where $\bar{v}_C$ is the molar volume of C, J_C^R is the chemical reaction rate of C ($moles/cm^2/sec$), X_{gb} is the thickness of the grain-boundary film, in which precipiation is occurring, and X_C is the thickness of the C layer. In processes such as this, the chemical potentials (or concentrations) of the diffusing species commonly approach a constant, steady-state value; consequently, equation (5) requires that the rate of precipitation of each component, i, by reaction exactly balance the net rate of supply by diffusion. Provided that diffusion through the layers of A and B is negligible,

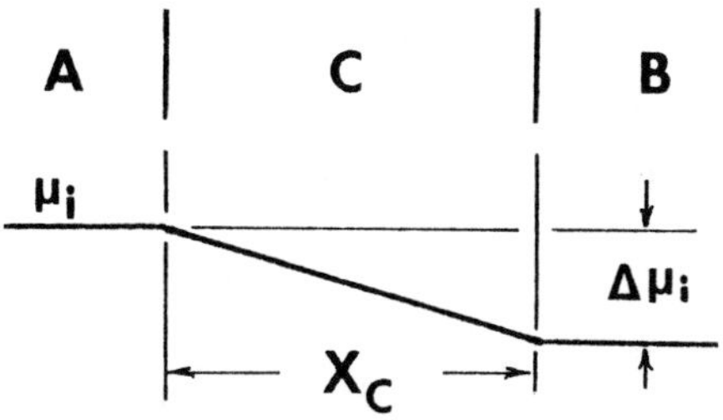

Figure 1. Variation in chemical potential of component i (μ_i) during growth of mineral C by reaction between layers of mineral A and mineral B. X_C is the thickness of the layer of C, $\Delta\mu_i$ is the difference in the value of μ_i in equilibrium with the assemblages AC and AB.

$$X_{gb}\, \nu_i^C\, J_C^R = J_i^D$$

at the zone boundaries, and

$$\frac{dX_C}{dt} = \bar{v}_C \frac{J_i^D}{\nu_i^C} .$$

Assuming that (i) the diffusing species are in equilibrium with the local mineral assemblage at all points along the diffusion path; (ii) reaction occurs only at zone boundaries; and (iii) L_{ii}^D coefficients are constant,

$$J_i^D = -L_{ii}^D \left(\frac{\Delta\mu_i}{X_C}\right) ,$$

and

$$\frac{dX_C}{dt} = \frac{\bar{v}_C\, L_{ii}^D\, (-\Delta\mu_i)}{\nu_i^C\, X_C} ,$$

where $\Delta\mu_i$ is the potential drop between the assemblages AC and BC. Integrating and solving for t gives the time required to grow a layer of C, which is X_C cm thick,

$$t = \frac{\nu_i^C\, X_C^2}{2L_{ii}^D\, \bar{v}_c\, \Delta\mu_i} . \qquad (6)$$

Application of the Kinetic Equations

As an illustration of how these equations can be used, consider the growth of the wollastonite rims formed between chert nodules and their marble matrix in a west Texas contact aureole (Joesten, 1974). The wollastonite rims grew at the expense of calcite and quartz by a process involving diffusion of CaO and SiO_2 through the growing wollastonite layer. The process can be represented schematically by a simple reaction cycle,

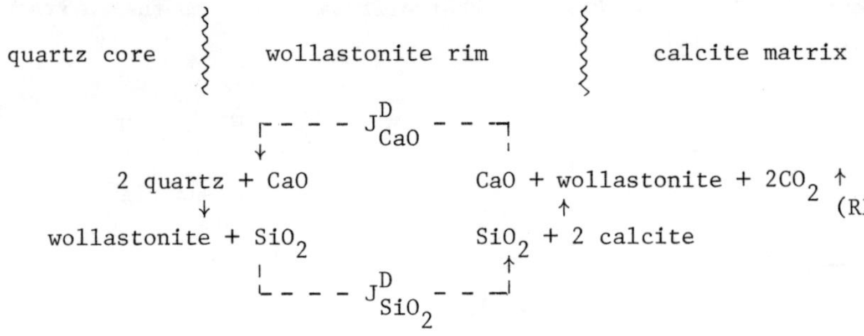

quartz core   } wollastonite rim   } calcite matrix

$$
\begin{array}{ccc}
& \overset{\ulcorner\;-\;-\;-\;J^D_{CaO}\;-\;-\;-\;\urcorner}{\downarrow} & \\[4pt]
2\ quartz + CaO & & CaO + wollastonite + 2CO_2 \uparrow \\
\downarrow & & \uparrow \qquad\qquad\qquad\qquad (R1)\\
wollastonite + SiO_2 & & SiO_2 + 2\ calcite \\[4pt]
& \underset{\llcorner\;-\;-\;-\;J^D_{SiO_2}\;-\;-\;-\;\lrcorner}{\uparrow} &
\end{array}
$$

in which diffusion of SiO_2 and CaO through the wollastonite rim (dashed
lines) is driven by the differences in the chemical potentials imposed
by the assemblages quartz-wollastonite and wollastonite-calcite. Diffusion almost certainly occurred through the network of grain boundaries
in the wollastonite layer, because lattice diffusion is much slower than
grain-boundary diffusion at metamorphic temperatures (Fisher, 1978, p.
1048).

Most processes, such as this, rapidly attain a steady state in
which chemical potentials at the reaction zone boundaries remain nearly
constant (assuming constant T), simply because the fluid film permeating
the grain-boundary network has negligible volume compared with the growing layer of wollastonite. The quantity of CaO and SiO_2 that must be
supplied to adjust the potentials in the grain-boundary film to the
steady state values is so much smaller than the quantity required to
grow a discernable layer of wollastonite that a steady state is achieved
very early in the growth process. Consequently, $\partial c_i/\partial t = 0$ in the grain-
boundary network, and the mass balance equation (5) requires that the
rate of precipitation at any point must exactly balance the net influx by
diffusion. For that reason, reaction (R1) was written with the zone
boundary reactions exactly balancing the diffusive fluxes. The coefficients in this reaction cycle depend upon the reference frame used to
measure the diffusion (*cf*. Brady, 1975, for a review of the possibilities;
also Chapter 6, this volume). In this example it is convenient to
measure the rate of diffusion relative to a reference frame moving with
the velocity of one of the components, say CaO. This permits us to consider the entire process in terms of a single flux, and leads to a simple
reaction cycle

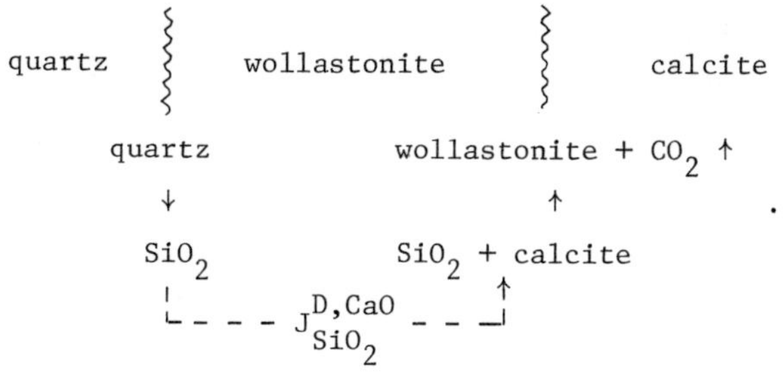

(R2)

More formally,

$$J^R_{qz}\ X_{gb} = J^{D,CaO}_{SiO_2} = J^R_{wo}\ X_{gb} \tag{7}$$

where X_{gb} refers to the thickness of the grain-boundary film within which precipitation is taking place, and serves to convert the expression for reaction rate from the unit volume basis implied by J^R_r to a unit cross-section basis, compatible with J^D_i. The subscripts qz and wo refer to the quartz dissolution and wollastonite precipitation reactions, respectively. $J^{D,CaO}_{SiO_2}$ represents the diffusive flux of SiO_2 measured on a CaO-fixed reference frame.

Substituting (1) and (3) into equation (7) and assuming linear chemical potential gradients,

$$L^R_{qz}\ A_{qz}\ X_{gb} = L^{D,CaO}_{SiO_2} \left(\frac{\Delta\mu_{SiO_2}}{\Delta x}\right) = L^R_{wo}\ A_{wo}\ X_{gb} \tag{7'}$$

where Δx is the thickness of the wollastonite zone. Rearranging, the left-hand equation gives

$$\frac{\Delta\mu_{SiO_2}}{A_{qz}} = \frac{L^R_{qz}\ X_{gb}\ \Delta x}{L^{D,CaO}_{SiO_2}} \ . \tag{7''}$$

We can now apply this relation to a nodule 49.4 m from the intrusive contact, with a 2.5 cm wollastonite zone (Δx) developed along a flat portion of the nodule, where the one-dimensional equations used here can be applied. The peak temperatures attained in the aureole are tightly constrained by five isograds, corresponding to the growth of wollastonite, tilleyite, rankinite, spurrite and larnite; interpolating, this nodule attained a peak temperature of approximately 830°C. Presumably, wollastonite grew only while the temperature of the nodule was above the equilibrium temperature of the reaction quartz + calcite →

wollastonite + CO_2 [600°C at 325 bars, the total pressure during metamorphism (Joesten, 1974)]. Reaction processes of this sort can be modeled approximately by assuming isothermal growth at a temperature equivalent to two-thirds of the maximum overstepping, or 753°C in this example. At this temperature, the relations in Table 1 give $L_{SiO_2}^{D,CaO}$ = 1.24×10^{-16} $mol^2/cal \cdot cm \cdot s$ and $X_{gb}L_{qz}^{R}$ = 1.43×10^{-9} $mol^2/cal \cdot cm^3 \cdot s$; substituting these values into equation (7") gives $\Delta\mu_{SiO_2}/A_{qz}^{R}$ = 2.88×10^7, implying that the local affinities were negligible compared with the total potential drop across the wollastonite zone. Consequently, the system was in local equilibrium, and we may use the values of the potentials in equilibrium with quartz-wollastonite and wollastonite-calcite to estimate the total potential drop across the wollastonite layer. Inserting thermochemical data from Robie *et al.* (1978) and estimates of $L_{SiO_2}^{D,CaO}$ from Table 1 into equation (6) gives an estimate of 5100 years for the time required to grow a 2.5 cm-thick wollastonite layer on the nodule we are considering. This figure is, of course, only an approximate value because it assumes growth at constant temperature, but more precise estimates based on detailed heat-flow calculations (Fisher and Joesten, unpublished data) give closely comparable values.

THE DISSIPATION FUNCTION AND ITS ROLE IN COMPLEX PROCESSES

In more complex processes involving several reactions, multicomponent diffusion, and perhaps heat flow all proceeding simultaneously, a more sophisticated analysis is often helpful. A natural starting point for a thermodynamic analysis of irreversible processes is the concept of entropy production. In discussions of the second law of thermodynamics, the total entropy change of a system is commonly written

$$dS = d_iS + d_eS \tag{8}$$

where d_eS represents the net flow of entropy into the system from its surroundings, and d_iS represents the entropy produced by processes occurring within the system. The second law of thermodynamics then states that

$$d_iS \geqslant 0$$

where the equality characterizes reversible processes, the inequality irreversible ones. In other words, d_iS is a measure of the degree of irreversibility of the processes occurring within the system. The starting point for a thermodynamic analysis of irreversible processes is to evaluate the term d_iS.

We begin by assuming that a system undergoing irreversible processes may be divided into parts which are sufficiently large that thermodynamic properties are well defined, but small enough that gradients in the intensive parameters are negligible. We then assume that the entropy variation in each of these subsystems may be expressed in terms of the same variables as in macroscopic systems at equilibrium,

$$Tds = du + Pdv - \sum_i \mu_i dn_i \tag{9}$$

where the extensive parameters are designated by lower-case letters to emphasize their local character. In the literature of irreversible thermodynamics, equation (9) is commonly referred to as an expression of the assumption of "local equilibrium." This phrase, however, has a very different meaning from the concept of local equilibrium as used in the petrologic literature. In petrology, local equilibrium means that at each point in a system all minerals and chemical species are in mutual equilibrium, so that local reaction affinities are everywhere negligible. Because of the equilibrium conditions relating the chemical potentials of individual species, the summation need include only the independent chemical components needed to characterize the system. It was in this sense that we used the concept of local equilibrium to evaluate the chemical potential drop across the wollastonite zone in the example discussed in the previous section. In irreversible thermodynamics, however, local equilibrium means only that equation (9) is assumed to hold when the summation is taken over all of the species in the system; spatial gradients in the intensive parameters are assumed to be on a local scale, but individual minerals or species need not be in equilibrium and reaction affinities need not be negligible, even locally.

Following Katchalsky and Curran (1967, p. 75-79), it is convenient to rewrite equation (9) in terms of the local concentrations of entropy, energy and matter, $s_v \equiv s/v$, $u_v \equiv u/v$, and $c_i \equiv n_i/v$, giving at constant v

$$Tds_v = du_v - \sum_i \mu_i dc_i \ . \tag{9'}$$

Dividing by dt while holding the spatial coordinates constant then gives

$$T \frac{\partial s_v}{\partial t} = \frac{\partial \mu_v}{\partial t} - \sum_i \mu_i \frac{\partial c_i}{\partial t} . \qquad (9'')$$

The left-hand side of equation (9") refers to the *total* entropy change at a point in a system. In order to proceed further we must distinguish the entropy change due to irreversible processes from that reflecting a net gain or loss of entropy from external sources. Because equation (8) can be applied to arbitrarily small volumes we may immediately write an equation giving the entropy balance at any point in a system,

$$\frac{\partial s_v}{\partial t} = \sigma - \frac{dJ_s}{dx} \qquad (10)$$

where σ is the local rate of entropy production due to irreversible processes. This equation is closely analogous to the mass balance equation (5), and simply expresses the fact that any difference between the local rate of entropy production and the net loss of entropy to the environment must produce a change in the local value of entropy. The rate of entropy production for the entire system may then be determined by evaluating σ and integrating over the volume of the system,

$$\frac{d_i S}{dt} = \int_v \sigma \ dV .$$

Inserting equation (10) into (9") and using the first law to express du_v we obtain after considerable manipulation (see Katchalsky and Curran, 1967, p. 76–80)

$$-T \frac{d}{dx} (J_s) + T\sigma = -T \frac{d}{dx} \left(\frac{J_q - \sum_i \mu_i J_i^D}{T} \right) +$$

$$\left(\frac{J_q - \sum_i \mu_i J_i^D}{T} \right) \frac{d}{dx} (-T) + \sum_i J_i^D \frac{d}{dx} (-\mu_i) + J_r^R A_r . \qquad (11)$$

Comparing the first terms on opposite sides of this equation, we see that

$$J_S = \frac{J_q - \sum_i \mu_i J_i^D}{T} . \qquad (12)$$

This relation shows that the entropy flux can be considered as consisting of two parts, one due to the transfer of heat by conduction, and one due to the transfer of energy by diffusion. Substituting this relation back into equation (11) gives

$$T\sigma = J_S \frac{d}{dx}(-T) + \sum_i J_i^D \frac{d}{dx}(-\mu_i) + J_r^R A_r \ . \qquad (13)$$

The quantity $T\sigma$, called the dissipation function by Lord Rayleigh, has the dimensions of free energy per unit time, and represents the local rate of dissipation of free energy as a result of irreversible processes within the system. Equation (13) indicates that the rate of dissipation can be expressed as a series of terms, each representing a flux multiplied by the force driving that flux. At equilibrium the forces and fluxes both equal zero, and the rate of free energy dissipation is of course zero.

Several alternative choices of fluxes and forces are possible in equation (13), so long as each flux multiplied by its conjugate force has the dimensions of free energy per unit time, and so long as the sum remains equal to $T\sigma$ (Katchalsky and Curran, 1967, p. 79-80). We have chosen the form of equations (12) and (13) so that diffusive fluxes and reaction rates are conjugate to the familiar forces of equations (1) and (3). As a consequence, we are forced to consider the flow of heat in terms of the entropy flux, J_S, rather than the conventional heat flux, J_q, of equation (2).

The dissipation function (13) provides the basis for a thermodynamic analysis of fluxes and forces in an evolving system. Some of the most useful results to emerge from such an analysis concern the values of the kinetic coefficients. Consider a system involving n irreversible processes, each proceeding at a rate J_i, driven by n forces, X_i, chosen so as to satisfy the relation

$$T\sigma = \sum_i^n J_i X_i \ . \qquad (13')$$

Numerous experiments have shown that most fluxes depend on more than one force (for example, the Soret effect, in which a diffusive flux is driven by a temperature gradient, or diffusion in multicomponent systems, in which the flux of any component can be driven by a gradient in the chemical potential of any other component). In general, we must expect that each flux will depend on all of the fluxes in the system; assuming linear rate laws,

$$J_i = L_{i1}X_1 + L_{i2}X_2 + \ldots L_{in}X_n \ . \qquad (14)$$

A full kinetic description of a system involving n fluxes therefore requires a total of n^2 kinetic coefficients, a somewhat daunting prospect in complex petrologic systems.

One of the major triumphs of irreversible thermodynamics has been to reduce the number of coefficients required to characterize the kinetics of an evolving system. One of the best known relations, originally due to Onsager (1931) is that the matrix of kinetic coefficients is symmetrical, so that

$$L_{ij} = L_{ji} \tag{15}$$

provided that the fluxes and forces are independent and chosen so as to satisfy equation (13). Naturally, corresponding cross-coefficients must have the same units; for example, the cross-coefficients coupling diffusion to heat flow and vice-versa both have the units mol/cm·sec·deg. For systems involving chemical reactions sufficiently close to equilibrium for the linear rate laws of equation (3) to hold, the validity of equation (15) can be demonstrated by straightforward thermodynamic arguments (Katchalsky and Curran, 1967, p. 91-97). For systems involving diffusion, however, the derivation of (15) is based on statistical mechanical arguments (i.e., on microscopic reversibility) and is in a sense somewhat extra-thermodynamic (Katchalsky and Curran, 1967, Ch. 15). The importance of equation (15) is two-fold. First, it immediately reduces the number of independent kinetic coefficients by nearly one-half (from n^2 to $(n^2+n)/2$). Secondly, it provides a simple experimental check on the validity of the theory, now confirmed by experimental determinations in a wide range of systems (Miller, 1969).

An additional limitation on the number of coefficients can be derived from symmetry considerations first pointed out by Curie (Katchalsky and Curran, 1967, p. 89). In isotropic systems, all of the kinetic coefficients must be invariant to transformations of the coordinate axes used to describe the system. This condition is clearly met by L_{ij} coefficients which are scalar quantities, such as those coupling vectorial fluxes (for example, diffusion) to vectorial forces (chemical potential gradients) or those relating scalar fluxes (for example, reaction rates) to scalar forces (affinities). However, if scalar fluxes are assumed to be cross-coupled to vectorial forces, the cross-coupling coefficient must be a vector. Any non-zero vector must necessarily depend on the orientation of the coordinate system used to describe the system, which is impossible in an isotropic system; consequently, there can be no cross-coupling of forces and fluxes of different tensorial order in isotropic systems. For example, if we assume that the fluid permeating the grain-boundary network of a metamorphic rock is isotropic, there can be no cross-coefficients coupling reaction rates to chemical potential gradients or coupling diffusion rates to reaction affinities.

In many systems, a still further reduction in the number of kinetic coefficients can be achieved by a careful choice of components. In systems involving diffusion, cross-coupling of diffusive fluxes will occur whenever the diffusing species contain more than one component. For example, $Si(OH)_4$ will diffuse in response to gradients in the chemical potential of either SiO_2 of H_2O, leading to cross-coupling between the components SiO_2 or H_2O (see Carman, 1968). Coupling of this sort can be largely avoided by choosing as components compositions which closely approximate the dominant diffusing species in the system. Coupling can also arise through interaction between electrically charged species or because of motion of the reference frame used to measure diffusion. However, the resulting cross-coefficients (L_{ij}) can be calculated from the diagonal coefficients (L_{ii}) using straightforward electrochemical considerations (Katchalsky and Curran, 1967, Ch. 11) and the rules governing transformation from one reference frame to another (Brady, 1975). By making judicious use of these approaches, it is frequently possible to reduce the number of independent kinetic coefficients to the number of independent fluxes in the system.

Finally, the dissipation function (13) provides useful constraints on the sign of the diagonal coefficients and the magnitude of whatever cross-coefficients remain. Consider a system involving two fluxes and two forces, for which the dissipation function may be written $T\sigma = J_1X_1 + J_2X_2 \geq 0$. Assuming linear kinetics, $J_1 = L_{11}X_1 + L_{12}X_2$ and $J_2 = L_{21}X_1 + L_{22}X_2$. Substituting these relations into the dissipation function and using the Onsager relations (15) gives

$$T\sigma = L_{11}X_1^2 + 2L_{12}X_1X_2 + L_{22}X_2^2 \geq 0 \ . \qquad (16)$$

Because either X_1 or X_2 can be made equal to zero, the inequality can hold only if $L_{11}X_1^2 \geq 0$ and if $L_{22}X_2^2 \geq 0$, requiring that the diagonal coefficients L_{11} and L_{22} both be positive. Furthermore, the properties of quadratic equations require that the inequality (16) can hold only if

$$L_{12}^2 \leqslant L_{11}L_{22} \tag{17}$$

(Katchalsky and Curran, 1967, p. 91), severely limiting the magnitude of the cross-coefficients.

Additionally, the dissipation function provides useful insight into the thermodynamic meaning of the steady state. Experience tells us that a system left to itself will evolve in such a way that it will eventually attain a state of equilibrium in which all the forces and all the fluxes vanish; the dissipation rate given by equation (13') is then zero. We may, however, impose certain constraints on a system so that it cannot reach equilibrium; for example, the mineral assemblages may impose a fixed gradient in the chemical potential of one component, while allowing the chemical potential gradient of another component to change freely. In such systems it is observed that the unconstrained potential will shift toward a steady value (the steady state), and that in the steady state the flux of the unconstrained component will vanish. To analyze a system of this type thermodynamically, consider a process involving two forces linearly related to two fluxes by constant L_{ij} coefficients; the dissipation function is then given by equation (16). Assume that X_2 is maintained at a constant value by some external constraint (for example, a gradient in chemical potential can be imposed by the local mineral assemblages) and differentiate (16) with respect to X_1, giving

$$\left(\frac{\partial T\sigma}{\partial X_1}\right)_{X_2} = 2L_{11}X_1 + 2L_{12}X_2 = 2J_1 \; .$$

In systems of this type, the steady state is characterized by the condition $J_1 = 0$, so that

$$\left(\frac{\partial T\sigma}{\partial X_1}\right)_{X_2} = 0 \; .$$

Because $T\sigma \geqslant 0$, the steady state must be characterized by a minimum rate of dissipation, or, as conventionally expressed in treatises on non-equilibrium thermodynamics, by a minimum rate of entropy production. In analyzing petrologic systems, the minimum dissipation principle provides an alternative criterion of the steady state which can be useful under certain circumstances (Fisher, 1973).

Further Applications

The relations discussed in the preceding section provide the basis for a more careful evaluation of the growth of the calc-silicate reaction zones discussed earlier. Our previous analysis assumed the rate of diffusion of SiO_2 through the growing wollastonite layer depended on the chemical potential gradient alone. We now see that we were justified in neglecting cross-coupling between diffusion and local chemical reactions; assuming that the fluid permeating the grain-boundary network can be considered isotropic, coupling between diffusion (a vectorial flux) and chemical affinity (a scalar force) is impossible by Curie's principle. However, both diffusion and heat or entropy flow are vectorial fluxes driven by vectorial forces and may be expected to be coupled. Consequently, the full expressions for the diffusion rate of SiO_2 ($J^{D,CaO}_{SiO_2}$, here simplified to J^D_{Si}) and the entropy flux (J_S) are

$$J^D_{Si} = L_{SiSi} \frac{d}{dx} (-\mu_{Si}) + L_{SiS} \frac{d}{dx} (-T)$$
$$J_S = L_{SSi} \frac{d}{dx} (-\mu_{Si}) + L_{SS} \frac{d}{dx} (-T) \quad . \qquad (18)$$

In order to evaluate the importance of cross-coupling on the diffusion of SiO_2 we must evaluate the relative magnitude of the terms representing the chemical and thermal contributions to J_{Si}. Following a procedure similar to that of Katchalsky and Curran (1967, p. 183),

$$L_{SiS} = s_T \, L_{SiSi} \, RT \qquad (19)$$

where s_T is the Soret coefficient. Typically, s_T is on the order of 10^{-3} deg^{-1}. Combining this value with estimates of L_{SiSi} from Table 1 for a temperature of 1026°K and a time when the wollastonite rim had reached a thickness of one cm gives $L_{SiS} = 2.52 \times 10^{-16}$ mol/cm/cm·s·deg. Inserting these values into equations (18),

$$\begin{aligned} J^D_{Si} &= (1.24 \times 10^{-6} \text{ mol}^2/\text{cal·cm·s})(3.49 \times 10^3 \text{ cal/mol·cm}) \\ &\quad + (2.52 \times 10^{-6} \text{ mol/cm·s·deg})(4 \times 10^{-2} \text{ deg/cm}) \\ &= 4.33 \times 10^{-13} \text{ mol/cm}^2\text{·s} + 1.01 \times 10^{-17} \text{ mol/cm}^2\text{·s} \;, \end{aligned}$$

suggesting that chemical diffusion (the first term) was approximately four orders of magnitude larger than thermal diffusion (the second term). This inference is strongly supported by the fact that the wollastonite rims are approximately concentric on the chert nodules. If diffusion had been significantly coupled to the temperature gradient, the wollastonite rim should have grown much thicker on the side of the nodule where chemical diffusion was aided by thermal diffusion than on the side where the two effects were opposed.*

Having established a reasonable value of L_{SiS}, we are now in a position to evaluate the possible effects of the chemical potential gradients on heat flow in the system. Using the Onsager relation (15), $L_{SSi} = L_{SiS}$; accordingly, we may substitute the value for L_{SiS} obtained above into the equation for J_S (18),

$$J_S = (2.52 \times 10^{-14}\ \text{mol/cm·s·deg})(3.49 \times 10^{3}\ \text{cal/mol·cm})$$
$$+ (6.82 \times 10^{-6}\ \text{cal/cm·s·deg}^2)(4 \times 10^{-2}\ \text{deg/cm})$$
$$= 8.79 \times 10^{-11}\ \text{cal/cm}^2\text{·s} + 2.73 \times 10^{-7}\ \text{cal/cm·s·deg}\ ,$$

showing that the heat flow driven by the chemical potential gradients in the system (the first term) is negligible in comparison to the heat flow driven by the temperature gradients (the second term). Consequently, heat flow patterns in the vicinity of the nodules can be analyzed to a very good approximation by conventional conductive heat transfer calculations.

IRREVERSIBLE THERMODYNAMICS, CHEMICAL REACTIONS AND PATTERN FORMATION IN PETROLOGY

Nature provides abundant examples of patterns, which arise from a combination of transport processes and chemical reactions. In this section, we shall first analyze the domain of validity of the linear rate law, as applied to chemical reactions. The process of pattern formation in petrology will be studied as a coupling of diffusion (or mass transport) and chemical reaction.

----

*Although thermal diffusion appears to have been negligible in this example, equation (17) contains an implied warning that the effect could be important in some systems. The value of L_{SiS} could in principle be as high as $L_{SiS} = \sqrt{L_{SiSi}L_{SS}}$. Values of L_{SS} are typically on the order of k/T, about 6.82×10^{-6} cal/cm·s·deg^2 in this example. Consequently, L_{SiS} could theoretically have been as high as 2.91×10^{-11} mol/cm·s·deg, about five orders of magnitude larger than the typical value used above. Had this occurred the rate of thermal diffusion could have exceeded the rate of chemical diffusion.

Linear Phenomenological Laws for Chemical Reactions

A general reaction can occur in either the forward or reverse directions. When the rates in either direction cannot be ignored, the *net* rate of the overall reaction is the difference between the forward and reverse rates. In Chapter 4 (*Transition State Theory*) it was shown that for any elementary reaction the net rate can be written as

$$R_{net} = R_+(1 - e^{\frac{\Delta G}{RT}}) \tag{20}$$

where R_+ is the forward rate of the reaction and ΔG is the Gibbs free energy change for the reaction as written. Near equilibrium ($\Delta G = 0$), the magnitude of ΔG is small relative to RT and equation (20) can be simplified to:

$$R_{net} = -\frac{R_+ \Delta G}{RT} \quad .$$

Defining the *affinity* of the chemical reaction, $A_r = -\Delta G$, then

$$R_{net} = \frac{R_+ A_r}{RT} \quad . \tag{21}$$

It is this linear relation between the rate and the affinity which was used in the previous section [*e.g.*, equation (13)]. However, as can be seen from equation (20), the linear law only holds near equilibrium, where $\Delta G = 0$. The treatment following equation (7") showed that, in some cases of geologic interest, A_r may indeed be close to zero. In other cases, however, the growth of a new crystal may occur under conditions far from equilibrium. If this is the case, the kinetics may follow a non-linear behavior. This type of behavior can give rise to patterns such as igneous and metamorphic banding or Liesegang rings, as we shall show below.

Before embarking on a treatment of non-linear kinetics, it is important to relate the linear domain of irreversible thermodynamics to data that are often obtained in experimental petrology. Let us consider the following dehydration reaction

$$\underset{muscovite}{KAl_3Si_3O_{10}(OH)_2} + \underset{quartz}{SiO_2} = \underset{andalusite}{Al_2SiO_5} + \underset{K\text{-}feldspar}{KAlSi_3O_8} + H_2O \tag{22}$$

such as may occur in the high grade metamorphism of pelitic rocks. Experiments may be carried out with crystals of muscovite, quartz, andalusite, and K-feldspar present in a capsule containing an accurately known amount of H_2O at fixed P_{fluid} and temperature. After a certain interval of time, Δt, the weight of the quartz in the capsule is carefully re-weighed. The change in the number of moles of quartz, Δn_{qtz}, divided by the time, Δt, $\Delta n_{qtz}/\Delta t$, gives the rate of the reaction (note that Δn_{qtz} may be positive or negative). If this experiment is repeated at different temperatures, holding all else constant, results such as those in Figure 2 are obtained (Kerrick, 1972). The net rate of the reaction is zero at the equilibrium temperature ($\Delta G = 0$), T_{eq}. Close to the equilibrium temperature, ΔG is small and equation (21) may be used:

$$\frac{\Delta n_{qtz}}{\Delta t} = \frac{R_+ A_r}{RT} . \tag{23}$$

ΔG can be written as $\Delta G = \Delta H - T\Delta S$. Furthermore, from the condition of equilibrium, $\Delta H - T_{eq}\Delta S = 0$. (Assume ΔH and ΔS do not vary with T in the neighborhood of T_{eq}). Combining these results, the affinity can be written as

$$A_r = -\Delta G = (T - T_{eq})\Delta S \tag{24}$$

a relation which is also used in Chapter 8. Therefore,

$$R_{net} = \frac{\Delta n_{qtz}}{\Delta t} = \frac{R_+ \Delta S}{RT_{eq}} (T - T_{eq}) \tag{25}$$

and the net rate depends linearly on the temperature, as the temperature is varied slightly to either side of the equilibrium temperature, T_{eq}. Hence, the slope of the $\Delta n_{qtz}/\Delta t$ versus T curve at T_{eq} can provide kinetic information. For example, the data in Figure 2 were obtained with Δt = 14 days. Therefore, the slope of the curve at T_{eq} is

$$\frac{dR_{net}}{dT} = -0.464 \ \mu g \ SiO_2/day \cdot K$$

or

$$\frac{dR_{net}}{dt} = -7.735 \times 10^{-9} \text{ moles } SiO_2/day \cdot K$$

$$= 7.735 \times 10^{-9} \text{ moles } H_2O/day \cdot K \ .$$

If the ΔS for the reaction at the prevalent conditions (note that this ΔS must be calculated using the standard thermodynamic equations and must include the entropy of the water) is around 100 Joule/mole·K, then combining equation (25), T_{eq} = 590°C and the above value of dR_{net}/dT yields

$$R_+ = 5.55 \times 10^{-7} \text{ moles } H_2O/day$$

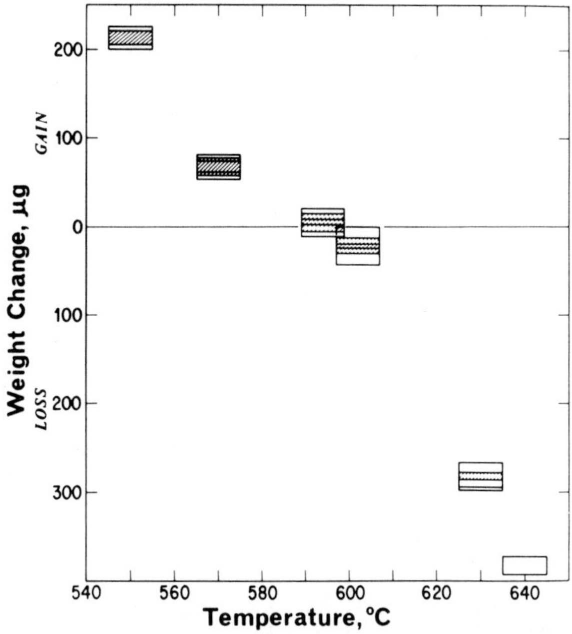

Figure 2. Experimental results of single crystal runs for the dehydration of muscovite at P_{fluid} = 3.5 kbar with X_{H_2O} = 0.5 and X_{CO_2} = 0.5. Height of rectangles gives weighing error; width of rectangles indicate temperature error. Overlap of rectangles is indicated by hachures. Duration of runs: 14 days.

for the forward rate of the dehydration reaction.

As discussed in Chapter 4, the net rates of overall reactions may sometimes follow slightly different equations:

$$R_{net} = R_{+}(1 - e^{\frac{n\,\Delta G}{RT}}) \qquad (26)$$

where n is some constant. The case n = 1 was discussed above. Simplifying equation (26) for small ΔG yields

$$R_{net} = \frac{R_{+}\,\Delta S\,n}{RT_{eq}}\,(T - T_{eq}) \;. \qquad (27)$$

The result is, thus, the same as equation (25), except for the factor n. For most applications the use of n = 1 is probably a safe bet. However, the determination of n must be carried out in applications of the method.

As ΔG increases or as T deviates further from T_{eq}, equations (20) or (26) predict a non-linear dependence of the net rate on the affinity and on temperature. To understand the temperature dependence, equation (20) can be rewritten as

$$R_{net} = R_{+}[1 - e^{\frac{\Delta H}{R}(\frac{1}{T} - \frac{1}{T_{eq}})}] \;. \qquad (28)$$

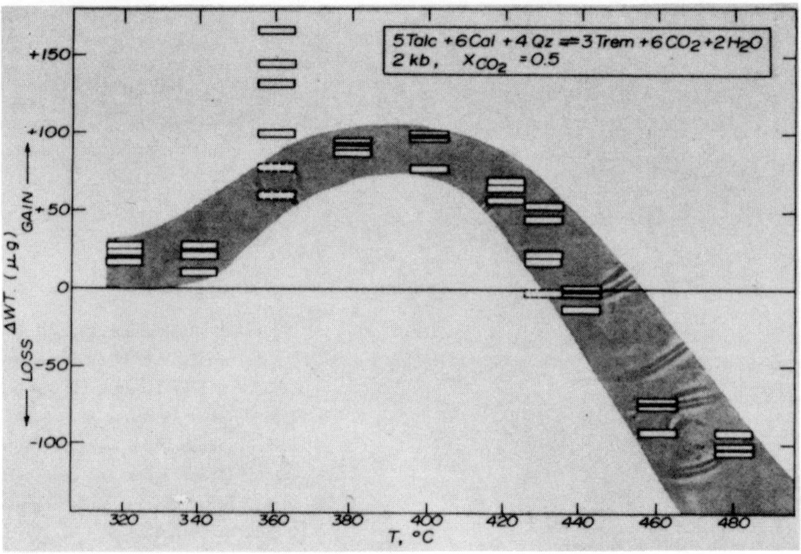

Figure 3. Experimental results of single crystal runs on the reaction shown in caption as P_f = 2 kbar, X_{CO_2}= 0.50. Dashed rectangles are from "tremolite-free" capsules (see below). Duration of runs was 21 days. Vertical and horizontal dimensions of rectangles in this weight change diagram give the uncertainties in weight change and temperature, respectively, of individual data points. Overlap of rectangles is indicated by black bands. Mineral abbreviations for figures: Cal = calcite; Diops = diopside; Dol = dolomite; Qz = ^{i}L quartz; Tr, Trem = tremolite; Fo = forsterite.

R_+, the forward rate of the reaction, is generally given by a rate constant times a function of the concentration of reactants and products (see Ch. 1), $f(c_i)$. These concentrations may, in fact, be specific areas of reactant minerals since heterogeneous reactions depend on surface areas (see Ch. 1). If the rate constant is written in the Arrhenius form, then equation (28) becomes:

$$R_{net} = A_+ e^{-\frac{E_+}{RT}} f(c_i)[1 - e^{\frac{\Delta H}{R}(\frac{1}{T} - \frac{1}{T_{eq}})}] . \qquad (29)$$

For small values of T, the term $\exp(-E_+/RT)$ begins to affect the behavior of R_{net}. Since the activation energy barrier, E_+, is generally greater than ΔH, the decrease in $\exp(-E_+/RT)$ is much greater than the increase in $\exp(\Delta H/RT)$ for T much less than T_{eq}; as a consequence, the net rate will tend to zero for small values of T. This type of behavior is shown by the reaction converting talc into tremolite in Figure 3 (Slaughter *et al.*, 1975).

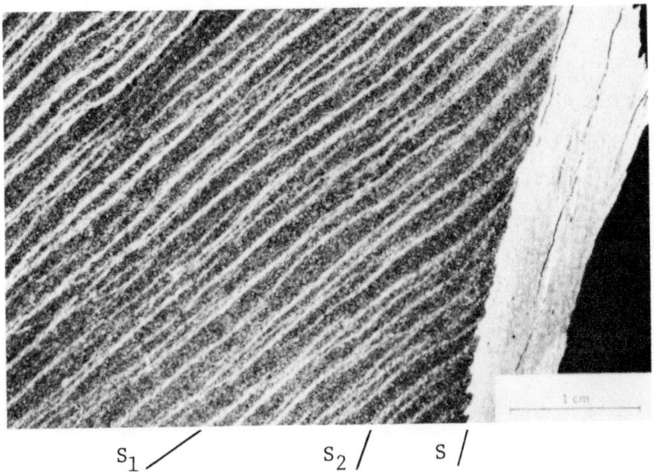

Figure 4. Differentiated layering (S_1) in a thin section of a specimen from the Upper Vicdessos Valley, Central Pyrenees, France. A bedding plane (S) can be seen at the right side of the photograph, separating a pelite (light area) from a quartz-rich bed (main part of photograph). The layering occurs in the quartz-rich bed; narrow layers, consisting almost exclusively of mica (light zones in the photograph), alternate with broader, quartz-rich zones. Mica in both zones is mostly aligned approximately parallel to the length of the zones, although the orientation pattern is complicated by a later crenulation cleavage (S_2) that is just visible in the photograph. (Nichols at 45°, Hobbs *et al.*, 1976).

Transport and Non-linear Irreversible Processes

In many natural reactions such as those discussed in the section on wollastonite growth, chemical potential gradients arise and drive diffusion fluxes. An amazing variety of petrologic processes involve coupled diffusion and chemical reactions. For example, the growth of crystals with a composition different from the bulk composition of a melt involves chemical reaction at the surface of the crystal and transport of the various components within the melt (see Ch. 8). Recently, Haase *et al.* (1980) have obtained oscillatory zoning in theoretical calculations of plagioclase growth from melt by use of non-linear kinetics. Another example is the precipitation of iron sulfide bands in sediments. This precipitation results from the diffusion of Fe^{2+}, liberated from iron-containing minerals, and the diffusion of S^{2-}, produced from the reduction reaction of sulfate, to a region where the chemical reaction forming the sulfide is occurring (Berner, 1969, 1980). The bands in some metamorphic rocks may result from dissolution of components in regions of stress and recrystallization in another region of lower stress, after suitable transport has occurred (Merino and Ortoleva, 1980). For example, Figure 4 (Hobbs *et al.*, 1976) shows a quartz-feldspar-mica schist, exhibiting bands consisting of alternating light quartz-feldspar layers and dark mica layers. These layers are at an angle to the bedding plane and cannot be sedimentary; they probably formed by a dissolution-diffusion-crystallization process. Vidale (1974) describes other observations of metamorphic differentiation layering. Finally, bands in igneous rocks are a well-known fact. Many layers have been ascribed to gravitational settling of crystals from melts or to flow patterns. However, it seems likely that some of the bands may be related to the kinetics of diffusion and chemical reactions. A recent discussion of layering in igneous rocks has been carried out by McBirney and Noyes (1979). For example, they ascribe possible diffusion-controlled Liesegang-type mechanisms to such features as orbicular structures in granites (Fig. 5) or the inch-scale layers in the Stillwater Complex of Montana (Fig. 6).

These examples clearly show the importance of pattern formation in nature. A detailed review of the kinetic literature on the topic of pattern formation and irreversible thermodynamics is beyond the scope of this chapter. For further discussion see the recent papers by Prigogine and Lefever (1978), Nicolis *et al.* (1978), Field and Noyes (1977), Noyes and Field (1977), Feinn *et al.* (1978), Lovett *et al.* (1978), Nicolis and Portnow (1973), Flicker and Ross (1974), as well as the classic paper by Stern (1954) and the books by Prigogine (1980) and Glasdorff and Prigogine (1971). While geochemists have proposed qualitative models to explain the occurrence of patterns in geological systems, further progress must stem from a quantification of these ideas within a kinetic framework. Therefore, the authors feel that an introduction to the basic concepts may be of interest to the geochemist or petrologist.

The rate equation for a species undergoing diffusion as well as chemical reaction is (see also Ch. 3):

$$\frac{\partial c}{\partial t} = D\frac{\partial^2 c}{\partial x^2} + f(c) \tag{30}$$

where D is the diffusion coefficient in the transporting medium (*e.g.*, aqueous or melt) and f(c) represents the chemical reaction rate law

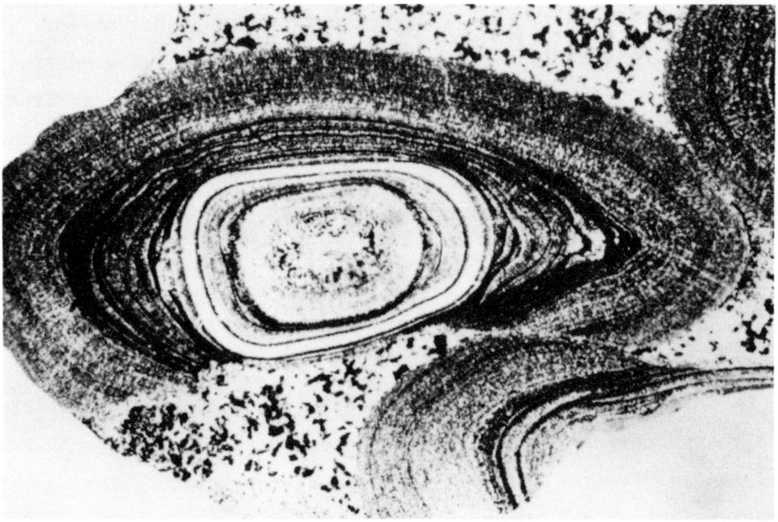

Figure 5. Photograph of an orbicular structure in a granite, exhibiting Liesegang bands. (McBirney and Noyes, 1979).

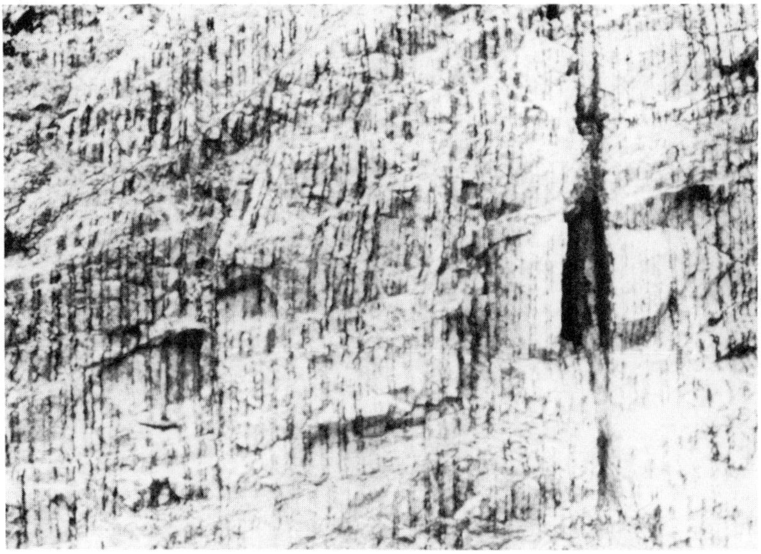

Figure 6. Inch-scale layers in the Stillwater Complex of Montana. Width of field is about 4 m. (McBirney and Noyes, 1979).

(as in Ch. 1 or Ch. 8). The classic theory of banding (Ostwald, 1897; Wagner, 1950; Prager, 1956) relies on the interaction between the diffusive fluxes of *two* components, which can react to form a precipitate. If the concentrations of the two components are c_1 and c_2, the kinetic equations become:

$$\frac{\partial c_1}{\partial t} = D_1 \frac{\partial^2 c_1}{\partial x^2} - f(c_1,c_2) \qquad (31a)$$

$$\frac{\partial c_2}{\partial t} = D_2 \frac{\partial^2 c_2}{\partial x^2} - f(c_1,c_2) \qquad (31b)$$

where the minus sign has been introduced because, in this case, $f(c_1,c_2)$ refers to the common rate of precipitation of the mineral composed of both 1 and 2. $f(c_1,c_2)$ may be a non-linear expression. The key kinetic assumption is that the concentration product, c_1c_2, must exceed the solubility product of the mineral by a certain amount of supersaturation before new nuclei can form and grow. After this supersaturation product is reached the mineral precipitates until the solubility product is reached in the solution. The following model can be used to illustrate the formation of bands by this mechanism. First, the length of the system is considered to be the unit length (i.e., x ranges from 0 to 1). Next, the initial condition is such that there is no amount of 1 or 2 in the system at t = 0:

$$c_1(x,0) = 0 \qquad 0 < x < 1$$

$$c_2(x,0) = 0 \qquad 0 < x < 1 .$$

In turn, the boundary conditions are

$$c_1(0,t) = 0.2 \qquad c_1(1,t) = 0$$

$$c_2(0,t) = 0 \qquad c_2(1,t) = 100 .$$

These boundary conditions ensure a flux of component 1 from left to right and a similar flux of component 2 from right to left, as time proceeds. Just as in the examples discussed previously, these conditions physically mimic the generation of the reacting components at different boundaries of a natural system. As the system evolves, each component begins to diffuse and the diffusion waves intermingle until the supersaturation product is reached at a particular locality (see Fig. 7). When this happens, the concentration of the reactants drops in the locality, thereby impoverishing the nearby regions of reactants. Therefore, no other precipitate can form in the vicinity, until the diffusion waves have a chance to proceed further away and reach the supersaturation product. Close to the region where the first band is precipitating, the concentration is kept low and this prohibits the neighboring region from reaching the supersaturation product. However, further away, the effect of the band weakens and the overall diffusion of the system allows the product c_1c_2 to increase again. In our example, the supersaturation condition was incorporated by allowing $f(c_1,c_2)$ to have the following form:

$$f(c_1,c_2) = 0 \qquad \text{if } c_1c_2 < 1.5$$

and no previous supersaturation has been reached;

$$f(c_1,c_2) = 10\, ln[\frac{c_1c_2}{0.25}] \qquad \text{if } c_1c_2 > 0.25$$

and supersaturation has been reached in the region.

Note that $f(c_1,c_2)$ is a non-linear function; f has also a "spatial memory" because it depends on the history of a locality, i.e., whether supersaturation has nucleated crystals or not. The equation for f assumes that $K_{sp} = 0.25$ and that the supersaturation product needed for crystal growth is 1.5. As can be seen in Figure 7, bands are indeed obtained as time goes on. Equations (36) were solved by finite difference in this example.

McBirney and Noyes (1979) have discussed a simple model for layering based on the variation of temperature in a cooling intrusion and the diffusion profiles in the melt induced by a crystallizing component. Although their model begins by assuming a growing crystal (unlike Fig. 7), the treatment predicts the formation of crystallization bands. In their model, the width of the bands is roughly given by D/V (similar to the D/V zones of crystal growth -- see Albarede and Bottinga, 1972), where D is the diffusion coefficient and V is the rate of advance of an isotherm deeper into the cooling melt. McBirney and Noyes note that D varies between 3×10^{-6} and 5×10^{-10} cm^2/sec and V between 3×10^{-7} and 1×10^{-8} cm/sec. Hence, the crystal spacing is between 10^{-3} cm and 3 meters, which encompasses the type of igneous bands usually observed.

McBirney and Noyes point to the diffusion coefficient in the melt as the primary control on the band spacing. This is also true of the Liesegang model just discussed. For example, increasing the rate of

Figure 7 legend on next page →

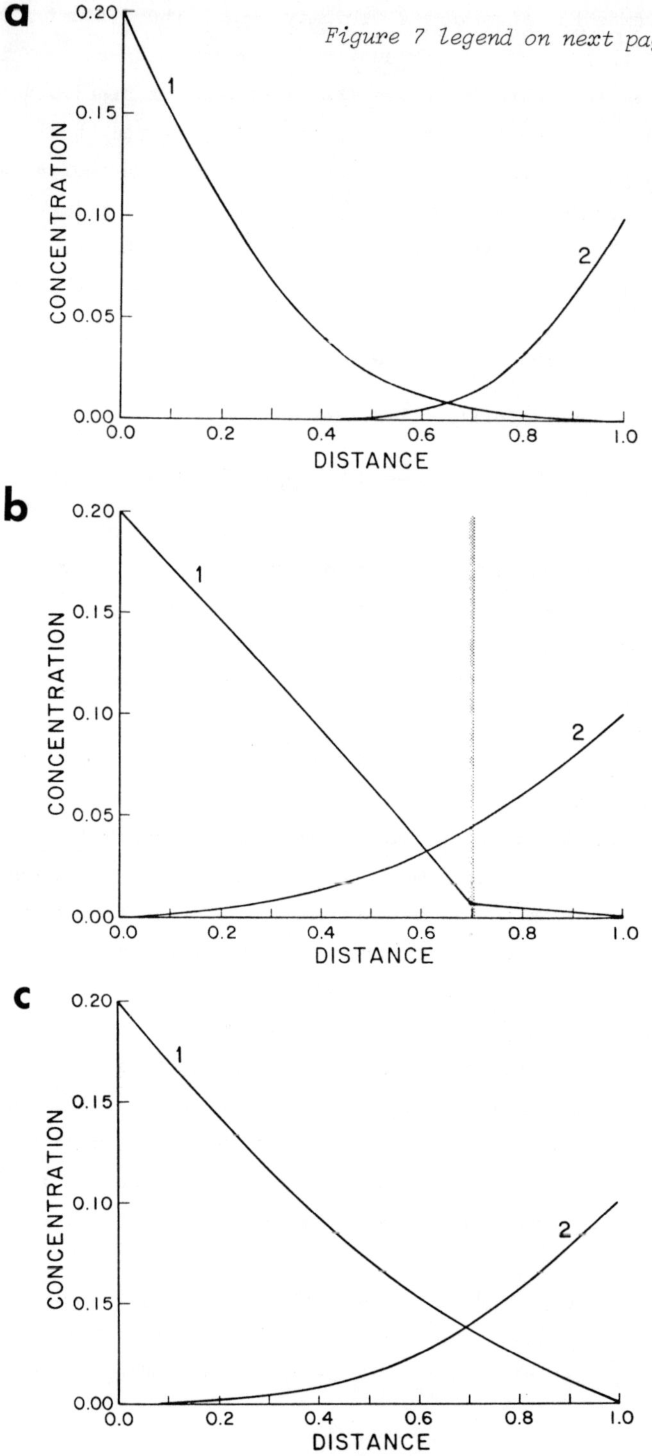

Figure 7. Time evolution of the usual double-diffusive model for Liesegang banding (see text). Note that the concentration of component 2 has been divided by 1000 to fit in the same plot. Initially $c_1 = 0.2$ at $x = 0$ and $c_2 = 100$ at $x = 1$ with $c_2 = 0$ in between. ($D_1 = 0.05$ and $D_2 = 0.02$ (in arbitrary distance and time units).

(a) Concentration profiles at $t = 1.0$. Note that the interdiffusion of 1 and 2 has not exceeded the supersaturation product and no precipitation has yet occurred.

(b) Concentration profiles at $t = 3.0$. Precipitation is imminent at $x = 0.7$.

(c) Concentration profiles at $t = 4.0$. A single precipitation band at $x = 0.7$ (started at $t = 3.12$) is growing (shown by vertical bar). The concentration of 1 (the minor component) has now decreased at $x = 0.7$.

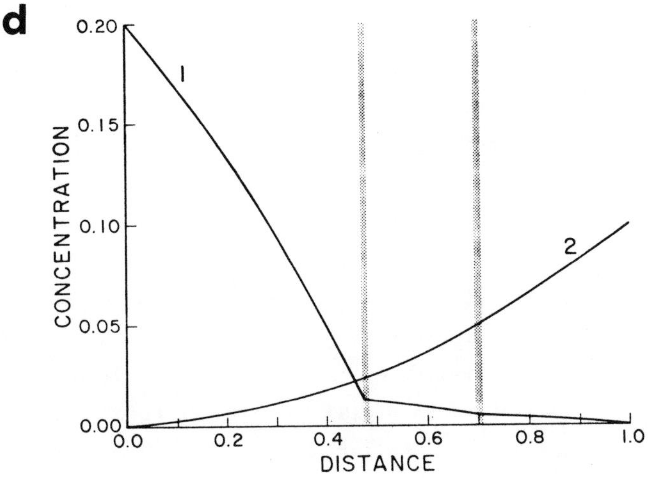

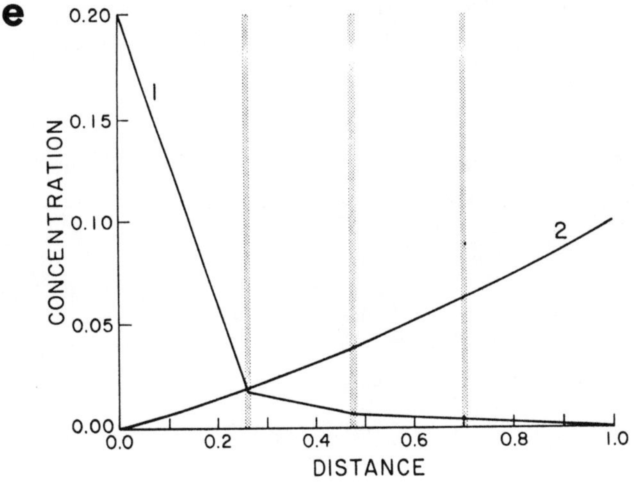

(d) Concentration profiles at $t = 5.0$. Two precipitation bands have now appeared at $x = 0.70$ and $x = 0.475$.

(e) Concentration profiles at $t = 10.0$. The three bands are at $x = 0.7$, 0.475 and 0.263. Note that the slopes of the concentration profiles on either side of the bands are not equal because there is now a sink term from the precipitation.

precipitation by nearly an order of magnitude to

$$f(c_1,c_2) = 50\ ln[\frac{c_1c_2}{0.25}]$$

does *not* change the location or the spacing of the bands. However, varying the diffusion coefficient of component 2, D_2, does change both the position and the spacing of the bands. For example, increasing D_2 from 0.02 to 0.1 shifts the initial bands to x = 0.325 and 0.175. Furthermore, the band spacing is now reduced to 0.150 from 0.225 in Figure 7. On the other hand, decreasing D_2 to 0.01 shifts the initial bands to the positions x = 0.775, 0.563 and 0.338, with spacings of 0.213 and 0.225. Note that unlike the case of McBirney and Noyes (which interrelates diffusion with temperature gradients), increasing D_2 tends to *decrease* the band spacing. This dependence of the bands on the value of the diffusion coefficient, D, is typical of Liesegang-type bands; on the other hand, the kinetics of reactions become very important in the next treatment of pattern formation.

The supersaturation theory relies on local boundary conditions to achieve banding. A more general method uses the properties of irreversible systems far from equilibrium. An introductory analysis to this type of pattern formation can give some powerful kinetic insights. Let us return to equation (30). For the time being, only *one* reactant is treated explicitly in equation (30) (all else is assumed constant). Normally, spatial inhomogeneities in chemical potential (caused by concentration gradients, anisotropic stresses or similar forces) are eliminated by the diffusion term in (30). If we begin with a homogeneous sample (no diffusion), the concentration will achieve steady state at c_o as long as

$$\frac{\partial c}{\partial t} = f(c_o) = 0 \ . \qquad (32)$$

For example, if the reaction refers to the dissolution/precipitation of silica, then the rate law is (see Ch. 1):

$$f(c_{H_4SiO_4}) = k(c_s - c_{H_4SiO_4}) \ .$$

Obviously, a homogeneous solution where $c_{H_4SiO_4} = c_s$ is at steady state ($\partial c/\partial t = 0$). At this point, no patterns have yet emerged. For patterns to emerge from a steady state system (or an equilibrium system if perturbed far enough) certain fluctuations must be amplified in the system.

To study the kinetic behavior of a system, which can vary both spatially and temporally, we assume that small fluctuations from steady state are a function of x and t:

$$c(x,t) = c_o + \alpha(x,t) \tag{33}$$

where α is the fluctuation. The next step is to treat the magnitude of α as small (relative to c_o) and insert (33) for c(x,t) in the rate law (30):

$$\frac{\partial(c_o + \alpha(x,t))}{\partial t} = D \frac{\partial^2(c_o + \alpha(x,t))}{\partial x^2} + f(c_o + \alpha(x,t)) \ . \tag{34}$$

Since α is small, the function f can be approximated by the first terms of a Taylor series:

$$f(c_o + \alpha) = f(c_o) + \left(\frac{\partial f}{\partial c}\right)_o \alpha \tag{35}$$

where $(\partial f/\partial c)_o$ is the derivative of f evaluated at c_o. Using equation (35) and $f(c_o) = 0$ in equation (34) yields:

$$\frac{\partial \alpha}{\partial t} = D \frac{\partial^2 \alpha}{\partial x^2} + (\partial f/\partial c)_o \ \alpha \ . \tag{36}$$

To solve equation (36) requires some mention of boundary conditions. Most systems (including geologic systems) can be treated as bounded systems. Let the scale of the system in the x-direction be given by the length, L. Furthermore, let us assume that the fluctuations are not arbitrarily imposed at the boundaries. Consequently, we will require that the fluctuations vanish at the boundaries:

$$\alpha(o,t) = \alpha(L,t) = 0 \ . \tag{37}$$

The solution of a diffusion problem in a bounded system is best handled by the method of separation of variables (see Crank, 1975). It can be checked that functions of the following type,

$$\alpha(x,t) = \alpha_o \ e^{w_n t} \ \sin\left(\frac{n\pi}{L} x\right) \tag{38}$$

are solutions to equation (36), which also satisfy the boundary condition (37). Inserting equation (38) into (36) yields:

$$\alpha_o \sin\left(\frac{n\pi x}{L}\right) w_n \ e^{w_n t} = -D \ \alpha_o \ e^{w_n t} \ \frac{n^2 \pi^2}{L^2} \sin\left(\frac{n\pi x}{L}\right)$$
$$+ (\partial f/\partial c)_o \ \alpha_o \ e^{w_n t} \ \sin\left(\frac{n\pi x}{L}\right) \ .$$

Dividing by $\alpha_o \sin(\frac{n\pi x}{L})\ e^{w_n t}$, we find that

$$w_n = (\partial f/\partial c)_o - D\,\frac{n^2\pi^2}{L^2} \quad . \tag{39}$$

Therefore, equation (38) is a solution of (36) as long as the frequency, w_n, obeys equation (39). At this point, the question of stability or pattern formation depends on the numerical properties of w_n. Imagine a system which maintains the concentration of c at the boundary at the steady state value, c_o. The initial condition may be any function, g(x), for $0 \leqslant x \leqslant L$. This initial function can be described by a Fourier series:

$$g(x) = \sum_n a_n \sin(\frac{n\pi x}{L}) + c_o \ . \tag{40}$$

Combining equations (38) and (40), the solution to the problem is then given by

$$c(x,t) = c_o + \sum_n a_n \sin(\frac{n\pi x}{L})\ e^{w_n t} \tag{41}$$

as long as the initial condition was not too far removed from c_o. What is the fate of the various spatial terms in equation (41) as the system evolves? The answer depends on the w_n in equation (39). If w_n is negative, the term containing $\sin(\frac{n\pi x}{L})$ will decay. If w_n is positive, the $\sin(\frac{n\pi x}{L})$ term will be amplified with time and a pattern with n peaks and valleys will result. Since the term with the biggest positive w_n will amplify the fastest, this term will dominate all other terms after long times and the final pattern will depend on the n for which w_n is largest and positive. If the rate law, f(c), is given by the usual linear term, $f(c) = k(c_s - c)$, then $(\partial f/\partial c)_o = -k$ is negative and thus w_n in equation (39) is negative for all values of n. In this case, the initial condition (40) decays until the system is homogeneous with $c = c_o$ everywhere. The only way to obtain a positive (unstable) frequency is for $(\partial f/\partial c)_o$ to be positive. Even when $(\partial f/\partial c)_o$ is positive, there will be a sufficiently large n* such that w_n will be negative for all n greater than n*. Hence, if $(\partial f/\partial c)_o$ is positive, only the w_n with $n < n^*$ will be unstable. Furthermore, the frequency with the largest positive value occurs at n = 1. This last observation leads to the result that even in the case that $(\partial f/\partial c)_o$ is positive, the only permanent spatial patterns are given by the sin $\frac{\pi x}{L}$ term and so there is no pattern formation of any consequence.

A very important result of our analysis is the requirement that $(\partial f/\partial c_o)$ be positive for the existence of unstable w_n. This result is the reason behind the often-quoted requirement of *autocatalysis* in all pattern-forming systems (*e.g.*, Prigogine, 1980, or Flicker and Ross, 1974). If $(\partial f/\partial c)_o$ is positive, then an increase in c will increase f(c) and hence increasing c will increase the rate of production of c by equation (30). Such a relation requires an autocatalytic step.

The same analysis can be carried out with more interesting results for a *two* species system. Now the equations become:

$$\frac{\partial c_1}{\partial t} = D_1 \frac{\partial^2 c_1}{\partial x^2} + f_1(c_1,c_2) \tag{42a}$$

$$\frac{\partial c_2}{\partial t} = D_2 \frac{\partial^2 c_2}{\partial x^2} + f_2(c_1,c_2) \ . \tag{42b}$$

Writing the fluctuations as

$$c_1(x,t) = c_1^o + \alpha_1(x,t) \tag{43a}$$

$$c_2(x,t) = c_2^o + \alpha_2(x,t) \tag{43b}$$

and inserting (43) into (42) yields the linear equations for α_i:

$$\frac{\partial \alpha_1}{\partial t} = D_1 \frac{\partial^2 \alpha_1}{\partial x^2} + (\partial f_1/\partial c_1)_o\, \alpha_1 + (\partial f_1/\partial c_2)_o\, \alpha_2 \tag{44a}$$

$$\frac{\partial \alpha_2}{\partial t} = D_2 \frac{\partial^2 \alpha_2}{\partial x^2} + (\partial f_2/\partial c_1)_o\, \alpha_1 + (\partial f_2/\partial c_2)_o\, \alpha_2 \tag{44b}$$

where the f_1 and f_2 functions were expanded in a Taylor series just as in the previous case. The solutions are again written in the form

$$\alpha_1 = \alpha_1^o\, e^{w_n t} \sin(\frac{n\pi x}{L}) \tag{45a}$$

$$\alpha_2 = \alpha_2^o\, e^{w_n t} \sin(\frac{n\pi x}{L}) \ . \tag{45b}$$

Inserting equations (45) for α_i into equations (44) gives the equations which the w_n must satisfy:

$$w_n \alpha_1^o = -D_1 \frac{n^2\pi^2}{L^2}\, \alpha_1^o + (\partial f_1/\partial c_1)_o \alpha_1^o + (\partial f_1/\partial c_2)_o \alpha_2^o$$

$$w_n \alpha_2^o = -D_2 \frac{n^2\pi^2}{L^2} \alpha_2^o + (\partial f_2/\partial c_1)_o \alpha_1^o + (\partial f_2/\partial c_2)_o \alpha_2^o .$$

If w_n must satisfy these equations for arbitrary α_1^o and α_2^o then the determinant of the coefficients must vanish, i.e., w_n satisfies the following equation

$$(a_{11} - D_1 \frac{n^2\pi^2}{L^2} - w_n)(a_{22} - D_2 \frac{n^2\pi^2}{L^2} - w_n) - a_{12}a_{21} = 0$$

where we have defined $a_{ij} = (\partial f_i/\partial c_j)_o$ for simplicity. This is a quadratic and the solution is

$$w_n = \frac{-b \pm [b^2 - 4c]^{1/2}}{2} \tag{46a}$$

where
$$b = (D_1 + D_2) \frac{n^2\pi^2}{L^2} - a_{11} - a_{22} \tag{46b}$$

$$c = (D_1 \frac{n^2\pi^2}{L^2} - a_{11})(D_2 \frac{n^2\pi^2}{L^2} - a_{22}) - a_{12}a_{21} . \tag{46c}$$

w_n in equation (46a) will be real and positive only if either of the following two conditions hold:

$$\text{(i)} \quad b < 0 \quad \text{and} \quad b^2 > 4c \tag{47a}$$

or
$$\text{(ii)} \quad b > 0 \quad \text{and} \quad c < 0 . \tag{47b}$$

b will be negative only if at least one of a_{11} or a_{22} is positive. If a_{11} and a_{22} are negative, then b is positive and w_n is positive only if c is negative. c could be negative if the product $a_{12}a_{21}$ was large and positive. But if a_{11} and a_{22} are negative, the largest positive w_n will once again occur when n = 1 and no spatial patterns will arise. The conclusion is that for the two-species case, some sort of autocatalysis (a_{11} or a_{22} must be positive) is still required for patterns (i.e., spatial instabilities) to occur.

Let us apply the results for the two-species case to an example (termed the *Brusselator*) treated by Nicolis *et al.* (1977). The equations of motion for the two species are

$$\frac{\partial c_1}{\partial t} = A + c_1^2 c_2 - (B + 1) c_1 + D_1 \frac{\partial^2 c_1}{\partial x^2} \tag{48}$$

$$\frac{\partial c_2}{\partial t} = B\, c_1 - c_1^2\, c_2 + D_2 \frac{\partial^2 c_2}{\partial x^2} \, . \tag{49}$$

In this case, A and B stand for the externally fixed concentrations of certain species. The steady state solution to the f_i functions:

$$f_1(c_1^o, c_2^o) = A + c_1^{o^2}\, c_2^o - (B + 1)c_1^o = 0$$

$$f_2(c_1^o, c_2^o) = B\, c_1^o - c_1^{o^2}\, c_2^o = 0$$

can be easily checked to be $c_1^o = A$ and $c_2^o = B/A$. Therefore,

$$a_{11} = (\partial f_1/\partial c_1)_o = 2c_1^o c_2^o - (B + 1) = B - 1$$

$$a_{12} = (\partial f_1/\partial c_2)_o = c_1^{o^2} = A^2$$

$$a_{21} = (\partial f_2/\partial c_1)_o = B - 2c_1^o c_2^o = -B$$

$$a_{22} = (\partial f_2/\partial c_2)_o = - c_1^{o2} = -A^2 \, .$$

These a_{ij} can be used in equation (46) to obtain the w_n. As a result, it can be shown than w_n will be unstable if the concentration B is such that

$$B > 1 + \frac{D_1}{D_2} A^2 + D_1 \frac{n^2\pi^2}{L^2} + \frac{A^2}{D^2 \frac{n^2\pi^2}{L^2}} \, .$$

Therefore, we can define the critical value of B for each n as

$$B_n^{crit} = 1 + \frac{D_1}{D_2} A^2 + D_1 \frac{n^2\pi^2}{L^2} + \frac{A^2}{D_2 \frac{n^2\pi^2}{L^2}} \, . \tag{50}$$

Figures 8 and 9 show the implications of equation (50). In the example, the concentration of A was taken to be 2 (in arbitrary units) and the diffusion coefficients taken as $D_1 = 0.02$ and $D_2 = 0.08$ (again in some arbitrary units) and $L = 1$. If we label the right side of equation (50) as B_n^{crit}, then the critical values of B for the different n are given in Table 2. Obviously, if B is less than 4.056 no B_n^{crit} will be less than B and all w_n will be negative. So if $B < 4.056$, the system should not develop any spatial patterns. On the other hand, suppose the concentration of B was 4.6. Looking at Table 2, we see that $B > B_n^{crit}$ for $n = 2$ and 3, in this case. Hence, spatial patterns are expected to

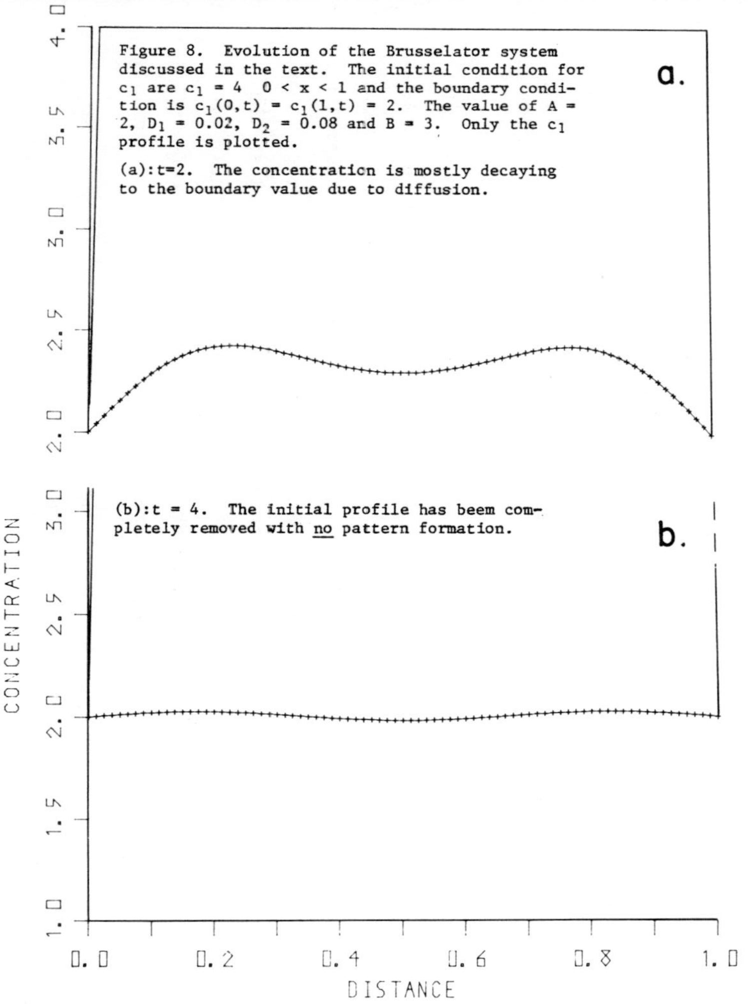

Figure 8. Evolution of the Brusselator system discussed in the text. The initial condition for c_1 are $c_1 = 4$ $0 < x < 1$ and the boundary condition is $c_1(0,t) = c_1(1,t) = 2$. The value of $A = 2$, $D_1 = 0.02$, $D_2 = 0.08$ and $B = 3$. Only the c_1 profile is plotted.

(a): $t=2$. The concentration is mostly decaying to the boundary value due to diffusion.

(b): $t = 4$. The initial profile has beem completely removed with <u>no</u> pattern formation.

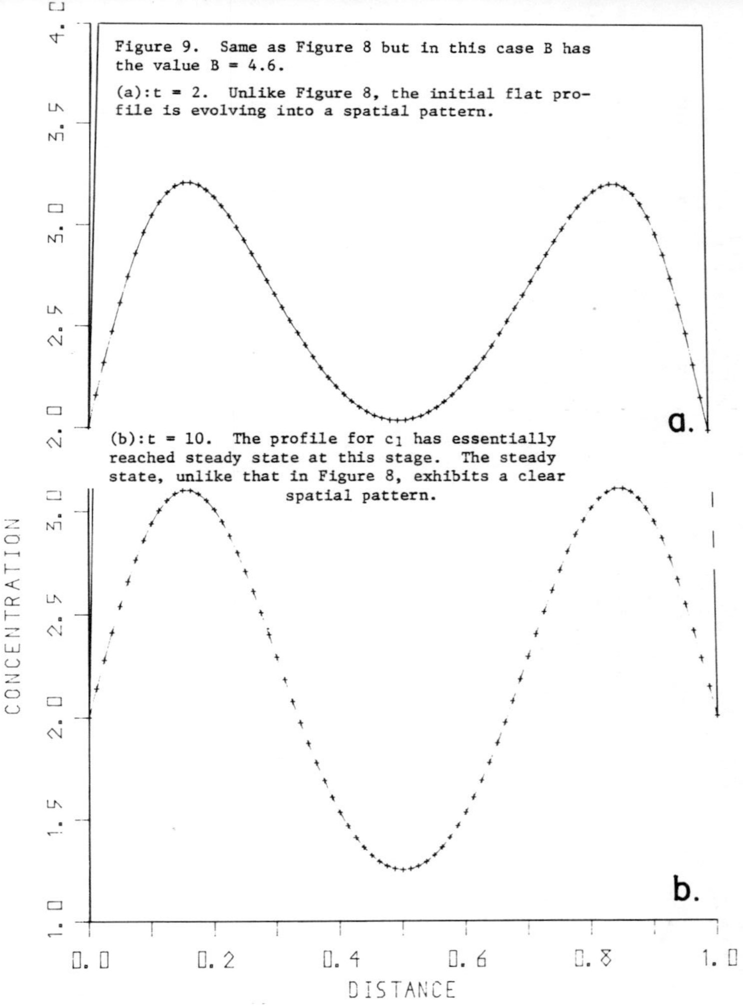

Figure 9. Same as Figure 8 but in this case B has the value $B = 4.6$.

(a): $t = 2$. Unlike Figure 8, the initial flat profile is evolving into a spatial pattern.

(b): $t = 10$. The profile for c_1 has essentially reached steady state at this stage. The steady state, unlike that in Figure 8, exhibits a clear spatial pattern.

TABLE 2

Stability Analysis - Critical Values of B

$D_1 = 0.02 \quad D_2 = 0.08$

$A = 2.0$

n	B_n^{crit}
1	7.263
2	4.056
3	4.339
4	5.475
5	7.137
6	9.247
7	11.776
8	14.712
9	18.051
10	21.790

develop which should look like $\sin(2\pi x)$ and $\sin(3\pi x)$ (x now goes from 0 to 1). Figures 8 and 9 bear this out nicely. The boundary condition for x and y is the corresponding steady state value (A and B/A, respectively). The initial condition is shown in the figures. Note that when B = 3, the initial fluctuation decays in a normal fashion until a homogeneous steady state is reached. However, for B = 4.6 a spatial pattern does develop, as predicted. Furthermore, the pattern looks like a $\sin(3\pi x)$ function ($0 \leqslant x \leqslant 1$). The first few w_n, in the case B = 4.6, are given by

$$w_1 = -0.693$$

$$w_2 = +0.365$$

$$w_3 = +0.195$$

$$w_4 = -0.714 \ .$$

All the other $w_n (n > 4)$ are negative (stable). Note that the largest positive w_n occurs at n = 2. However, the pattern in Figure 9 does not resemble $\sin(2\pi x)$. The problem is that the *initial* condition was symmetric; hence, only odd integers n in $\sin(n\pi x)$ can contribute to the problem. This is why the final pattern resembled the pattern from the other n with a positive w_n, i.e., n = 3. The important point is not only that the onset of patterns can be predicted but the *type* of pattern as well.

There are some interesting conclusions that can be carried over to petrologic applications. For example, analysis of either equation (51) or (55) shows that if L is very small, the possibility of obtaining positive w_n is also very small. Therefore, all other things being equal, banding should occur preferentially where the transport of the two (or more) components takes place over a large distance. Physically, this result stresses the dominance of diffusion over short distances, thereby overcoming any fluctuations by homogenization. Another general result is that if the maximum instability (biggest positive w_n) corresponds to n*, the spatial term $\sin \frac{n^*\pi x}{L}$ will dominate the evolution of the system. This term will produce bands that are spaced a distance $\frac{2L}{n^*}$ apart. If an estimate of L (the size of the system) can be made from field evidence (*e.g.*, the field relations between a reducing horizon as a source of Fe^{2+} and a source of O_2 in the formation of iron oxide bands), the spacing of the bands can then provide information on n*. In turn, knowing n* and using the general equation (51), knowledge can be obtained of the chemical coupling in the system (the a_{ij}). Alternatively, a reasonable guess for n*, based on some chemical reaction models, can be used with the field data on the bands to obtain an estimate of L, the size of the reacting system at the time of formation of the bands.

CONCLUSION

The formulation of reaction rates and transport rates from the point of view of entropy production can help elucidate petrologic processes. The theory of irreversible thermodynamics relates fluxes of heat or mass linearly to thermodynamic forces (temperature gradients or chemical potential gradients) near equilibrium. One of the tasks of the petrologist interested in kinetics is to find the appropriate phenomenological coefficients relating fluxes and forces. In turn, these phenomenological coefficients can be used to estimate the rates of growth of mineral zones during metamorphism or diagenesis.

A consequence of the general theory of irreversible thermodynamics is that each flux may be influenced by any of the thermodynamic forces. This type of coupling leads to such effects as thermal diffusion, the Soret effect and uphill diffusion. The theory, however, constrains the number and size of the coupling phenomenological coefficients, L_{ij}. Estimates of the L_{ij} enable a calculation of the importance of these coupling effects.

The linear theory of irreversible thermodynamics relates the rates of reactions to the free energy difference of the reaction. This relation can be used along with experimental data to obtain the individual rates of reactions in petrology.

The behavior of systems far from equilibrium is shown to be different from that near equilibrium. In particular, the formation of spatial patterns so common in petrology, is intimately related to the action of a system, when far from equilibrium. The analysis of Liesegang type models as well as the spatial patterns arising from autocatalytic effects can provide useful insights into the formation of differentiated layering in petrology.

Albarede, F. and Bottinga, Y. (1962) Kinetic disequilibrium in trace element partitioning between phenocrysts and host lava. Geochim. Cosmochim. Acta, 36, 141-156.

Berner, R. A. (1969) Migration of iron and sulfur within anaerobic sediments during early diagensis. Amer. J. Sci., 267, 19-42.

——— (1980) *Early Diagenesis: A Theoretical Approach.* Princeton Univ. Press, Princeton, N. J., 241 p.

Bowen, N. L. (1913) The melting phenomena of the plagioclase feldspars. Amer. J. Sci., Series IV, 36, 577-599.

Brady, J. B. (1975) Reference frames and diffusion coefficients. Amer. J. Sci., 275, 954-983.

Carman, P. C. (1968) Intrinsic mobilities and independent fluxes in multicomponent isothermal diffusion, I. Simple darken systems. J. Phys. Chemistry, 72, 1707-1712.

——— (1968) Intrinsic mobilities and independent fluxes in multicomponent isothermal diffusion, II. Complex darken systems. J. Phys. Chemistry, 72, 1713-1721.

Crank, J. (1975) *The Mathematics of Diffusion,* 2nd ed., Clarendon Press, Oxford, 414 p.

de Groot, S. R. and Mazur, P. (1962) *Nonequilibrium Thermodynamics.* North Holland Publishing Company, Amsterdam, 510 p.

Feinn, D., Ortoleva, P., Scalf, W., Schmidt, S., Wolff, M. (1978) Spontaneous pattern formation in precipitating systems. J. Chem. Phys., 69, 27-39.

Field, R. J. and Noyes, R. M. (1977) Mechanisms of chemical oscillators: Conceptual bases. Acc. Chem. Res., 10, 214-221.

Fisher, G. W. (1977) *Nonequilibrium Thermodynamics in Metamorphism. Thermodynamics in Geology.* 381-403.

——— (1978) Rate laws in metamorphism. Geochim. Cosmochim. Acta, 42, 1035-1050.

Flicker, M. and Ross, J. (1974) Mechanism of chemical instability for periodic precipitation phenomena. J. Chem. Phys., 60, 3458-3465.

Glansdorff, P. and Prigogine, I. (1971) *Thermodynamic Theory of Structure, Stability and Fluctuations.* Wiley-Interscience, New York, 306 p.

Goldschmidt, V. M. (1911) Die Kontaktmetamorphose in Kristianiagedeit. Oslo: Vidensk. Skr. I, Math.-Nat. Kl., No. 11.

Haase, C. A., Chadam, J., Feinn, D. Ortoleva, P. (1980) Oscillatory zoning in plagioclase feldspar. Science, 209, 272-275.

Hobbs, B. E., Means, W. D., Williams, P. F. (1976) *An Outline of Structural Geology.* John Wiley and Sons, New York, 571 p.

Joesten, R. (1977) Evolution of mineral assemblage zoning in diffusion metasomatism. Geochim. Cosmochim. Acta, 41, 649-670.

Katchalsky, A. and Curran, P. F. (1967) *Nonequilibrium Thermodynamics in Biophysics.* Harvard Univ. Press, Cambridge, MA, 248 p.

Kerrick, D. M. (1972) Experimental determinations of misconite + quartz stability with $P_{H_2O} < P_{total}$. Amer. J. Sci., 272, 946-958.

Lovett, R., Ortoleva, P., Ross, J. (1978) Kinetic instability in first order phase transitions. J. Chem. Phys., 69, 947-955.

McBirney, A. R. and Noyes, R. M. (1979) Crystallization and layering of the Skaergaard Intrusion. J. Petrology, 20, 487-554.

Merino, E. and Ortoleva, P. (1980) Temporal development of fabric in uniaxially stressed polycrystalline media - a theory. Contrib. Mineral. Petrol., 71, 429.

Miller, P. G. (1969) The experimental verification of the Onsager reciprocal relations. In *Transport Phenomena in Fluids.* H. J. M. Hanley, ed., Dekker, New York, 377-432.

Nicolis, G., Erneux, T., Herschkowitz-Kaufman, M. (1978) Pattern formation in reacting and diffusing systems. Adv. Chem. Phys., 263-315.

——— and Portnow, J. (1973) Chemical oscillations. Chem. Reviews, 73, 365-384.

Noyes, R. M. and Field, R. J. (1977) Mechanisms of chemical oscillators: experimental examples. Acc. Chem. Res., 10, 273-280.

Onsager, L. (1931) Reciprocal relations in irreversible processes II. Phys. Rev., 38, 2265-2279.

Ostwald, W. (1897) *Lehrbuch der Allegemeinen Chemie.* Engleman, Leipzig.

Prager, S. (1956) Periodic precipitation. J. Chem. Phys. 25, 279-283.

Prigogine, I. (1967) *Introduction to Thermodynamics of Irreversible Processes.* Wiley Interscience, New York, 119 p.

———— (1980) *From Being to Becoming*. W. H. Freeman Co., San Francisco, 272 p.

———— and Lefever, R. (1978) Coupling between diffusion and chemical reactions. Adv. Chem. Phys., 1-17.

Robie, R. A., Hemingway, B. S., Fisher, J. R. (1978) Thermodynamic properties of minerals and related substances at 298.15°K and 1 bar (10^5 pascals) pressure and at higher temperatures. U. S. Geol. Surv. Bull. 1452, Washington, D. C.

Stern, K. H. (1954) The Liesegang phenomenon. Chem. Reviews, 54, 79-99.

Slaughter, J., Kerrick, D. M., Wall, V. J. (1975) Experimental and thermodynamic study of equilibria in the system CaO-MgO-SiO_2-H_2O-CO_2. Amer. J. Sci., 275, 143-162.

Vidale, R. (1974) Metamorphic differentiation layering in pelitic rocks of Dutchess County, New York. In *Geochemical Transport and Kinetics*, Hofmann, A. W. *et al.*, (eds.), Carnegie Institution of Washington Publication 634, 273-286.

Wagner, C. (1950) Mathematical analysis of the formation of periodic precipitations. J. Colloid Sci., 5, 85-97.

Chapter 6

DIFFUSION in ELECTROLYTE MIXTURES

David E. Anderson

SECTION 1: INTRODUCTION

Electrolytes, whether solid or liquid, are mixtures in which at least one of the components is partially or completely dissociated into ions. Diffusion, in olivine, for example, is usually envisaged as the migration of Fe^{2+} and Mg^{2+} ions through a fixed anion lattice (SiO_4^{4-}); thus, from the point of view of diffusion theory and thermodynamics, the olivine behaves as a strong electrolyte. Other examples are silicate melts and aqueous electrolytes such as NaCl-KCl-H_2O. Despite the obvious differences in their physical nature, the analysis of diffusion in these various phases has a common theoretical basis.

The movement of ions through an electrolyte in response to an applied electrical field is termed ionic conductance. The random mixing of ions or undissociated molecules (for example, water molecules in aqueous systems), in the absence of applied fields, is termed diffusion. We shall be concerned only with diffusion in isothermal, isobaric systems in the absence of all external fields. Furthermore, it is assumed that the mixture as a whole is composed of some combination of electrically neutral (molecular) components. In these circumstances, electrical equilibrium will be maintained at all points and at all times during diffusion: Any tendency for cations and anions to separate into regions with a net electrical charge would create very strong electrical potentials that would immediately cause ionic conduction to restore local charge equilibrium. Some idea of the very large potential associated with the isolation of a very small quantity of a particular ion can be gained from the calculations of Guggenheim (1967, p. 298). Thus, even in the absence of an applied electric potential, there is a virtual electric potential -- the diffusion potential -- that constrains the independent migration of cations and anion in electrolytes.

It follows that any set of equations used to describe the flux of ions within an electrolyte mixture must always be solved simultaneously in a way that sets the electric current to zero. The importance of the diffusion potential cannot be overemphasized.

It is often convenient to analyze diffusion in electrolytes in terms of molecular components such as NaCl or $Fe_3Al_2Si_3O_{12}$, even though

NOTATION USED IN CHAPTER 6

A	Symbol for the anion in (5-17).
A_i	Polynomial fitting coefficient in (5-23).
A^a_{ik}	Element of the matrix defined by (4-5).
B^a_{ik}	Element of the matrix used to transform fluxes from reference frame a to reference frame b; (4-5, 4-7).
C_i	Symbol for a cation in (5-17).
D^a_i	Diffusion coefficient of the *ith* molecular component in a binary mixture (cm^2/sec); (2-2).
D^a	Mutual diffusion coefficient in a binary solution; (2-4).
D^*	Tracer diffusion coefficient; (3-8).
D^*_c	Cation tracer diffusion coefficient at infinite dilution in an aqueous electrolyte; (8-3).
D^*_a	Anion....
D^o_i	Diffusion coefficient at infinite dilution defined by (8-1).
D^a_{ik}	Independent diffusion coefficient relating the flux of the *ith* salt to the concentration gradient of the *kth* salt; (2-10).
$\tilde{D}^a_{ik}$	Dependent diffusion coefficient....; (2-9).
$\bar{D}^a_{ik}$	Independent diffusion coefficient....; (4-19).
F	The Faraday 96,493 coulombs/equiv.
G_{ik}	Thermodynamic factor defined by (6-6).
I^o	Electrical current.
J^a_i	Molar flux of the *ith* salt in reference frame a (moles/cm^2sec).
L^a_{ik}	Independent thermodynamic diffusion coefficient relating the flux of the *ith* salt to the chemical potential gradient of the *kth* salt ($moles^2$/joule cm sec); (2-18).
$\tilde{L}^a_{ik}$	Dependent thermodynamic diffusion coefficient....; (2-16).
M_i	Molecular weight of the *ith* component (g/mole).
R	The Gas Constant (8.3144 joules/mole deg).
S_{ik}	Non-singular (modal) matrix used to diagonalize the matrix D^a_{ik}; (6-20).
T	Temperature in degrees Kelvin.
$\bar{V}_i$	Partial molal volume of component i (salt or solvent) (ml/mole).
X^a_i	Independent chemical potential gradient of salt i; (5-3).
a	Superscript. Reference frame a (a = o,m,n,v).
a_i	Activity of the *ith* salt; (5-16).
$\bar{a}_i$	Normalized weighing factor for reference frame a; (4-2).
$\bar{b}_i$	...reference frame b.
c_i	Molar concentration of the *ith* molecular component (moles/liter).
c_{ic}	Molar concentration of the *ith* cation.
c_{ia}	...anion.
$\bar{c}_i$	Transformed (pseudo-) concentrations defined by (6-29).
f^*_i	Activity coefficient on a mole-fraction concentration scale.

g	Molar Gibbs function.
j^a_i	Molar flux of the *ith* ion in an electrolyte mixture (moles/cm^2sec).
ℓ^a_{ij}	Ionic transport relating the flux of the *ith* ion to the chemical potential gradient of the *jth* ion ($moles^2$/joule cm sec); (7-6).
m	Superscript. Mass-fixed reference frame.
m_1	Molality of solute in a binary aqueous electrolyte (moles/kilogram solvent).
n	Superscript. Molecular reference frame.
n_i	Number of atoms per unit area of lattice plane i (i = 1,2); (3-5).
$\bar{n}$	Number of atoms per unit volume; (3-7).
o	Superscript. Solvent-(lattice-)fixed reference frame.
r_{ic}	Stoichiometric coefficient for the *ith* cation in the *ith* salt; (5-17).
r_{ia}	...anion....
v^a	Reference velocity for reference frame a (cm/sec); (4-2).
v_f	Net drift velocity of atoms; (3-8).
y	Distance (cm).
y_i	Concentration gradient of the *ith* molecular component (moles/cm; $\partial c_i/\partial y$).
$\bar{y}_i$	Transformed concentration gradient defined by (6-18).
y^*_i	Molar mean ion activity coefficient of the *ith* solute; (5-18).
x^a_i	Independent chemical potential gradient of the *ith* ion; (7-3).
x_i	Mole fraction of *ith* component.
z_i	Signed valence of the *ith* cation.
z_a	Signed valence of the anion.
Λ^a	Characteristic root (eigenvalue) of the matrix D^a_{ik}; (2-24, 6-20).
Γ	Jump frequency; (3-5).
Φ	Osmotic coefficient relating solvent activity to molal solute concentrations.
γ^*_i	Molal activity coefficient of the *ith* salt.
δ_{ij}	Kronecker function (δ_{ij} = 1, i = j; δ_{ij} = 0, i ≠ j).
η	Boltzmann-Matano variable in (3-7).
λ	Lattice spacing parallel to y.
λ^o_i	Equivalent conductance of the *ith* ion at infinite dilution -- the limiting ionic conductance.
μ_i	Chemical potential of the *ith* salt (joules/mole).
μ^o_i	Chemical potential of the *ith* salt in the standard state.
μ_{ic}	Chemical potential of a cation; (7-1).
μ_{ia}	Chemical potential of an anion; (7-1).
μ_{ij}	Chemical potential derivatives defined by (2-21).
ρ_i	Partial density of the *ith* salt (g/ml).
ρ^o	Density of pure water at 25°C (g/cm^3).
σ	Entropy production (entropy produced per unit volume per unit time).
ϕ	Electrical potential (volts).
ω_i	Mass fractions for the *ith* molecular component ($\rho_i/\Sigma\rho_i$).

these components may be completely dissociated in the mixture. The choice of molecular components obviates all of the problems associated with the diffusion potential and allows the structure of diffusion theory to be presented in a simple form (*e.g.*, de Groot and Mazur, 1962, Ch. 11). Some of the most useful quantities that can be measured, however, are related to ionic species (*e.g.*, tracer diffusion coefficients). And, any consideration of diffusion mechanisms or construction of kinetic models must ultimately deal with the ionic nature of the mixture. In practice, both ionic and molecular fluxes are employed. Sections 1 to 6 are concerned with the fluxes of molecular components; ionic fluxes are considered in Sections 7 and 8.

The discussion will be restricted to diffusion in a single phase -- the kind of problem encountered in liquids and single crystals. The simplest class of single phase electrolytes are those with a common anion; examples include $NaCl-CaCl_2-H_2O$ and, for cation exchange, silicates such as the aluminous garnets (cations -- Fe^{2+}, Mg^{2+}, Mn^{2+}, etc.; anion -- $Al_2Si_3O_{12}^{6-}$) and the alkali feldspars (cations -- K^+, Na^+, etc.; anion -- $AlSi_3O_8^{1-}$). Only mixtures of this kind will be considered here. The treatment of diffusion in a system with multiple anions as well as cations proceeds along the same lines, but requires an even more complicated and cumbersome notation. Anderson and Buckley (1974), Brady (1975b), Christensen (1977), Fisher (1977), Lasaga (1979b), and Lasaga *et al.* (1977) have discussed various aspects of the problem of diffusional exchange between coexisting phases.

Amongst the aims of the theoretical studies of diffusion in the last half century have been: to clarify the exact meaning of Fick's first law as applied to systems of two or more diffusing components; to determine the nature of the quantity measured in a diffusion experiment; to optimize the collection of diffusion data, especially in multicomponent systems; to simplify the solution of the partial differential equations [i.e., equations (2-7), (2-8), (2-22) and (2-23) below]. These objectives are very important for the study of diffusion in geological systems which commonly contain a large number of possible components for diffusion and, as for example in the case of silicates, are often difficult to study experimentally (*e.g.*, slow diffusion rates; lack of readily-made stoichiometric phases; difficulties in forming good diffusion couples). In these circumstances simple, yet reasonably accurate, solutions to the differential equations and estimation or extrapolation of diffusion coefficients becomes important.

A useful, but by no means, exclusive starting point for a study of the general principles of diffusion, diffusion in electrolytes and the application of theory to natural systems may be found in Anderson and Graf (1976), Brady (1975a,b), Cooper (1974), Crank (1956), Darken (1948), Fisher (1977), Kirkaldy (1970), Kirkwood *et al.* (1960), Lasaga (1979a,b), Manning (1968b, 1974), Miller (1966, 1967a,b), Moynihan (1971), de Groot and Mazur (1962) and Gupta and Cooper (1971).

Many of the examples cited below are taken from aqueous electrolytes, since, thanks in no small part to the recent efforts of D. G. Miller and J. A. Rard, we now have for the first time accurate thermodynamic and diffusion data over a wide range of compositions in some systems of immediate importance to geochemistry (Rard and Miller, 1979a,b, 1980, 1981).

There will be certain conventions used in this chapter. Unless otherwise noted, all concentrations and concentration-dependent quantities are expressed in molar units (moles/liter). For convenience, all equations are written for one-dimensional diffusion parallel to the y-axis. (The more common x-axis is not used to avoid confusion with mole fractions later on.)

Throughout, we shall consider a system of n diffusing components; in this case, only (n-1) of the components form an independent set. Matrix notation is applied *only* to sets of independent thermodynamic or diffusional quantities. Thus the symbol [ ] *always* denotes an (n-1)x (n-1) matrix; the symbols $[\ ]^{-1}$ and $[\ ]^{T}$ are used for the inverse and transpose of such a matrix. An (n-1) column vector is represented by ( ).*

SECTION 2: GENERAL STRUCTURE OF DIFFUSION THEORY FOR MOLECULAR COMPONENTS

In this section, the general equations needed to describe the diffusion of molecular components are summarized. This summary is intended to emphasize the underlying simplicity and unity of the theory.

----

*All of the matrix methods used here are covered by Hildebrand (1952) or Wilkinson (1965).

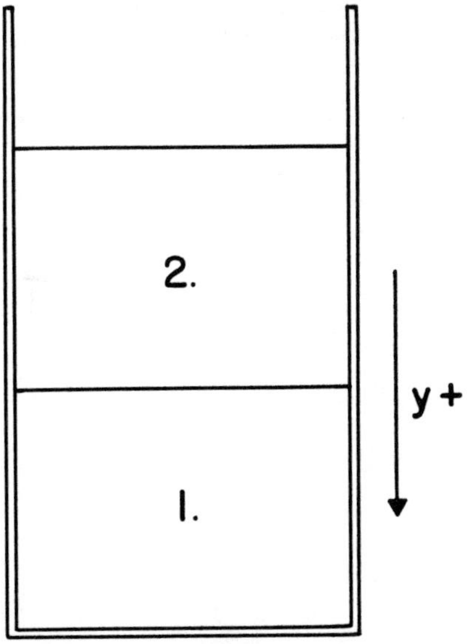

Figure 1. Free diffusion cell for two liquids 1 and 2. Changes in volume during mixing are marked by movement of the upper (free) surface of liquid 2. For (3-16), the lower boundary of the couple is fixed by the cell wall.

Later sections deal with some of the many complications that arise from the application of the theory to experimental and natural systems.

A flux is defined as the amount of matter flowing across a plane of unit area for unit time. Therefore, it is important to define clearly the nature of the reference plane involved in the flux. The mixing of components, which results from diffusion fluxes, is commonly accompanied by a change of volume. In the case of a liquid system in an open vertical cell (Fig. 1), volume changes, and the consequent bulk flow of liquid, may be detected by movement of the meniscus. It is convenient in this situation to choose a reference plane for fluxes, that moves with the local center of volume. Since there is no net transfer of volume across this plane,

$$\bar{V}_1 J_1^v + \bar{V}_2 J_2^v + \ldots + \bar{V}_n J_n^v = 0 \qquad (i = 1, 2, \ldots, n) \tag{2-1}$$

where $\bar{V}_i$ and J_i^v are the partial molar volumes and diffusion fluxes of the n diffusing components, respectively. The volume-fixed reference plane is denoted by the superscript v.

For the simplest system, n = 2, Fick's first law states that for each component (Crank, 1956, p. 2)

$$J_1^v = -D_1^v(\partial c_1/\partial y) \qquad J_2^v = -D_2^v(\partial c_2/\partial y) \tag{2-2}$$

where D_i^v is the diffusion coefficient of component i (i = 1 or 2). The inclusion of the minus sign by Fick was meant to ensure that a positive flux would correspond to a negative gradient of c_i. Some usual units are: J_v^i - moles/cm^2/sec; D_i^v - cm^2/sec; c_i - moles/cm^3; y - cm. This particular set of units and concentration scale is one of a number of possible choices, but it is conventional to pick a combination of units for J_i^v, c_i and y that gives D_i^v the dimensions (length)^2 x (time)$^{-1}$. Another common choice is based on mass (gram) concentrations (g/cm^3) and mass fluxes ($g/cm^2/sec$).

If c_i has units of moles/ℓ and $\bar{V}_i$ has units of ml/mole, it follows that

$$c_1\bar{V}_1 + c_2\bar{V}_2 + \ldots + c_n\bar{V}_n = 1000.027 \ . \qquad (2\text{-}3)$$

Differentiating (2-3) yields $\Sigma\bar{V}_i(\partial c_i/\partial y) = 0$. Therefore, using this last equation with n = 2 and equation (2-2) in equation (2-1) requires that

$$D_1^V = D_2^V \equiv D^V \ . \qquad (2\text{-}4)$$

It follows that only one equation

$$J_i^V = -D^V(\partial c_i/\partial y) \qquad (i = 1 \text{ or } 2) \qquad (2\text{-}5)$$

is needed to describe diffusion in a binary system. D^V is often referred to as a mutual diffusion coefficient. Kirkwood *et al.* (1960) have given a simple mathematical analysis of diffusion in the liquid cell depicted in Figure 1. Their analysis fixes quite clearly the concepts involved and explicitly confirms that there can be no bulk flow if the $\bar{V}_i$ are constant (Onsager, 1945).

Conservation of mass requires that in the absence of sources and sinks (Crank, 1956, p. 3; t = time)

$$\partial c_i/\partial t = -\partial J_i^V/\partial y \qquad (i = 1 \text{ or } 2) \ . \qquad (2\text{-}6)$$

Substitution from equation (2-5) for J_i^V leads to the partial differential equations

$$\partial c_i/\partial t = D^V(\partial^2 c_i/\partial y^2) \qquad (i = 1 \text{ or } 2) \qquad (2\text{-}7)$$

if D^V is not a function of concentration, and

$$\partial c_i/\partial t = \partial(D^V \partial c_i/\partial y)/\partial y \qquad (i = 1 \text{ or } 2) \qquad (2\text{-}8)$$

if D^V is concentration dependent.

Fick's first law was deduced by analogy with Fourier's law of heat conduction, and, as a consequence, related one flux to one gradient by a single proportionality constant (Tyrrel, 1964, has described the content and historical setting of Fick's original paper published in 1855). The way, therefore, in which Fick's first law is to be applied to a mixture with multiple gradients is not immediately obvious. In a binary system a flux (or gradient) of one component in one direction must be matched by a flux (or gradient) of the second component in the other direction [equation (2-1)]. Thus, although Fick's first law may be applied separately to each component [equation (2-2)], it is possible to define a common (mutual) diffusion coefficient for the two components.

Consider, however, a system such as $CaCl_2$-NaCl-H_2O where the two electrolytes are completely dissociated. A flux of Ca^{2+} ions in one direction must be matched by some combination of Cl^{1-} and Na^{1+} fluxes -- a combination that depends on the nature of the concentration gradients present and the intrinsic mobility of Na^{1+} and Cl^{1-} ions. Consequently, the flux of Ca^{2+} is coupled, through the diffusion potential, to the migration of Na^{1+} and Cl^{1-}. The fluxes of the molecular components $CaCl_2$ and NaCl will, of course, reflect the ultimate combination of ionic fluxes determined by the different mobilities of the three ions and the diffusion potential.

Onsager (1945), following his earlier work on nonequilibrium processes (Onsager, 1931a,b), proposed an extended form of Fick's first law. Onsager described the flux of component i in a multicomponent solution ($n > 2$) by

$$J_i^v = -\sum_{k=1}^{n} \tilde{D}_{ik}^v(\partial c_k/\partial y) \qquad (i = 1, 2, \ldots, n) \,. \qquad (2\text{-}9)$$

The last equation suggests that the flux of i is a linear function of each of the n concentration gradients. By introducing equations (2-1) and (2-3), one flux and one concentration gradient, respectively, may be eliminated from equation (2-9), so that (Onsager, 1945)*

$$J_i^v = -\sum_{k=1}^{n-1} D_{ik}^v y_k \qquad (i = 1, 2, \ldots, n-1) \,, \qquad (2\text{-}10)$$

where y_k is a convenient abbreviation for $(\partial c_k/\partial y)$; which component is eliminated is arbitrary. Equation (2-10) is a matrix equation that may be written as

$$(J^v) = -[D^v](y) \qquad (2\text{-}11)$$

or, for n = 3,

----

*Note that the matrices $\tilde{D}_{ik}^v$ in equation (2-9) and D_{ik}^v in equation (2-10) are different. Although dependent quantities are generally eliminated, the dependent set has some useful properties (Cullinan, 1965).

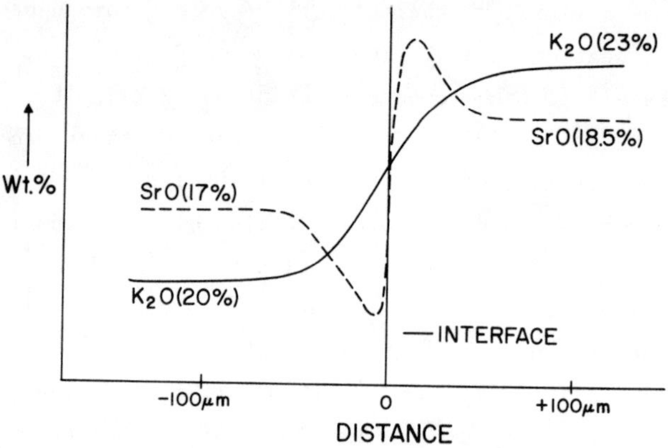

Figure 2. Redrawn concentration-distance curves for K_2O and SrO measured in glasses of the system K_2O-SrO-SiO_2 by Varshneya and Cooper (1962, their Fig. 3). Initial compositions are shown in weight percent at the undisturbed ends of the profiles.

$$\begin{pmatrix} J_1^v \\ J_2^v \end{pmatrix} = - \begin{bmatrix} D_{11}^v & D_{12}^v \\ D_{21}^v & D_{22}^v \end{bmatrix} \begin{pmatrix} y_1 \\ y_2 \end{pmatrix} . \qquad (2\text{-}12)$$

The magnitudes of the D_{ik}^v (and D^v) depend on the choice of concentration units for J_i^v and c_i in equation (2-9), even though the dimensions of D_{ik}^v are comprised only of length and time. Molar- and mass-based coefficients are related by (Miller, 1959)

$$D_{ik}(\text{molar}) = D_{ik}(\text{mass}) \times (M_k/M_i) . \qquad (2\text{-}13)$$

Only the off-diagonal coefficients (D_{ik}, $i \neq k$) are modified by this particular conversion. In a binary system, D^v will have the same numerical value on both scales. Conversions between other scales may alter all coefficients.

The significance of equation (2-10) may be illustrated by reference to Figure 2. The redistribution of K_2O during annealing is in accord with our intuitive notions of diffusion: homogenization by the migration of a component down its own concentration gradient. The concentration profile of SrO, however, presents a different situation, since the flux and the concentration gradient of SrO have the same sign, and SrO appears to be diffusing up its own concentration gradient ("uphill" diffusion).

With K_2O = 1 and SrO = 2, the two independent flux equations (the system in Fig. 2 has n = 3) are

$$J_1^V(K_2O) = -\underset{(1.20)}{D_{11}^V} y_1 - \underset{(0.03)}{D_{12}^V} y_2 \qquad J_2^V(SrO) = -\underset{(-1.36)}{D_{21}^V} y_1 - \underset{(0.013)}{D_{22}^V} y_2 \ . \qquad (2\text{-}14)$$

The values of the molar coefficients ($cm^2/sec \times 10^{10}$), recalculated from Table 2 of Varshneya and Cooper (1972), are shown in parentheses.

D_{12}^V has the same sign as D_{11}^V and is very much smaller (<0.2%). Thus, the contribution of the term $D_{12}^V y_2$ to the flux of K_2O must be small. In contrast, D_{21}^V is larger than D_{22}^V and of opposite sign and the term $D_{21}^V y_1$ is sufficiently large to reverse the flux of SrO. Each of the components K_2O, SrO and SiO_2 may be made to diffuse up its own concentration gradient by choosing suitable terminal compositions for the diffusion couple (Varshneya and Cooper, 1972). Other obvious examples of uphill diffusion have been illustrated by Darken* (1949), Chandhok *et al.* (1962) and Vignes and Sabatier (1969).

The minus sign included in Fick's first law ensures that the mutual (binary) diffusion coefficient D^V be positive. On the other hand, the signs of the D_{ik}^V are not fixed in a multicomponent system, although thermodynamics imposes conditions on certain combinations of the coefficients (Section 6).

The addition of (nonequilibrium) thermodynamics to diffusion theory has greatly clarified the meaning of Fick's first law in binary and multicomponent solutions (Onsager, 1945). At constant temperature and pressure, the conditions for chemical equilibrium in a system of n molecular components are (Gibbs, 1961)

$$\mu_i^\alpha = \mu_i^\beta = \mu_i^\gamma = \ldots = \mu_i^\sigma \qquad (i = 1, 2, \ldots, n) \qquad (2\text{-}15)$$

where μ_i is the chemical potential of i (joules/mole), and the superscripts $\alpha\ \beta\ \gamma \ldots \sigma$ may be taken to denote either separate phases or domains of local equilibrium within a single phase. The approach to equilibrium involves the elimination of chemical potential gradients within a single phase and among different phases for each component. Therefore, it is natural to construct a set of linear equations (Onsager and Fuoss, 1932; Onsager, 1945), analogous to equation (2-9), relating fluxes to chemical potential gradients, namely

----

*Darken's experiment reveals quite vividly the way in which an initially homogeneous component may be temporarily redistributed -- in this case to produce a discontinuity -- by uphill diffusion.

$$J_i^v = - \sum_{k=1}^{n} \tilde{L}_{ik}^v (\partial\mu_k / \partial y) \qquad (i = 1, 2, \ldots, n) \ . \qquad (2\text{-}16)$$

The coefficients $\tilde{L}_{ik}^v$ (moles^2/joule cm sec) are referred to as thermo-dynamic diffusion coefficients, phenomenological coefficients or con-ductance coefficients.

With the assumption -- in most cases involving only diffusion, the very reasonable assumption -- that local equilibrium is maintained throughout the diffusion zone, the Gibbs-Duhem equation (Prigogine and Defay, 1954)

$$\sum_{i=1}^{n} c_i d\mu_i = 0 \qquad (T, \ P \ \text{constant}) \qquad (2\text{-}17)$$

may be used, along with equation (2-1), to remove one gradient and one flux from (2-16). Thus [Section 5, equations (5-2) and (5-3)],

$$J_i^v = \sum_{k=1}^{n-1} L_{ik}^v X_k^v \qquad (i = 1, 2, \ldots, n\text{-}1) \qquad (2\text{-}18)$$

or
$$(J^v) = [L^v](X^v) \ , \qquad (2\text{-}19)$$

where X_k^v is a set of n-1 independent chemical potential gradients that may be obtained from equation (5-3). Linear equations similar to equa-tion (2-18) that join independent fluxes and thermodynamic forces (X) are often called phenomenological equations in nonequilibrium thermo-dynamics.

The empirical equations (2-10) and the phenomenological equations (2-18) give different but related descriptions of diffusion. Chemical potentials are functions of each of the (n-1) independent concentrations in an n component system, or

$$\mu_i = \mu_i(c_1, \ c_2, \ c_3, \ \ldots, \ c_{n-1}) \qquad (i = 1, 2, \ldots, n\text{-}1) \ . \qquad (2\text{-}20)$$

Consequently, the gradients in (2-10) and (2-18) are related by

$$[\mu] \equiv \mu_{ij} \equiv (\partial\mu_i / \partial c_j)_{T,P,c_k} \qquad (i,j = 1, 2, \ldots, n\text{-}1)$$

$$\partial\mu_i / \partial y = \sum_{j=1}^{n-1} \mu_{ij} (\partial c_j / \partial y) \qquad (i = 1, 2, \ldots, n\text{-}1) \ . \qquad (2\text{-}21)$$

Thermodynamics defines chemical equilibrium in terms of chemical potentials and only indirectly in terms of concentrations; an equation of state must be invoked to calculate equilibrium values of c_i for a given set of equilibrium values of μ_i. Chemical potential gradients follow concentration gradients in binary mixtures, but in multicomponent mixtures, equation (2-21) establishes a more complex and less obvious relationship between the gradients of μ_i and c_i. The terms μ_{ij} $(i \neq j)$ are only zero in a mixture of nonelectrolytes if i mixes ideally; in an electrolyte mixture with a common anion, such terms can never be zero (Miller, 1959, equation 44).

The empirical equations, (2-10), are framed in terms of directly measurable concentration gradients and lead to a set of differential equations analogous to (2-7) and (2-8):

$$\partial c_i/\partial t = \sum_{k=1}^{n-1} D_{ik}^{V}(\partial^2 c_k/\partial y^2) \qquad (D_{ik}^{V} \text{ const.; } i = 1, 2, \ldots, n-1) \quad (2\text{-}22)$$

and

$$\partial c_i/\partial t = \sum_{k=1}^{n-1} \partial(D_{ik}^{V}\partial c_k/\partial y)/\partial y \quad (D_{ik}^{V} \text{ not const.; } i = 1, 2, \ldots, n-1). \quad (2\text{-}23)$$

These equations form the basis for the analysis of diffusion in the laboratory and in natural systems.

The phenomenological equations depend on quantities (μ_i) that are not measurable,* but whose behavior is prescribed by thermodynamics. And as pointed out by Fisher (1977), in complex natural systems, it may be easier to determine or to fix variations of chemical potentials than concentrations during the evolution of a natural system.

The fluxes (J^V) and the forces (X^V), and therefore the coefficients [L^V], may be linked to the production of entropy (see Chapter 5) during a natural (irreversible) process (Onsager, 1931a,b; see also Section 5). The expression for the entropy production provides (1) a sound basis for the examination of the theoretical structure of diffusion and (2) a connection between isothermal, isobaric diffusion and other irreversible processes (for example, ionic conduction; Miller, 1966, 1967a,b; Eckman *et al.*, 1978).

Onsager (1945) noted that it should be possible to diagonalize [D^V] and thereby reduce equation (2-10) to an equivalent set of binary

---

*Chemical potentials are always calculated, via an equation of state, from some directly measurable quantity. The simplest equation of state, μ_i = const. + $RT\ln c_i$, assumes ideal mixing and is derived ultimately from the equation of state for a perfect gas (Denbigh, 1966).

equations analogous to equation (2-5): equations in which each component appears to diffuse only on its own concentration gradient. The diagonalization of $[D^V]$, which is not generally symmetric, depends on the existence of a non-singular matrix, [S], such that the transformation

$$[S]^{-1}[D^V][S] = [\Lambda] \tag{2-24}$$

produces a diagonal matrix $\Lambda_{ik} = \delta_{ik}\Lambda_k$ (i = 1, 2, ..., n-1). Subsequent studies of (1) the thermodynamic properties of the L-coefficients that follow from the production of entropy during diffusion and (2) the additional requirements for stability of a mixture during diffusion have demonstrated that [S] always exists (Cullinan, 1965; Gupta and Cooper, 1971; Kirkaldy, 1959, 1970; Kirkaldy *et al.*, 1963a; Sundelöf and Södervi, 1964).

Provided that the D^V_{ik} are not concentration dependent, the equivalent binary equations are (Onsager, 1945; Kirkaldy *et al.*, 1963a; Cullinan, 1965)

$$[S]^{-1}(J^V) = [\Lambda][S]^{-1}(y) \tag{2-25}$$

or

$$\bar{J}^V_i = \Lambda_i(\bar{y}_i) \qquad (i = 1, 2, \ldots, n-1) \tag{2-26}$$

where $\bar{J}^V_i$ and $\bar{y}_i$ are elements of the vectors $[S]^{-1}(J^V)$ and $[S]^{-1}(y)$, respectively. Λ_i, which is the *ith* characteristic root or eigenvalue of $[D^V]$, is an analogue of the mutual diffusion coefficient in equation (2-5). Thus, each component in the multicomponent solution may be treated separately as a member of a pseudo-binary mixture. Equation (2-26) leads to a simple set of partial differential equations analogous to (2-7). Further details are given in Section 6.

SECTION 3: MEASURABLE QUANTITIES

Although a wide variety of experimental techniques and conditions are employed, only two kinds of diffusion coefficients are measured: the tracer and chemical diffusion coefficients. Diffusion coefficients may be a function of temperature, direction, concentration, pressure and time. The pressure dependence is small in liquids and solids and the time dependence is usually ignored in experiments, although both may be important in some geological applications.

In some crystals, the dependence of the diffusion coefficient on concentration may stem not only from the usual chemical components but also from point defects. The defects may be considered as very dilute components and we may write chemical equations and calculate equilibrium constants for them as we would for the other chemical components (Swalin, 1962; van Gool, 1966; Manning, 1968; see Chapter 7 by Lasaga). The list of defects that may occur in any compound is very long, but few defects are important to diffusion and usually one particular defect will exert a dominant influence at a given temperature and pressure.

The most importanct defects are vacancies, since an exchange between an ion and vacancy provides the simplest mechanism for diffusion (i.e., the mechanism with the lowest activation energy; for an introduction to diffusion mechanism, see Manning, 1974). Although the concentration of vacancies is very small, a minor change in their concentration may significantly alter diffusion rates. In ideally stoichiometric compounds, vacancies on cation sites must be accompanied by vacancies on anion sites or interstitial cations to preserve electrical neutrality; in nonstoichiometric crystals, cation vacancies may be associated with aliovalent impurities or excess anions (Swalin, 1962; see Chapter 7).

An excess of anions (oxygen) may be formed by the reaction

$$\tfrac{1}{2}O_2(\text{gas}) = O(\text{crystal}) + V \qquad (3\text{-}1)$$

where V is a neutral cation vacancy. The neutral vacancies, by acting as acceptors, may ionize once

$$V = V^- + e^+ \qquad (3\text{-}2)$$

or twice

$$V^- = V^= + e^+ \qquad (3\text{-}3)$$

and so on, where e^+ is an electron hole in the valence band of the ion donating electrons. In the oxides of the transition metals, the formation of the charged vacancies is associated with oxidation of the metal; for example,

$$Fe^{3+} = Fe^{2+} + e^+ \; . \qquad (3\text{-}4)$$

Buening and Buseck (1973) have demonstrated experimentally the probable existence of this kind of dependency in the system Fe_2SiO_4-Mg_2SiO_4; their results are expressed as a function of the partial pressure of oxygen, since, from equation (3-1), the equilibrium concentration of vacancies (or the ratio of Fe^{2+} to Fe^{3+}) may be controlled by buffering the partial pressure of oxygen (see also Chapter 7 and Swalin, 1962).

The above method of forming ionized vacancies is common in oxides of the transition metals and is reflected in the conductance and diffusion properties of these materials (Swalin, 1962). To what degree it is important in Fe silicates other than olivine remains to be seen.

Buening and Buseck (1973) have also shown that the mutual diffusion coefficient for Fe-Mg exchange in olivine is largest parallel to the *c*-axis and smallest parallel to *b* -- a result that may be readily explained by an examination of the crystal structure. In crystals, the diffusion coefficients are the elements of a second rank tensor linking the two vectors J_i and y_i. Unless diffusion is one-dimensional, they must be handled as a tensor quantity.

The tracer coefficient is normally determined by the penetration of a tracer ion into a homogeneous phase. Homogeneity in solids means not only a uniform chemical composition (including point defects) but the absence of fast diffusion paths such as grain boundaries and fractures. The flux (J_{12}) of atoms between two planes (1 and 2) a distance λ apart in a crystal is (Manning, 1968, 1974):

$$J_{12} = n_1 T_{12} \tag{3-5}$$

and the reverse flux is

$$J_{21} = n_2 T_{21} \tag{3-6}$$

where n_1 and n_2 are the number of atoms of the diffusing species on planes 1 and 2, respectively, and T_{12} and T_{21} are the frequencies (sec^{-1}) for the forward and reverse jumps. Given that $n = \lambda\bar{n}$, where $\bar{n}$ is the total number of atoms per unit volume, it may be shown (Manning, 1974) that the net flux is

$$J = J_{12} - J_{21} = -\frac{1}{2}\lambda^2(T_{12} + T_{21})(\partial\bar{n}/\partial y) + \lambda(T_{12} - T_{21})\bar{n} \tag{3-7}$$

or

$$J = -D^*(\partial\bar{n}/\partial y) + v_F\bar{n} \ . \tag{3-8}$$

Thus, the tracer diffusion coefficient D^* is proportional to the average jump frequency ($T_{12} + T_{21}$). The term v_F, which equals $\frac{1}{2}(T_{12} - T_{21})$, has the dimensions length/time; when T_{12} is not equal to T_{21}, it represents the net velocity with which atoms drift to the left or right. "Forces" which may cause T_{12} to differ from T_{21} include an electrical field, a

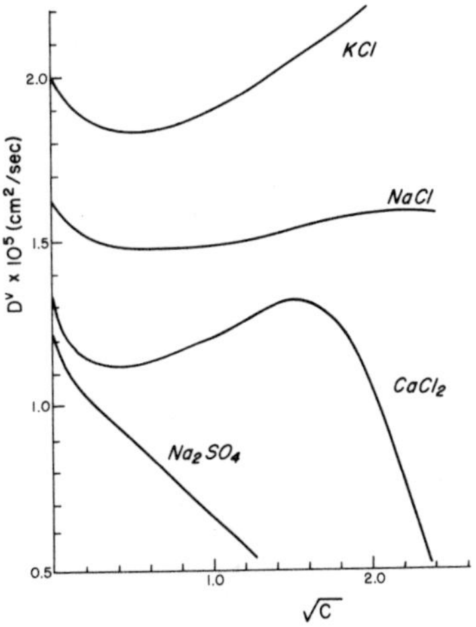

Figure 3. Plot of the experimental values of the mutual diffusion coefficients D^v versus the square root of concentration (molar) for binary aqueous solutions of Na_2SO_4, $CaCl_2$, NaCl and KCl (25°C; data from Rard and Miller, 1979a,b and 1980). The most concentrated solutions ($CaCl_2$ and NaCl are approximately 6 molar).

temperature gradient or the non-ideal portion of a chemical potential gradient (Manning, 1974, Table 1).

Chemical diffusion coefficients, that is, coefficients measured in a concentration gradient, are normally concentration dependent. The exact form of the dependency is variable (Fig. 3) and may change with temperature. In a binary mixture, D^V is found from a solution of equation (2-8) for a couple of the form shown in Figure 1. The experiment is arranged so that 1 and 2 are initially homogeneous and there is a sharp discontinuity in components 1 and 2 at the interface at time t = o. Equation (2-8) may be reduced to a homogeneous, first-order equation by the introduction of a new variable $\eta = y/\sqrt{t}$ (the Boltzmann-Matano solution; Matano, 1933; Crank, 1956, p. 232). For an infinite couple* (Fig. 4a), the boundary conditions (where c is either component 1 or 2)

$$c(+y,o) = c(+\infty,t) = c^o \tag{3-7}$$

and

$$c(-y,o) = c(-\infty,t) = c^1 \tag{3-8}$$

may be expressed in terms of the new variable as

$$c(\eta = +\infty) = c^o \tag{3-9}$$

and

$$c(\eta = -\infty) = c^1 . \tag{3-10}$$

With the new variable, eqaution (2-8) becomes

$$\frac{\partial c}{\partial \eta} \cdot \frac{\partial \eta}{\partial t} = \frac{\partial \eta}{\partial y} \frac{\partial}{\partial \eta} D \frac{\partial c}{\partial \eta} \cdot \frac{\partial \eta}{\partial y} \tag{3-11}$$

or

$$-\frac{\eta}{2} \frac{\partial c}{\partial \eta} = \frac{\partial}{\partial \eta} D \frac{\partial c}{\partial \eta} \tag{3-12}$$

---

*That is two phases, initially separated by a sharp plane, in which the compositions at the ends of the couple remain undisturbed by diffusion (*e.g.*, Fig. 2). The term free diffusion, usually applied to liquids, denotes an infinite couple. Some other important configurations are treated by Crank (1956), Kirkaldy (1970) and Dunlop *et al.* (1972).

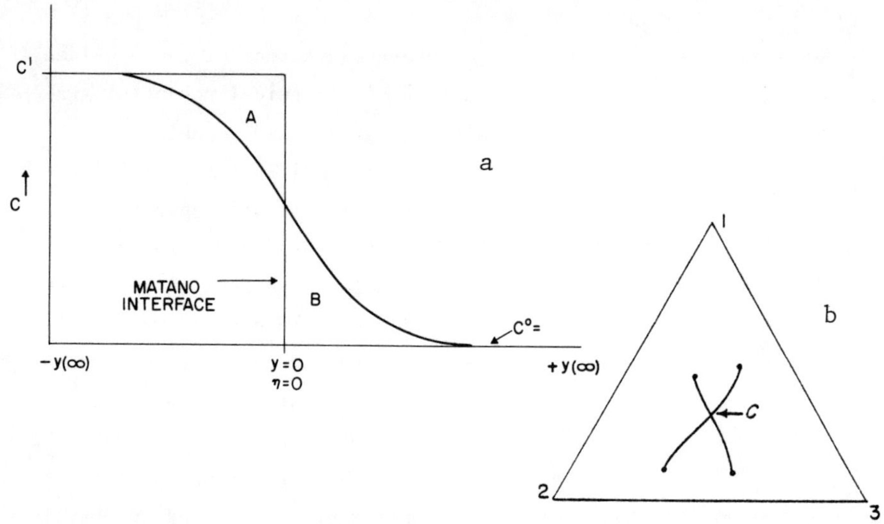

Figure 4. (a) Concentration curve for one component in a binary diffusion couple versus the Boltzmann variable $\eta(y/\sqrt{t})$. Equation (3-14) sets the areas A and B equal and defines η = o (y = o) at the Matano interface.
(b) Diffusion paths for two experimental couples in a ternary system (components 1, 2 and 3). Equation (3-19) determines the values of D^v_{11}, D^v_{12}, D^v_{21} and D^v_{22} uniquely at the point of intersection of the paths (C).

Integration of equation (3-12) produces

$$\int_{c_o}^{c} \frac{\eta}{2}\, dc = -D \left.\frac{dc}{d\eta}\right|_c , \tag{3-13}$$

where $c^o < c < c^1$. Since, for an infinite couple D $dc/d\eta$ = o at $c = c^1$

$$\int_{c^o}^{c^1} \eta dc = \int_{c^o}^{c^1} y dc = 0 . \tag{3-14}$$

Thus, y = η = o (the "Matano" interface) must be chosen so that the areas A and B are equal in Figure 4a. Rearrangement of equation (3-13) gives

$$D(c) = -\frac{1}{2}\frac{d\eta}{dc}\int_{c_o}^{c} \eta dc = -\frac{1}{2t}\frac{dy}{dc}\int_{c^o}^{c} y dc , \tag{3-15}$$

where $\eta = y/\sqrt{t}$ has been re-introduced.

After the Matano interface has been located, the integral and the gradient $d\eta/dc$ (dy/dc) may be evaluated graphically (if the $\bar{V}_i$ are constant, the Matano interface is the initial interface or weld). Precise data (on dc/dy) are better handled by numerical methods (Dunlop *et al.*, 1972), since graphical methods are very prone to errors at either end of the c-y curve. Less precise data (*e.g.*, microprobe data) may be treated by a method suggested by Hall (1953); his method allows linear regression methods to be used to improve the estimate of D^V and programming of much of the computation.

Equation (3-15) is only exact if the concentration differences across the couple are so small (ideally, vanishingly small) that $\bar{V}_1$ and $\bar{V}_2$ may be taken as constant. On the other hand, the accuracy of D^V is determined chiefly by the accuracy with which the concentration profile can be measured -- large concentrations usually favor better measurement (the method of chemical analysis is, of course, all important). The way out of this dilemma is to compute the error in D^V caused by varying the partial molar volumes, that is, the error due to bulk flow. Historically, the study of this problem in metals arose out of experiments performed by Kirkendall (1942) and Smigelskas and Kirkendall (1947) on the shift of inert markers during diffusion -- a phenomenon sometimes referred to as the Kirkendall shift.

As confirmed vividly by the experiments of Correa da Silva and Mehl (1951), bulk flow is not confined to liquid systems. The bulk flow, as noted earlier, is driven by volume changes associated with non-ideal mixing; in crystalline materials the immediate cause is changes in unit cell dimensions with composition. In some crystalline materials, an additional component associated with a flux of vacancies may also be present, and indeed, may be the major component.

Kirkwood *et al.* (1960) have shown that the flux J_1^c measured across a plane fixed a distance y above the base of the cell in Figure 1 is

$$J_1^c = -D^V(\partial c_1/\partial y) - c_1 \int_{+\infty}^{y} \frac{\partial \bar{V}_1}{\partial c_1} \frac{D^V}{(1-c_1\bar{V}_1)} \left[\frac{\partial c_1}{\partial y}\right]^2 dy \qquad (3\text{-}16)$$

(the integration assumes that the positive direction of y is down and that the initial composition is unchanged at the base of the cell, $+\infty$). The flux J_1^c combines (a) a diffusive flux given by the first term on the right-hand side and (b) a convective flux (integral term) proportional to the accumulated volume changes below y. Equation (3-16) establishes, in what is a typical experimental situation, that the quantity measured is D^V plus an error fixed by the integral term. The integral is zero if $\bar{V}_1$ is constant and diminishes rapidly as $\partial c_1/\partial y$ becomes smaller.

Duda and Vrentas (1965) have derived a formula analogous to (3-16) in terms of the variable η. The resulting equation, which is more readily applicable to experiments, contains a Boltzmann-Matano integral [equation (3-15) above] plus a second integral in y that disappears, when volume changes are zero.

Prager (1953), Crank (1956), Baluffi (1960), Wagner (1969) and Brady (1975a) have also formulated expressions for determining D^V. Brady, following Hartley and Crank (1949), re-defines concentration and distance units in crystals to obtain a simple and readily applicable form of the Boltzmann-Matano integral. All of these methods require either partial molar volumes or density versus concentration data. Brady (1975a) has given a particularly lucid treatment of the problem in crystalline materials.

Experimental methods for measuring diffusion coefficients in gases, liquids and solids (with the emphasis on liquids) have been extensively reviewed by Dunlop *et al.* (1972). Rard and Miller (1979a,b; 1980, 1981) have measured diffusion coefficients and densities from dilution to near saturation in aqueous solutions of NaCl, $CaCl_2$, $MgCl_2$, Na_2SO_4, $MgSO_4$, $BaCl_2$ and KCl.

Freer (1980) has published an up-dated compilation of diffusion data in oxides (which includes a brief review of some of the newer experimental techniques) and has almost completed a comparable compilation of data for silicates.

The Boltzmann-Matano analysis may be applied to a ternary system, if the D^v_{ik} are concentration dependent. For the boundary conditions (infinite couple)

$$c_i(+y,o) = c_i(+\infty,t) = c_i^o \qquad (3\text{-}17)$$

and

$$c_i(-y,o) = c_i(-\infty,t) = c_i^1 \qquad (3\text{-}18)$$

the substitution of $\eta = y/\sqrt{t}$ into equation (2-23) and integration gives

$$\int_{c_i^o}^{c_i} \eta dc_i = -\sum_{k=1}^{n-1} D^v_{ik} \left.\frac{dc_k}{d\eta}\right|_{c_1} \qquad (3\text{-}19)$$

and

$$\int_{c_i^o}^{c_i^1} \eta dc_i = 0 \qquad (3\text{-}20)$$

for the Matano interface (which must be the same for all components). The four independent diffusion coefficients for a ternary system may be computed from two diffusion couples with different terminal compositions and one common composition in their diffusion zones. The coefficients are only determined at that single, common composition (Fig. 4b).

When the $\bar{V}_i$ are nearly constant, there is an obvious advantage in forming long diffusion paths with multiple intersections (Vignes and Sabatier, 1969). On the other hand, for short diffusion paths, it is reasonable to assume that D^V_{ij} are concentration independent; equation (2-22) may then be solved exactly without reference to the awkward Boltzmann-Matano analysis (Fujita and Gosting, 1956; Kirkaldy, 1959). Kirkaldy *et al.* (1963b) and Kirkaldy (1970) have reviewed various methods for optimizing the determining of diffusion coefficients.

Whatever method is used, the burden of mapping the variation of the D^V_{ik} over a substantial compositional range in a ternary system is obvious; the burden would be even greater in systems where the D^V_{ik} are dependent on p_{O_2} as well as the concentration of the major components (and, of course, also T). The measurement of the concentration gradients in aqueous electrolytes is itself difficult, but Rayleigh interferometry may be made to yield very accurate results (Dunlop *et al.*, 1972). The diaphragm cell (*e.g.*, Wendt and Shamin, 1970) is easier to use, but the resulting diffusion coefficients are not as accurate.

The total number of measurements completed in ternary systems (metals, glasses and aqueous electrolytes) is small (Burchard and Toor, 1962; Dayanda and Grace, 1965; Dunlop 1957a,b; Dunlop and Gosting, 1955; Fujita and Gosting, 1956, 1960; Kim *et al.*, 1973; Revzin, 1972; Schuck and Toor, 1963; Sucov and Gorman, 1965; Varshneya and Cooper, 1972; Vignes and Sabatier, 1969; Wendt, 1962; Wendt and Shamin, 1970; Zeibold and Ogilvie, 1967). Only Vignes and Sabatier and Ziebold and Ogilvie have mapped the coefficients across a whole isotherm. There have apparently been no measurements made in ionic crystals or in quaternary systems of any kind. Gupta and Cooper (1971; Cooper, 1974) have suggested that it may be possible to obtain all the elements of $[D^V]$ from a single experiment; this method, however, has not been tested yet.

SECTION 4: REFERENCE FRAMES

The volume-fixed reference frame specified by equation (2-1) is one of a number of possible choices. Studies of diffusion and conductance in aqueous electrolytes and their inter-relationship, for example, commonly employ fluxes measured with respect to the local velocity of water molecules (solvent-fixed frame). Lattice-fixed reference frames, which have the same transformation properties as solvent-fixed frames, may be very important to the treatment of diffusion in silicates (Brady, 1975a; Anderson, 1976).

A comprehensive treatment of reference frames, following on the work of Hooyman (1956), has been completed by de Groot and Mazur (1962, Ch. 11). Although less comprehensive, the analysis of Kirkwood *et al.* (1960) sets out the same principles in a simple and less abstract way.

Once mechanical equilibrium has been reached in the diffusion zone (Prigogine, 1947; de Groot and Mazur, 1962, p. 43), diffusion fluxes may be defined by

$$J_i^a = c_i(v_i - v^a) \qquad (i = 1, 2, \ldots, n) \ . \tag{4-1}$$

The velocity of i parallel to y, v_i, is measured with respect to a fixed (laboratory) reference frame. v^a is a weighted reference velocity defined by (Hooyman, 1956)

$$v^a = \sum_{i-1}^{n} \bar{a}_i v_i \quad \text{and} \quad \sum_{i=1}^{n} \bar{a}_i = 1 \ , \tag{4-2}$$

where the superscript a denotes a reference frame corresponding to the choice of $\bar{a}_i$. Thus, the term $c_i v^a$ in equation (4-1) subtracts a component from the total flux $c_i v_i$ of i equal to the bulk flow of i. Different choices of $\bar{a}_i$ correspond to different ways of specifying the bulk flow.

Multiplication of equation (4-1) by $\bar{a}_i/c_i$, followed by summation gives

$$\sum_{i=1}^{n} (\bar{a}_i/c_i)J_i^a = \sum_{i=1}^{n} \bar{a}_i v_i - v^a \sum_{i=1}^{n} \bar{a}_i = 0 \ , \tag{4-3}$$

an expression that is frequently taken to define reference frames. The substitution of $\bar{a}_i = c_i\bar{V}_i/1000$ immediately reproduces equation (2-1).

A lattice-fixed or solvent-fixed frame, in which v^a is taken as the velocity of one of the components k, is obtained by putting $\bar{a}_i = \delta_{ik}$

(i.e., $J_k^o = 0$). In electrolyte solutions with no neutral solvent, k will usually have to be an ionic species. A convenient choice in silicates is often the anion (*e.g.*, $Al_2Si_3O_{12}^{6-}$ in garnet; Section 1). The resulting ionic fluxes must be combined according to the scheme set out in Section 7, in order to get molecular fluxes in a lattice-fixed frame (Anderson, 1976).

Setting $\bar{a}_i$ equal to the mass fractions ω_i or the mole fractions x_i, will define a mass-fixed or molecular reference frame. With the above choices, equation (4-3) becomes

$$\sum_{i=1}^{n} \bar{V}_i \bar{J}_i^v = \sum_{i=1}^{n} M_i J_i^m = J_k^o = \sum_{i=1}^{n} J_i^n = 0 \qquad (4\text{-}4)$$

where the superscripts v, m, o and n denote the volume-fixed, mass-fixed, solvent (lattice)-fixed and molecular frames, respectively. Weighing factors for mass fluxes have been tabulated by de Groot and Mazur (1962, p. 240).

Fluxes may be transformed from reference frame b [defined by weighing factors $\bar{b}_i$ in equation (4-2)] to reference frame a by (de Groot and Mazur, 1962):

$$(J^a) = [B^{ab}](J^b) \ . \qquad (4\text{-}5)$$

The elements of the $(n-1)^2$ matrix $[B^{ab}]$ needed to transform the (n-1) vector (J^b) are derived from the expression (Hooyman, 1956)

$$J_i^a = J_i^b + c_i u \qquad (i = 1, 2, \ldots, n) \qquad (4\text{-}6)$$

where according to equation (4-1), u is given by $u = v^b - v^a$. With this equation for u and equations (4-5) and (4-6), the elements B_{ik}^{ab} are given by (de Groot and Mazur, 1962)

$$B_{ik}^{ab} = \delta_{ik} + (\bar{a}_n \bar{b}_k / \bar{b}_n - \bar{a}_k)(c_i / c_k) \qquad (i,k = 1, 2, \ldots, n-1) \ . \qquad (4\text{-}7)$$

Pre-multiplication of (4-5) by the inverse matrix, $[B^{ab}]^{-1}$ gives

$$[B^{ab}]^{-1}(J^a) = (J^b) \ . \qquad (4\text{-}8)$$

But

$$(J^b) = [B^{ba}](J^a) \qquad (4\text{-}9)$$

and hence

$$[B^{ab}]^{-1} = [B^{ba}] \qquad (4\text{-}10)$$

so that the inverse matrix exists and its elements may be found by interchanging a and b in equation (4-7).

TABLE 1

	$NaCl - H_2O$		$CaCl_2 - H_2O$	
c_{SALT}	0.0016	5.0678	0.0017	5.57
$\bar{V}_{SALT}$	17.006	23.256	16.798	43.062
$D^V \times 10^5$	1.576	1.585	1.251	0.5213
$D^o \times 10^5$	1.576	1.796 (13%)	1.251	0.6858 (31%)
$D^m \times 10^5$	1.576	1.348 (15%)	1.251	0.3889 (25%)
$D^n \times 10^5$	1.576	1.629 (3%)	1.251	0.610 (17%)

The values of D^V are from Rard and Miller (1979a); partial molar volumes were calculated from density versus concentration data using the equations of Dunlop and Gosting (1959) and data from the above authors. The numbers in parentheses give the percentage difference $(D^V - D^a) / D^V$.

Diffusion coefficients are transformed by

$$[D^a] = [B^{ab}][D^b] \tag{4-11}$$

since, with equations (4-5) and (2-11)

$$(J^a) = [B^{ab}](J^b) = -[B^{ab}][D^b](y) = -[D^a](y) \ . \tag{4-12}$$

The mutual diffusion coefficients, D^o, D^m and D^n, in a binary system are related to the measurable coefficients D^V by (with two as the dependent component)

$$D^a = (\bar{a}_2/\bar{b}_2)D^V = D_1^a = D_2^a \tag{4-13}$$

or

$$\text{(a)} \quad D^o = (1000/c_2\bar{V}_2)\ D^V \tag{4-14}$$

$$\text{(b)} \quad D^m = (\omega_2/c_2\bar{V}_2)D = (M_2/\bar{V}_2)D^V \tag{4-14}$$

$$\text{(c)} \quad D^n = (x_2/c_2\bar{V}_2)D^V$$

where ρ and c are the total density and total concentration of the solution.

Calculated values of D^o, D^m and D^n for the binary aqueous electrolytes $NaCl-H_2O$ and $CaCl_2-H_2O$ are shown in Table 1 for a dilute and a concentrated solution. As expected, there are negligible differences among the various coefficients in the dilute solutions -- at infinite dilution all reference frames coincide with a fixed frame. The partial molar volumes are much larger in the concentrated $CaCl_2$ solution than the $NaCl_2$ solution. Consequently, the differences between D^v and D^o, D^m and D^n are larger, especially for the important and often used transformation $D^v \rightarrow D^o$.

For a ternary system (dependent component three)

$$D_{ij}^o = \sum_{k=1}^{2} B_{ik}^{ov} D_{kj}^v \qquad (i,j = 1,2) \tag{4-15}$$

and

$$B_{ik}^{ov} = \delta_{ik} + (c_i \bar{V}_k / c_3 \bar{V}_3) . \tag{4-16}$$

Miller (1965) has listed D_{ij}^v and D_{ij}^o for the system $NaCl$-KCl-H_2O at 25°C. The percentage differences between D_{ij}^o and D_{ij}^v in the most concentrated solution [three molar; $c_1(NaCl) = 1.5$; $c_2(KCl) = 1.5$] are: D_{11} - 4.8%; D_{12} - 99%; D_{21} - 21%; D_{22} - 5.5%. The volume changes of mixing are relatively small in this system and the on-diagonal coefficients are little altered by the transformation. The off-diagonal coefficients are small (D_{12}^v/D_{11}^v = 7.2%; D_{21}^v/D_{22}^v = 17.7%) and are drastically changed by the transformation. As D_{ik}^v/D_{ii}^v $(i \neq k) \rightarrow 1$, however, the percentage change will diminish.

The proper identification of reference frames is crucial in the measurement, estimation or theoretical manipulation of diffusion coefficients. The effects of ignoring reference frames in the solution of the differential equations (2-7), (2-8), (2-22) or (2-23) is clearly proportional to the variation of partial molar volumes with concentration. The differential equations are normally solved in the same volume-fixed frame in which measurement is made, but many estimation procedures (Section 8) produce coefficients in a solvent-fixed frame.

Starting with equations (4-9) and (4-11), it is possible to cast the flux equations in a self-consistent form such that $[\bar{D}^m]$, $[\bar{D}^n]$ and $[\bar{D}^o]$ (or in general, $[\bar{D}^a]$ where a = m, n or o) may be derived from $[D^v]$ by a similarity transformation -- a result that is important in Section 6. We have for reference frame a (a = m, n or o), from equations (4-5) and (2-11),

$$[B^{av}](J^v) = (J^a) = -[B^{av}][D^v][B^{av}]^{-1}[B^{av}](y) \tag{4-17}$$

where the unit matrix $[B^{av}]^{-1}[B^{av}]$ has been included in the right-hand side of the equation. The term $[B^{av}](y)$ may be reduced as follows (de Groot and Mazur, 1962, p. 258), using equation (4-7)

$$\left([B^{av}](y)\right)_i = \sum_{j=1}^{n-1} B_{ij}^{av} y_j = y_i + \frac{c_i \bar{a}_n}{\bar{b}_n} \sum_{j=1}^{n} \frac{\bar{b}_j}{c_j} y_j - c_i \sum_{j=1}^{n} \frac{\bar{a}_j}{c_j}$$

$$(i = 1, 2, \ldots, n-1) \; . \qquad (4\text{-}18)$$

The second term disappears with the substitution of $\bar{b}_j = c_j \bar{V}_j$ [equation (2-3)]. The third term is zero when $\bar{a}_j = x_j$ ($\sum_{j=1}^{n} y_j = 0$) and $\bar{a}_j = \omega_j$. For a solvent-fixed frame, equation (4-18) becomes $y_i - (c_i/c_n)y_n$. Thus, we may write the flux equations in the form

$$J_i^a = -\sum_{j=1}^{n-1} \bar{D}_{ij}^a y_j \qquad (i = 1, 2, \ldots, n-1) \qquad (4\text{-}19)$$

for a = n or m, and

$$J_i^a = \sum_{j=1}^{n-1} \bar{D}_{ij}^a \left(y_j - \frac{c_j}{c_n} y_n\right) \quad (i = 1, 2, \ldots, n-1) \qquad (4\text{-}20)$$

when a = o. Combining equations (4-17) and (4-19) or (4-20),

$$[\bar{D}^a] = [B^{av}][D^V][B^{av}]^{-1} \; . \qquad (4\text{-}21)$$

Pre-multiplication of equation (4-21) by $[B^{av}]^{-1}$ and post-multiplication by $[B^{av}]$ gives

$$[D^V] = [B^{av}]^{-1}[\bar{D}^a][B^{av}] \qquad (4\text{-}22)$$

which proves that $[\bar{D}^a]$ and $[D^V]$ are similar matrices when the diffusion laws are written in the form of equation (4-17) or, specifically, (4-19) and (4-20).

SECTION 5: TRANSFORMATION AND CALCULATION OF L-COEFFICIENTS

During the course of an irreversible process, the rate of entropy production (the amount of entropy produced per unit volume per unit time) is an invariant: The value of σ cannot be determined by the particular scheme introduced to describe the process. Onsager (1931a,b) showed that σ may be calculated (if local equilibrium is maintained) from a linear combination of fluxes and forces. For diffusion at constant T and P (Onsager, 1945; de Groot and Mazur, 1962)

$$T\sigma = -\sum_{i=1}^{n-1} J_i^a(\partial\mu_i/\partial y) - J_n^a(\partial\mu_n/\partial y) =$$

$$= \sum_{i=1}^{n-1} J_i^a[1 + \sum_{k=1}^{n-1} (c_k\bar{a}_i/c_i\bar{a}_n)]\partial\mu_k/\partial y \tag{5-1}$$

where equation (4-3) and the Gibbs-Duhem relationship, (2-17), have been used to eliminate J_n^a and $(\partial\mu_n/\partial y)$, respectively. By defining (de Groot and Mazur, 1962)

$$A_{ik}^a = \delta_{ik} + (c_k\bar{a}_i/c_i\bar{a}_n) \qquad (i,k = 1, 2, \ldots, n-1) \tag{5-2}$$

and

$$X_i^a = -\sum_{k=1}^{n-1} A_{ik}^a(\partial\mu_k/\partial y) \qquad (i = 1, 2, \ldots, n-1) \tag{5-3}$$

equation (5-1) may be shortened to

$$T\sigma = \sum_{i=1}^{n-1} J_i^a X_i^a = (J^a)(X^a) \ . \tag{5-4}$$

The invariance of σ means that, with equation (4-8),

$$T\sigma = (J^b)(X^b) = \big([B^{ab}][J^b]\big)(X^a) = (J^b[B^{ab}]^T(X^a) \tag{5-5}$$

and hence,

$$(X^b) = [B^{ab}]^T(X^a) \ . \tag{5-6}$$

It follows from equation (4-5) that

$$(J^a) = [B^{ab}](J^b) = [B^{ab}][L^b](X^b) = [B^{ab}][L^b][B^{ab}]^T(X^a) \tag{5-7}$$

and, therefore

$$[L^a] = [B^{ab}][L^b][B^{ab}]^T \tag{5-8}$$

where equation (2-19) has been used for both (J^a) and (J^b).

Onsager (1931a,b) showed that with the "proper" choice of fluxes and forces $[L^a]$ is symmetric;

$$[L^a] = [L^a]^T \qquad (L_{ik}^a = L_{ki}^a, \ i\neq k; \ i,k = 1, 2, \ldots, n-1) \ . \tag{5-9}$$

The expression, (5-4), for the entropy production may be constructed from a straightforward and reasonable extension of the first and second laws of thermodynamics to continuous processes (de Groot and Mazur, 1962, Ch. 2,3). The deduction of the linear relations, (2-19), from (5-4) is equally straightforward, although the domain over which the assumption of linearity is reasonable can only be determined by experience (the same is true of Fick's law). Unfortunately, in Onsager's analysis, the "proper" choice of flux and forces and the subsequent deduction of the symmetry relations, (5-9), is based in part on microscopic arguments (the symmetry relations are deduced from the reversibility of microscopic fluctuations about the equilibrium state of thermodynamic parameters). In contrast, the thermodynamic fluxes and forces are macroscopic quantities, that, often, may be chosen in a variety of ways. The choice of fluxes and forces, the symmetry relations and the transformation properties of the L-coefficients have remained controversial; a clear discussion of the problem, with special reference to the experimental verification of relations (5-9), has been given by Miller (1974; see also the discussion following his paper).

It is fair to say for isothermal, isobaric diffusion that in the range in which the phenomenological equations (2-19) are linear, the L-coefficients are symmetric whenever they link independent fluxes and independent thermodynamic forces of the form of equation (5-3). Moreover, their symmetry is preserved by the transformations (5-8), a set of transformations that follow from the invariance of $T\sigma$. In (5-8), if $[L^b]$ is symmetric, then so is $[L^a]$, as can easily be verified by writing out the terms of (5-9) for n = 3.

The coefficients L^v_{ik} are especially important in many diffusion studies (*e.g.*, Miller, 1966, 1967a). The calculation of the L^v_{ik} from measured values of D^v_{ik} is best carried in two steps: $D^v_{ik} \rightarrow D^o_{ik} \rightarrow L^o_{ik}$. The equations needed for the first step are (4-15) and (4-16). We have, in a solvent-fixed reference frame, from equations (2-10) and (2-18)

$$(J^o) = [D^o](y) = [L^o](X^o) \ . \qquad (5\text{-}10)$$

But, from (5-3),

$$(X^o) = [A^o](\partial\mu/\partial y) \qquad (5\text{-}11)$$

where $(\partial\mu/\partial y)$ is a vector with elements $(\partial\mu_i/\partial y)$. Chemical potential gradients $(\partial\mu/\partial y)$ and concentration gradients (y) are related by equation

$$(\partial\mu/\partial y) = [\mu](y) \ . \qquad (5\text{-}12)$$

Thus, from the last three equations,

$$[D^o] = [L^o][A^o][\mu] \ . \qquad (5\text{-}13)$$

Successive post-multiplication of both sides of equation (5-13) by $[\mu]^{-1}$ and $[A^o]^{-1}$ gives

$$[L^o] = [D^o][\mu]^{-1}[A^o]^{-1} \ . \qquad (5\text{-}14)$$

That $[A^o]$ is a unit matrix is easily proved by the substitution of $\bar{a}_i = \delta_{ik}$ $(i = 1, 2, \ldots, n-1)$ into equation (5-2). Thus, equation (5-14) simplifies to

$$[L^o] = [D^o][\mu]^{-1} \ . \qquad (5\text{-}15)$$

To calculate L^o_{ik}, the differentials $\partial\mu_i/\partial c_j$ that make up $[\mu]$ must be evaluated. The chemical potential of a salt i (*e.g.*, NaCl, Fe_2SiO_4) is (Robinson and Stokes, 1959, Ch. 2)

$$\mu_i = \mu_i^o + RT\mathit{ln}a_i \qquad (5\text{-}16)$$

where μ_i^o is a constant (T,P constant) and a_i is the activity of i. The dissociation reaction of the *ith* electrolyte may be represented by

$$(C_i)_{r_{ic}}A_{r_{ia}} = r_{ic}C_i^{z_i} + r_{ia}A^{z_a} \ . \qquad (5\text{-}17)$$

The stoichiometric coefficients (r_{ic} and r_{ia}) and the signed valences (z_i and z_a) for the dissociation of $CaCl_2$ are: $r_{ic} = 1$; $r_{ia} = 2$; $z_i = +2$; $z_a = -1$. The activity of i is then (Robinson and Stokes, 1959; $r_i = r_{ic} + r_{ia}$).

$$a_i = c_{ic}^{r_{ic}} c_{ia}^{r_{ic}} (y_i^*)^{r_i} \qquad (5\text{-}18)$$

where y_i^* is the mean ion activity coefficient of i on a molar concentration scale. With the additional relations

$$c_{ic} = r_{ic}c_i \qquad (5\text{-}19)$$

and (for a mixture with a common anion)

$$c_a = \Sigma r_{ia}c_i \ , \qquad (5\text{-}20)$$

substitution of equation (5-18) and (5-16) and differentiation gives (Miller, 1959)

$$\mu_{ij} = RT \frac{r_{ic}\delta_{ij}}{c_j} + \frac{r_{ia}r_{ja}}{\Sigma r_{ia}c_i} + r_i \frac{\partial \mathit{ln}y_i^*}{\partial c_j} \ . \qquad (5\text{-}21)$$

From this point on the discussion will be confined to binary and ternary aqueous electrolytes, although the principles involved apply to all electrolyte mixtures.

For a binary aqueous electrolyte, equation (5-21) reduces to (electrolyte = component 1)

$$\mu_{11} = \frac{r_1}{c_1}(1 + c_1 d\ln y_i^*/dc_1)RT \ . \tag{5-22}$$

Clearly, y_1^* must be known as a function of c_1 to find μ_{11}.

Mean ion molal activity coefficients (γ_1^*) are obtained from measurements of the osmotic coefficient Φ of the solution at different molalities (m_1) and integration of the Gibbs-Duhem equation (*cf.* Rard and Miller, 1980 for Na_2SO_4 and $MgSO_4$ solutions; Rard *et al.*, 1977 for $CaCl_2$ solutions). The coefficients may be fitted to a polynomial series containing a Debye-Hückel constant (A):

$$\Phi = 1 - (A/3)\sqrt{m_1} + \Sigma A_i m_1^{r_i} \ . \tag{5-23}$$

Here r_i are powers to which the least squares coefficients A_i are to be raised (and not, as elsewhere in this text, stoichiometric coefficients). Then,

$$\ln\gamma_1^* = -A\sqrt{m} + \sum_i A_i \frac{r_i+1}{r_i} m_1^{r_i} \ , \tag{5-24}$$

an expression that is easily differentiated with respect to m_1. Kirkwood and Oppenheim (1961) have set out the principles needed to convert activity coefficients from one concentration scale to another. The conversion of γ_1^* to y_1^* is accomplished by (Miller, 1959)

$$y_1^* = (1000\rho_2^o/c_2M_2)\gamma_1^* \tag{5-25}$$

(ρ_2^o = density of pure water). With equation (5-25), the term in parentheses in (5-22) may be evaluated as follows (Miller, 1966, Appendix 2):

$$\frac{d\ln y_1^*}{dc_1} = \frac{\bar{V}_1}{c_2\bar{V}_2} + \frac{dm_1}{dc_1}\frac{d\ln\gamma_1^*}{dm_1} \tag{5-26}$$

which, with

$$\frac{dm_1}{dc_1} = \frac{m_1}{c_1}\left(1 + \frac{c_1\bar{V}_1}{c_2\bar{V}_2}\right) \tag{5-27}$$

becomes

$$(1 + c_1\frac{d\ln y_1^*}{dc_1}) = 1 + \frac{c_1\bar{V}_1}{c_2\bar{V}_2}(1 + m_1 d\ln\gamma_1^*/dm_1) \ . \tag{5-28}$$

For a binary solution, equation (5-14) simplifies to

$$L^o = D^o/\mu_{11} = D^o c_1/(1 + c_1 d\ln y_1^*/dc_1)RTr_1 \ . \tag{5-29}$$

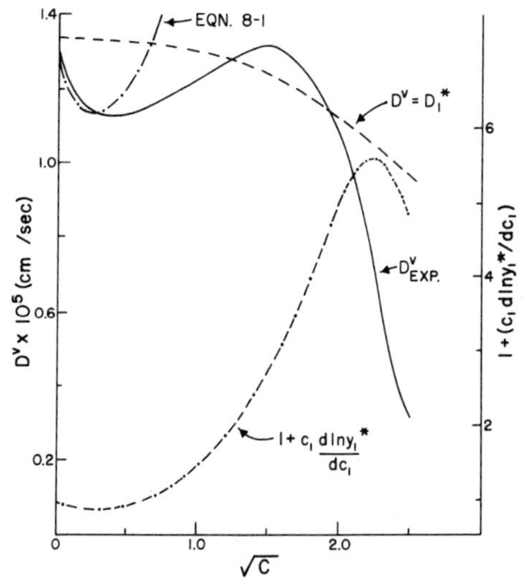

Figure 5. Estimation of D^V in the system $CaCl_2$-H_2O by the Nernst-Hartley equation 8-1 and the approximation $D^V = D_1^*$. Experimental values of D^V are from Rard and Miller (1979a). Values of the thermodynamic factor for $CaCl_2$, $1 + c_1\, d\ln y_1/dc_1$, are calculated from Rard *et al.*, (1977).

A plot of the thermodynamic factor ($1 + c_1 d\ln y_1/dc_1$) versus c_1 for $CaCl_2$ is given in Figure 5.

The osmotic coefficients and the activity coefficients in a ternary mixture are found from an expression for the excess Gibbs function developed by Scatchard (1961). The expression combines the contributions from the two limiting binary salt-water systems and mixing terms. The result is a lengthy polynomial series comprised of Debye-Hückel terms and three sets of fitting coefficients (*e.g.*, Rush and Robinson, 1968). Although algebraically complex, the expression for $\ln\gamma_i^*$ is easily differentiated to obtain $d\ln\gamma_i^*/dm_j$; the equation needed for the conversion to $d\ln y_i^*/dc_j$ has been derived by Miller (1959). The Debye-Hückel constants and fitting coefficients for about 30 ternary (aqueous) systems have been compiled by Rush (1969) and Hamer and Wu (1972).

SECTION 6: THE DIAGONALIZATION OF $[D^V]$

The proof that $[D^V]$ may always be diagonalized depends on a demonstration that $[D^n]$ may be constructed from two real, symmetric and positive matrices $[L^n]$ and $[G]$ (Kirkaldy, 1959, 1970; Kirkaldy *et al.*, 1963a; Cullinan, 1965). The reason for beginning with $[D^n]$ rather than $[D^V]$ will become apparent below [equation (6-11)]. The derivations below, although they follow those of Cullinan (1965), are not dependent on special combinations of reference frames and concentration units and may be applied to either solids or liquids.

The substitution of

$$(J^n) = [L^n](X^n) \quad (6\text{-}1)$$

into the expression for the entropy production [see equation (5-1)]

$$T\sigma = (J^n)(X^n) \quad (6\text{-}2)$$

yields a symmetric quadratic form (Hildebrand, 1952, p. 36)

$$T\sigma = (X^n)^T[L^n](X^n) \ . \quad (6\text{-}3)$$

By virtue of the second law of thermodynamics, the entropy production is zero at equilibrium and positive for all natural processes. Thus,

$$(X^n)^T[L^n](X^n) \geq 0 \ . \quad (6\text{-}4)$$

The conditions for equation (6-4) to be positive, definite are (Hildebrand, 1965, p. 52; Wilkinson, 1965, p. 28-30)

$$\sum_{i=1}^{n-1} L^n_{ii} > 0; \qquad \sum_{i,k=1}^{n} L^n_{ii}L^n_{kk} - L^n_{ik}L^n_{ki} > 0; \qquad \det|L_{ik}| > 0 \quad (6\text{-}5)$$

where det stands for the determinant of the matrix. When $n = 3$, the second conditions requires, since $L^n_{12} = L^n_{21}$ [equation (5-9)], that $L^n_{11}L^n_{22} > 0$. Then since by the first condition $L^n_{11} + L^n_{22} > 0$, L^n_{11} and L^n_{22} must each be positive. On the other hand, L^n_{12} $(=L^n_{21})$ may be positive or negative. All reported values of L^n_{11} and L^n_{22} are indeed positive (Miller, 1959, 1965; Vignes and Sabatier, 1969; Revzin, 1972).

The second matrix [G] is connected with the molar Gibbs function g and has as elements

$$G_{ik} = \frac{\partial^2 g}{\partial c_i \partial c_k} \qquad (i,k = 1, 2, \ldots, n-1) \quad (6\text{-}6)$$

so that it is obviously symmetric. de Groot and Mazur (1962, p. 246-247) have shown that

$$\frac{\partial^2 g}{\partial c_i \partial c_k} = \sum_{j=1}^{n-1} (\delta_{ij} + c_j/c_n)\mu_{jk} \qquad (i,k = 1, 2, \ldots, n-1) \ . \quad (6\text{-}7)$$

The conditions that a mixture remain stable (i.e., be at the "bottom of a well") during diffusion can also be expressed in terms of G_{ik} (Prigogine and Defay, 1954; de Groot and Mazur, 1962, Appendix 2): i.e.,

$$\sum_{i=1}^{n-1} \sum_{k=1}^{n-1} dc_i G_{ik} dc_k \geq 0 \ . \quad (6\text{-}8)$$

Substitution for (X^n) in (6-1) from equation (5-3) gives

$$(J^n) = -[L^n][A^n](\partial\mu/\partial y) \qquad (6\text{-}9)$$

or, with equation (2-21)

$$(J^n) = -[L^n][A^n][\mu](y) \ . \qquad (6\text{-}10)$$

For a molecular reference frame $\bar{a}_i = x_i$, and hence, using equations (5-2) and (6-7)

$$[A^n][\mu] = \sum_{j=1}^{n-1} A^n_{ij}\mu_{jk} \qquad (i,k = 1, 2, \ldots, n-1)$$

$$= \sum_{j}^{n-1} (\delta_{ik} + c_k/c_n)\mu_{jk} = G_{ik} \ . \qquad (6\text{-}11)$$

Note that equation (6-11) is only true for a molecular reference frame if molar-based quantities are used, and is the basis for the form of equation (6-1). For mass-based quantities, the equation analogous to (6-11) holds for a mass-fixed reference frame.

Substitution of equation (6-11) into (6-10) yields

$$(J^n) = -[L^n][G](y) \qquad (6\text{-}12)$$

and, by comparison with equation (4-19),

$$[\bar{D}^n] = [L^n][G] \qquad (6\text{-}13)$$

and

$$(X) = [G](y) \ . \qquad (6\text{-}14)$$

By combining the transpose of (6-13) with the symmetry relations $[G] = [G]^T$, we obtain

$$(X^n)^T = (y)^T[G] \qquad (6\text{-}15)$$

and finally, with equations (6-15) and (6-1)

$$(y)^T[G][L^n][G](y) \geq 0 \ . \qquad (6\text{-}16)$$

The symmetry of $[G]$ and $[L^n]$ ensures the symmetry of the combination $[G][L^n][G]$.

If the stability conditions (6-8) are restated as

$$(y)^T[G](y) \geq 0 \ , \qquad (6\text{-}17)$$

then equations (6-16) and (6-17) are quadratic forms in (y) with associated real, symmetric matrices $[L]$ and $[G]$, both of which are positive definite. New variables $\bar{y}$ can be introduced by a transformation (Hildebrand, 1962, p. 48; Korn and Korn, 1961, p. 371)

$$(y) = [S](\bar{y}) \tag{6-18}$$

such that the quadratic forms in equations (6-16) and (6-17) simplify to a sum of squares (i.e., no $\bar{y}_i\bar{y}_j$ terms) of the $\bar{y}_i$. Methods for obtaining the transformation matrix, [S], are given below. As a result of this transformation, relation (6-16) is reduced to

$$(y)^T[G][L^n][G](y) = (\bar{y})^T[\Lambda^n](\bar{y}) \tag{6-19}$$

where $[\Lambda^n]$ is a diagonal matrix composed of the characteristic roots of $[L^n][G]$ (see Korn and Korn, 1961, p. 371).* Because the matrix $[G][L^n][G]$ associated with the quadratic form (6-19) is real and symmetric, the characteristic roots are real, and, if (6-19) is to be positive, all positive ($\Lambda_i^n > 0$, i = 1, 2, ..., n-1; Hildebrand, 1965, p. 54). But by virtue of (6-13), the Λ_i^n are also the characteristic roots of $\bar{D}^n$. And furthermore, since similar matrices have the same characteristic equation and therefore the same characteristic roots, the Λ_i^n are the characteristic roots of $[D^V]$ (and, of course, the similar matrices $[\bar{D}^m]$ and $[\bar{D}^o]$; Hildebrand, 1965, p. 54).

In practice, the diagonalization of $[D^V]$ is carried out by the similarity transformation (Hildebrand, 1952, p. 42; Wilkinson, 1965, p. 6)

$$[S]^{-1}[D^V][S] = [\Lambda] \tag{6-20}$$

which because [S] is nonsingular [equation (6-18)] is always possible. When n = 3, Λ_1 and Λ_2 are the roots of the determinantal equation

$$\begin{vmatrix} D_{11}^V-\Lambda & D_{12}^V \\ D_{21}^V & D_{22}^V-\Lambda \end{vmatrix} = 0 \tag{6-21}$$

or

$$(D_{11}^V-\Lambda)(D_{22}^V-\Lambda) - D_{12}^V D_{21}^V = 0 \,. \tag{6-22}$$

The two roots of equation (6-22) are (Kirkaldy, 1970)

$$\Lambda_1,\Lambda_2 = \tfrac{1}{2}\left[(D_{11}^V + D_{22}^V) \pm [(D_{11}^V + D_{22}^V)^2 - 4(D_{11}^V D_{22}^V - D_{12}^V D_{21}^V)]^{\frac{1}{2}}\right] \tag{6-23}$$

For Λ_1 and Λ_2 each to be real and positive requires that

---

*Korn, G. A. and Korn, T. M. (1961) *Mathematical Handbook for Scientists and Engineers*. McGraw Hill, New York.

$$D^V_{11} + D^V_{22} > 0$$

$$D^V_{11}D^V_{22} - D^V_{12}D^V_{21} > 0 \qquad (6\text{-}24)$$

$$(D^V_{11} + D^V_{22})^2 \geq 4(D^V_{11}D^V_{22} - D^V_{12}D^V_{21}) \;;$$

the first two conditions ensure that the roots are positive and the last condition that they are real. Gupta and Cooper (1971) have stated the equivalent conditions for a quaternary system.

The data collected so far in ternary systems are consistent with equations (6-24) (see list of references near the end of Section 3).

Kirkaldy (1970) has also re-stated the conditions (6-25) on the assumption that D^V_{11} and D^V_{22} are individually positive, whereas equation (6-25) requires only that their sum be positive. D^V_{11} is positive in all the systems studied thus far, D^V_{22} is positive in all of these systems except for $K_2O\text{-}SrO\text{-}SiO_2$ (Varshneya and Cooper, 1972). Many of the systems also violate an ancillary condition that follows from Kirkaldy's assumption, namely that D^V_{12} and D^V_{21} should generally have the same sign. It appears that equation (6-24) is the only general statement that may be derived from equation (6-19).

The k*th* column of [S] may be found by solving the equations (Hildebrand, 1965; Hollingsworth, 1967)

$$\sum_{j=1}^{n-1} (D^V_{ij} - \delta_{ij}\Lambda_k)S_{jk} = 0 \,. \qquad (6\text{-}25)$$

For example, the first column (k = 1) is given by (n = 3)

$$(D^V_{11} - \Lambda_1)S_{11} + D^V_{12}S_{21} = 0$$

$$D^V_{21}S_{11} + (D^V_{22} - \Lambda_1)S_{21} = 0 \qquad (6\text{-}26)$$

where Λ_1 is obtained from equation (6-22). Each column of $[S]$-S_{1k}, S_{2k}, ..., $S_{(n-1)k}$- is a characteristic vector (eigenvector).

If $[D^V]$ is constant, that is, if $[D^V]$ and (y) commute, then premultiplication of equation (2-22) by $[S]^{-1}$ and inclusion of the unit matrix $[S][S]^{-1}$ on the right-hand side yields

$$[S]^{-1}\frac{\partial(c)}{\partial t} = [S]^{-1}[D^V][S][S]^{-1}\frac{\partial^2(c)}{\partial y^2} \qquad (6\text{-}27)$$

or, with equation (6-20),

$$\frac{\partial \bar{c}_i}{\partial t} = \Lambda_i(\partial^2 \bar{c}_i/\partial y^2) \ . \qquad (6\text{-}28)$$

Here (c) is a vector with components, c_1, c_2, ..., c_{n-1}, and

$$\bar{c}_i = \sum_{k=1}^{n-1} S_{ik}^{-1} c_k \ . \qquad (6\text{-}29)$$

Once the new concentrations $\bar{c}_i$ have been calculated, equation (6-28) is analogous in every way to (2-7) -- Fick's second law for diffusion where D^V is constant -- and may be solved in the same way (Gupta and Cooper, 1971; Cullinan, 1965; Toor, 1964a,b). When D^V is constant, (2-7), and hence equation (6-28), usually have exact solutions; in contrast, where D^V is concentration dependent, the partial differential equations (2-8) and (2-23) must generally be solved by numerical techniques. Thus, the diagonalization of D^V and the uncoupling of equation (2-23) greatly simplifies the treatment of diffusion processes. Λ_i is a mutual diffusion coefficient for a mixture of i and the dependent component n; the multicomponent mixture may be treated as a set of unrelated (uncoupled) binary mixtures.

Lasaga (1979b) has discussed the properties of the roots Λ_i and the application of the method to exchange processes in silicates (Toor, 1964b also discusses diffusion and chemical exchange). Cooper and Gupta (1971) and Christensen (1977) have applied the methods above to diffusion-controlled growth in ceramic systems. Loomis (1978a,b) has attempted to simulate the effect of diffusion on compositional zoning in garnets with a program based on equation (6-28).

As shown by Gupta and Cooper (1971), Varshneya and Cooper (1972), and Cooper (1974), normalized eigenvectors may be used as basis vectors for an examination of the geometric properties of diffusion paths.

Where $[D^V]$ is concentration dependent and therefore $[D^V]$ and (y) do not commute, the uncoupling of equation (2-23) is impossible.

SECTION 7: IONIC FLUXES

The chemical potentials of cations μ_{ic} and the anion* (μ_a) are defined by the relationship

----

*Note, once again, that we are only treating mixtures with one anion.

$$\mu_i = r_{ic}\mu_{ic} + r_{ia}\mu_a , \tag{7-1}$$

since the chemical potential of an isolated ion is not a physically measurable quantity (Guggenheim, 1967; Denbigh, 1966). It is permissible to write equations containing the terms $d\mu_{ic}/dy$ and $d\mu_a/dy$ if the total system of equations are eventually combined in a way that produces the form (7-1). In other words, the set of equations must reflect local equilibrium in the solution; arguments concerning the diffusion potential (Section 1) apply equally to fluxes and thermodynamic forces, and for exactly the same reasons.

The simplest and most elegant method of dealing with ionic quantities is that laid out by Miller (1967a,b). Although developed primarily for aqueous electrolytes, Miller's equations and methods may be adapted quite easily to the treatment of crystals and ionic melts.

The entropy production for diffusion of ionic components is, in a solvent-fixed reference frame,

$$T\sigma = \sum_{i=1}^{m} j_i^o x_i^o \tag{7-2}$$

where the summation is over the m ionic species in the mixture. The thermodynamic forces are

$$x_i^o = -[\partial\mu_{ic}/\partial y + z_i F(\partial\phi/\partial y)] \tag{7-3}$$

for the m-1 cations, and

$$x_m^o = -[\partial\mu_m/\partial y + z_m F(\partial\phi/\partial y)] \tag{7-4}$$

for the common anion. Here ϕ is an electrical potential in volts and F is the Faraday constant. By analogy with equation (5-16)

$$\mu_{ic} = \mu_{ic}^o + RT\,ln\,y_{ic}c_{ic} \tag{7-5}$$

(Robinson and Stokes, 1959; a similar equation may be written for the anion).

The linear relations analogous to equation (5-4) are

$$j_i^o = \sum_{j=1}^{m} \ell_{ij}^o x_j^o \qquad (i = 1, 2, \ldots, m) \tag{7-6}$$

where the j_i^o are ionic fluxes in a solvent-fixed reference frame. It should be noted that through equations (7-4) and (7-5) the fluxes are specified in terms of three gradients: (1) $RT\,d\,ln\,c_{ic}/dy$; (2) $RT\,d\,ln\,y_{ic}/dy$, a term that allows for non-ideal mixing of the ions; (3) $d\phi/dy$, a term that takes account of the electrical interactions of the ions.

Setting the flux of water to zero in an aqueous electrolyte establishes a solvent-fixed reference frame and an independent set of components for equation (7-6). As a result, the matrix of ℓ^o_{ij} coefficients is symmetric (Miller, 1967a).

The number of ionic components in a crystal can often be reduced by considering diffusion on sub-lattices (Howard and Lidiard, 1964; Anderson, 1976). For example, the exchange of Fe^{2+}, Mn^{2+} and Mg^{2+} in the ternary mixture $(Fe,Mn,Mg)_3Al_2Si_3O_{12}$ takes place on a "cation" sub-lattice defined by the array of dodecahedral sites in the garnet. The tetrahedral (Si), octahedral (Al) and oxygen sites may be combined to form a single "anion" sub-lattice. Taking the flux of the anion $Al_2Si_3O_{12}^{6-}$ as zero defines a lattice-fixed reference frame and allows equation (7-6) to be applied to the diffusion of the three remaining cations [thereby reducing the number of ℓ^o_{ij} coefficients in equation (6-5) from 36 to 9]. Miller (1966, footnote 18) has argued that the matrix ℓ^o_{ij}, as defined by equation (7-6), should be symmetric, even though the fluxes and forces are not strictly independent [see equation (7-7)]. Elementary kinetic calculations (*e.g.*, Anderson and Graf, 1978) do yield a symmetric matrix.

During diffusion, the total electrical current I^o in the solvent (lattice)-fixed frame must be zero, i.e.,

$$I^o = \sum_{k=1}^{m} z_k F j^o_k = 0 \; . \qquad (7\text{-}7)$$

The application of equation (6-6) to diffusion in quasi-crystalline or crystalline materials takes a simple form when the unit cell parameters remain constant (either because of ideal mixing or small variations in concentrations). All of the sub-lattices are coupled electrically; if, however, the number of lattice sites and ions (including, if present, ionized defects) on a particular sub-lattice are conserved during diffusion, then equation (7-7) will automatically ensure electrical neutrality. Where unit cell parameters vary substantially, it may be necessary initially to re-define concentrations and distance units (Birchenall *et al.*, 1948; Kirkaldy, 1957; Anderson and Buckley, 1974; Anderson, 1976; Brady, 1975a).

It is convenient at this point to introduce a common variable, μ_i', for the chemical potentials of all m ions (cations *and* anion); thus, equations (7-3) and (7-4) may be combined to give

$$x_i^o = -[\partial\mu_i'/\partial y + z_i F(\partial\phi/\partial y)] \qquad (i = 1, 2, \ldots, m) \ . \qquad (7\text{-}8)$$

Solutions of (7-6) that are consistent with equation (7-7) have been worked out by Miller (1967a, p. 618) for aqueous electrolytes. Substitution of equation (7-8) into (7-6), followed by insertion of the resulting expression for j_i^o into equation (7-7), enables a calculation of the term $\partial\phi/\partial y$. Knowing $\partial\phi/\partial y$, x_i^o in equation (7-8) can be calculated and therefore j_i^o in equation (7-6) can also be computed. For ternary solution with a common anion (i = 1,2,3) the result is

$$j_i^o = -\frac{\sum_{j=1}^{3}\sum_{k=1}^{3}\sum_{\ell=1}^{3} \ell_{ij}^o \ell_{k\ell}^o z_k [z_\ell(\partial\mu_j'/\partial y) - z_j(\partial\mu_\ell'/\partial y)]}{\sum_{k=1}^{3}\sum_{\ell=1}^{3} z_k \ell_{k\ell}^o z_\ell} \ . \qquad (7\text{-}9)$$

Charge conservation (i.e., overall the mixture must be made up of molecular components, Section 1) requires that

$$r_{ic} z_i + r_{ia} z_a = 0 \qquad (i = 1 \text{ or } 2) \ . \qquad (7\text{-}10)$$

The important point is that with the last equation and (7-1), it may be easily shown that the terms $(z_\ell \partial\mu_j'/\partial y - z_j \partial\mu_\ell'/\partial y)$ for ionic components in equation (7-8) that are not zero (ℓ = j) are equal to $(z_\ell/r_{ic})\partial\mu_i/\partial y$; the chemical potential gradients of molecular components.

Taking the anion as component 3, for example,

$$z_{1c}(\partial\mu_{2c}/\partial y) - z_2(\partial\mu_{1c}/\partial y) = \frac{z_1}{r_{2c}}\,\partial\mu_2/\partial y - \frac{z_2}{r_{1c}}\,\partial\mu_1/\partial y \ . \qquad (7\text{-}11)$$

Thus equation (7-6) successfully combines the fluxes and thermodynamic forces in molecular combinations, for (Miller, 1967a)

$$J_i^o = \frac{j_i}{r_{ic}} = -\frac{\sum_{j=1}^{2}\sum_{k=1}^{3}\sum_{\ell=1}^{3} \frac{z_k z_\ell}{r_{ic} r_{jc}} (\ell_{ij}^o \ell_{k\ell}^o - \ell_{i\ell}^o \ell_{kj}^o)\frac{\partial\mu_j}{\partial y}}{\sum_{k=1}^{3}\sum_{\ell=1}^{3} z_k \ell_{k\ell}^o z_\ell} \qquad (7\text{-}12)$$

Comparison with

$$J_i^o = -\sum_{j=1}^{2} L_{ij}^o \partial\mu_j/\partial y \tag{7-13}$$

[i.e., equation (2-18) for a solvent-fixed frame] gives

$$L_{ij}^o = \frac{\sum_{k=1}^{3}\sum_{\ell=1}^{3} \frac{z_k z_\ell}{r_{ic} r_{jc}} (\ell_{ij}^o \ell_{k\ell}^o - \ell_{i\ell}^o \ell_{kj}^o)}{\sum_{k=1}^{3}\sum_{\ell=1}^{3} z_k \ell_{k\ell}^o z_\ell} \tag{7-14}$$

Miller's solution of equation (7-6) is easily modified to deal with diffusion via ionized vacancies -- the mechanism suggested for Fe-Mg exchange in olivine by Buening and Buseck (1973). Assuming that the anion sub-lattice is the dependent component m, the fluxes on the cation sub-lattice are

$$j_i^o = -\sum_{k=1}^{m-1} \ell_{ik}^o [(\partial\mu_{kc}/\partial y - \partial\mu_{v=}/\partial y) + z_k F(\partial\phi/\partial y)] \tag{7-15}$$

where $z_{v=}$ and $\mu_{v=}$ are the signed valence and chemical potential of the ionized vacancy (Anderson, 1976). The total electrical current on the cation sub-lattice is

$$I^o = 0 = \sum_{i=1}^{m-1} z_{ic} F j_{ic}^o + z_{v=} F j_{v=}^o \tag{7-16}$$

and, hence

$$j_{ic}^o = \frac{\sum_{j=1}^{m-1}\sum_{\ell=1}^{m-1}\sum_{k=1}^{m-1} (z_{kc}-z_{v=}) \ell_{ij}^o \ell_{k\ell}^o z_{\ell c} \frac{\partial(\mu_{jc}-\mu_{v=})}{\partial y} - z_{jc}\frac{\partial(\mu_{jc}-\mu_{v=})}{\partial y}}{\sum_{k=1}^{m-1}\sum_{\ell=1}^{m-1} (z_{kc}-z_{v=}) \ell_{k\ell}^o z_{\ell c}} . \tag{7-17}$$

If diffusion occurs solely by interchange of cations (without vacancies), $z_{v=}$ equal zero and equation (7-17) reduces to (7-12).

Exchange processes in silicates often take place among ions of the same valence (*e.g.*, Fe^{2+}-Mg^{2+}-Mn^{2+} or K^+-Na^+). In these circumstances

$$z_{1c} = z_{2c} = \ldots = z_{(m-1)c} \tag{7-18}$$

and

$$r_{1c} = r_{2c} = \ldots = r_{(m-1)c} \tag{7-19}$$

and the terms $(\partial\mu_{v=}/\partial y)$ and $(z_{kc}-z_{v=})$ cancel out. With equations (7-1) and (7-9), (7-17) becomes

$$z_\ell(\partial\mu_{jc}/\partial y) - z_j(\partial\mu_{\ell c}/\partial y) = (\partial\mu_j/\partial y - \partial\mu_\ell/\partial y) \tag{7-20}$$

and

$$J_i^o = \frac{\sum_{j=1}^{m-1} \sum_{k=1}^{m-1} \sum_{\ell=1}^{m-1} \ell_{ij}^o \ell_{k\ell}^o z_\ell (\partial\mu_j/\partial y - \partial\mu_\ell/\partial y)}{r_{ic} \sum_{k=1}^{m-1} \sum_{=1}^{m-1} \ell_{k\ell} z_\ell} . \quad (7\text{-}21)$$

The ℓ_{ij}^o are fundamental transport quantities, since they directly reflect the interactions among ions during diffusion and conduction. Unfortunately, they are also difficult to obtain; Miller (1966, 1967a) has derived the equations needed to calculate the ℓ_{ij}^o from the equivalent conductance, transference numbers and L-coefficients in binary and ternary aqueous electrolytes. There appear to be no data available on ℓ-coefficients in crystalline or fused electrolytes.

The significance of the coefficients ℓ_{ij}^o/N_i (where N, the equivalent concentration, is equal to $r_{ic} z_i c_i$) may be summarized, very briefly, as follows (Miller, 1966):

(1) The on-diagonal coefficients ℓ_{ii}^o/N_i are chiefly a measure of the mobility an ion would have in the absence of cation-anion interactions. The limiting value at infinite dilution is (by extrapolation) a constant property of each ion, no matter what other ions are present. In binary solutions, the calculated values (all chloride solutions) vary slowly with ionic strength (Miller, 1966, Figs. 1 and 2).

(2) The off-diagonal coefficients mainly represent cation-anion interactions, which appear to be coulombic in origin. Although zero at infinite dilution, the coefficients increase rapidly with ionic strength in the measured binary systems (Miller, 1966, Fig. 3). Clearly, from the nature of equations (7-15) and (5-13), it is not possible to interpret the D_{ij}^o or L_{ij}^o in terms of ionic interactions, since they represent complex combinations of ℓ_{ii}^o and ℓ_{ij}^o $(i \neq j)$.

SECTION 8: ESTIMATION OF DIFFUSION COEFFICIENTS

Because of the burden of experimental work in binary, and especially multicomponent systems, methods for estimating diffusion coefficients have long remained attractive. The two must useful quantities are the tracer diffusion coefficient, which may be determined in either solids or liquids, and, in aqueous electrolytes, the limiting ionic conductance (λ_i^o). The latter is the equivalent conductance extrapolated to infinite dilution; values of λ_i^o at 293.15°K have been tabulated by Robinson and Stokes (1959) and various handbooks of physico-chemical data. Both the tracer diffusion coefficient (Section 3) and the limiting ionic conductance (Robinson and Stokes, 1959, Ch. 6) measure the random motion of an ion in the absence of concentration gradients [e.g., D*, as defined by equation (3-7), is proportional to the average jump frequency]. Thus, the problem, simply stated, is to predict the concentration dependence of the D_{ik}^v at various pressures and temperatures; for D*, this is equivalent to predicting the compositional dependence of T_{12} and T_{21} (Section 3); since diffusion is a short-range phenomenon, the reasoning behind equation (3-7) may be, and has been, applied fairly successfully to liquids.

Methods for computing diffusion coefficients in aqueous electrolytes from limiting ionic conductances or tracer diffusion coefficients have been published by Albright and Mills (1965), Anderson and Graf (1978), Lane and Kirkaldy (1965, 1966), Lasaga (1979), Miller (1967a) and Wendt (1965). The forerunner of the models that depend on λ_i^o, applicable only to binary solutions, is the Nernst-Hartley equation (Hartley, 1931; Robinson and Stokes, 1959, p. 286) which states (salt = 1, H_2O = 2, cation = c, anion = a)

$$D^o = \frac{RT}{F^2} \frac{z_c + |z_a|}{|z_c + z_a|} \frac{\lambda_c^o \lambda_a^o}{\lambda_c^o + \lambda_a^o} (1 + c_1 d\ln y_1^*/dc_1) \qquad (8-1)$$

or

$$D^o = D_1^o (1 + c_1 + d\ln y_1^*/dc_1) \qquad (8-2)$$

where D_1^o is equal to the first three terms on the right-hand side of (8-1). Note that D_1^o is the value of D^o at infinite dilution [i.e., when the thermodynamic term $c_1 d\ln y_1^*/dc_1$ is equal to zero in equation (8-1)]. Note that (8-1) specifies a coefficient in a solvent-fixed frame (Miller, 1966), a distinction that is often not made, but which may become important in more concentrated solutions, if D^o in equation (8-1) is compared with D^v. Note, that as before, D^o is a mutual diffusion coefficient.

The tracer diffusion coefficients D_c^* (cation) and D_a^* (anion) at infinite dilution are related to λ_c^o and λ_a^o by (Robinson and Stokes, 1959, p. 317)

$$D_c^* = \frac{RT\lambda_c^o}{|z_c|F^2} \quad \text{and} \quad D_a^* = \frac{RT\lambda_a^o}{|z_a|F^2} . \tag{8-3}$$

Hence, with equations (8-1) and (8-2),

$$D_1^* = \frac{RT}{F^2} \frac{(z_c + |z_a|)D_c^* D_a^*}{z_c D_c^* + z_a D_a^*} = D_1^o . \tag{8-4}$$

Li and Gregory (1974) have listed values of D_c^* and D_a^*.

A simple estimate may be made by setting D^o equal to D_1^o or D_1^* (i.e., by ignoring the term $c_1 d\ln y_1^*/dc_1$; Li and Gregory, 1974; Lasaga, 1979a). Thus, on a plot of D^o versus c_1 is approximated by a straight line or on a plot of D^V versus c_1 by a slightly curved line (Fig. 5). Reasonable estimates will emerge wherever the total concentration dependence of D^o is small. As noted by Li and Gregory, and is obvious from an inspection of Figure 3, the maximum difference will be small in a NaCl solution [the curve for D^o, if plotted on Figure 3, would fall slightly above the curve for D^V, since $C_2\bar{V}_2 < 1000$ in equation (4-14a) at all concentrations].

The differences between predicted values of D^V [calculated from D^o via equation (4-14a)] and experimental coefficients are more significant in $CaCl_2$ solutions (Fig. 5). In a 0.25 molar solution ($c_1 = 0.5$) the difference is about 20 percent and in a 5.57 molar solution about 95 percent ($D_1^o = 1.334 \times 10^{-5}$ cm^2/sec). The differences will increase steadily in solutions such as Na_2SO_4 (Fig. 3) and Mg_2SO_4, $ZnSO_4$ and $BaCl_2$ (Rard and Miller, 1979b, 1980).

The complete Nernst-Hartley equation (8-1) gives very poor results in $CaCl_2$ solutions above a concentration of 0.06 molar, eventually reaching differences of 400-500 percent in the most concentrated solutions (Fig. 5). As pointed out by Miller, a whole term proportional to ℓ_{12}^o/N is missing from the Nernst-Hartley equation and, since this term is only small very near infinite dilution, the equation can be expected to do poorly in general.

The models of Lane and Kirkaldy (1965) and Wendt (1965) also fare poorly in concentrated solutions; the modified Lane and Kirkaldy model put forward by Anderson and Graf (1978) does much better and tracks the curve D^o versus c_1, quite accurately above 1.5 molar concentration

(Anderson and Graf; in preparation).* At lower concentrations, and in NaCl solutions, their model yields results comparable to the simplified Nernst-Hartley equation (i.e., $D^o = D_1^o$).

Miller (1967a) has tested various models against data at five compositions (0.5 to 3.0 molar total concentration) in the system NaCl-KCl-H_2O and one composition (0.45 molar total concentration) in each of the systems LiCl-KCl-H_2O and LiCl-NaCl-H_2O. By far the best results are produced by Miller's own model (LN), which depends on relationships between the coefficients ℓ_{ik}^o/N in the ternary system and the two limiting salt-water binaries. The Lane and Kirkaldy (1965, 1966) and Wendt (1965) yield inferior, but fairly reasonable, estimates in the less concentrated solutions. All of Miller's estimates are within six percent of the measure coefficients; the Lane and Kirkaldy and Wendt models give differences of up to 25 percent in the coefficients in the 3 molar NaCl-KCl solution. The model of Anderson and Graf (1978) gives better estimates than either of these models and approaches the accuracy of model LN of Miller.

The general application of Miller's model LN will require sufficient data to determine the ℓ_{ij}^o in the associated binary systems; a highly desirable situation, but one that demands a great deal of experimental conductance, diffusion and thermodynamic data. The models of Lane and Kirkaldy (including Anderson and Graf, 1978) and Wendt are, at heart, multi-dimensional extensions of the Nernst-Hartley equation and, as such, depend on limiting ionic conductances and thermodynamic data. In many respects, the system NaCl-KCl-H_2O -- the only system in which there are reliable measurements over a significant range of compositions -- does not provide a very stringest test of any of these models. The coefficients D_{11}^v and D_{22}^v do not vary by more than 13 percent over the compositional range of the data, and the coefficients D_{12}^v and D_{21}^v are small compared to D_{11}^v and D_{22}^v (1-30 percent; Miller, 1965).**

----

*A new method developed by D. L. Graf (Graf, D. L., Anderson, D. E. and Woodhouse, J. B.; submitted to J. Soln. Chem.) gives better than 2% accuracy from infinite dilution to saturation in binary solutions.

**More experimental data are badly needed in binary and, especially, ternary mixtures. The effects of, for example, partial dissociation and of ion-pairing (Lasaga, 1979a) have not been rigorously tested in multicomponent solutions.

Darken (1948) derived the equation

$$D^V = (x_2D_1^* + x_1D_2^*)(1 + d\ln f_1/d\ln x_1) \qquad (8\text{-}5)$$

to calculate the mutual diffusion coefficient from tracer diffusion coefficients in binary metal alloys (components 1 and 2). The general similarity to the Nernst-Hartley equation is obvious. Here, x_i are mole fractions and f_1 is an activity coefficient on a mole fraction scale. Darken's equation is inexact (for a summary, see Howard and Lidiard, 1964, or Manning, 1968) and cannot be applied to ionic crystals in the above form.

Brady (1975), Anderson (1976), and Lasaga (1979) have proposed models that might be applied to ionic crystals (or, by extension ionic melts), none of which has been directly tested. The formulation of analogues of Darken's equations for ionic materials must deal with two problems which have no counterpart in metals.

Firstly, the tracer diffusion coefficient measures the migration of ions and therefore fluxes and chemical potential gradients (or concentration gradients) must be combined to produce a zero electrical potential. When the $\bar{V}_i$ are relatively constant, and diffusion occurs on only one sub-lattice, the methods set out in Section 7 may be employed. The problem is more complex, however, when diffusion on one sub-lattice is coupled to diffusion on another sub-lattice. The exchange of NaSi-CaAl in plagioclase is one example; there is also a possibility that coupling might take place between an ionized vacancy on one sub-lattice and electron holes on another (this possibility was considered, and rejected by Buening and Buseck, 1973, for diffusion in olivine). A re-definition of ionic components, after the manner suggested by Brady (1975b), to combine the sub-lattices may be one way to circumvent this problem.

Secondly, many systems that appear to be binary are in fact ternary systems at the microscopic level: The exchange of Fe_2SiO_4 and Mg_2SiO_4 in olivine, for example, involves the migration of Fe^{2+}, Mg^{2+} and V^{2-} (vacancies). The vacancies must be included in the microscopic flux equations [*e.g.*, equations (7-6)] if the dependence of the mutual diffusion coefficient D^V on P_{O_2} is to be retrieved, as well as the concentration dependence on Fe_2SiO_4 or Mg_2SiO_4 (Anderson, 1976). In metals, the chemical potential of a vacancy is zero at equilibrium, but as is clear from the equilibrium reactions (3-2) and (3-3), the chemical potential of an ionized vacancy is not zero. Rather,

$$\mu_V = \mu_{V^=} + 2\mu_{e+} \ . \qquad (8\text{-}6)$$

And, as has been emphasized in Section 7, only the linear combination of $\mu_{V^=}$ and μ_{e+} has any meaning; there are, therefore, no grounds for setting $\mu_{V^=}$ equal to zero (or, for that matter, any other value).

COMMENTS

The treatment of practical problems involving diffusion reduces to solving one or more differential equations. A wide variety of analytical and numerical solutions already exist for binary mixtures (Crank, 1956).* Where it is possible, the uncoupling of the multicomponent flux equations allows the same solutions to be applied component by component to a mixture containing any number of salts. A limited number of solutions have also been found for coupled flux equations (Kirkaldy, 1970).

Cation exchange processes -- whether they involve exchange between a pore fluid and a mineral, a crystal and a melt, or a pair of crystals in a metamorphic rock -- depend on rates of migration of ions in each of the coexisting phases. In general, we may expect that the rate of diffusion in one of the phases will govern the overall exchange rate. Thus, one of the problems addressed here, the construction of a self-consistent set of flux equations for ionic components in a particular phase, is a necessary first step in finding the simplest, appropriate differential equations.

By way of summary, the following points are important.

(1) The formulation of the phenomenological equations (2-18) for diffusion, by way of the concepts of entropy production and local equilibrium, constitutes a direct and logical extension of classical thermodynamics to the analysis of continuous processes.

(2) Construction of phenomenological equations for ionic components defines a simple and physically realistic method for determining the number of diffusing components.

----

*Especially the revised edition published in 1975.

(3) The methods set out by Miller (1967a) permit ionic fluxes and ionic chemical potential gradients to be combined in a way that ensures local electrical equilibrium throughout the diffusion zone.

(4) In electrolyte mixtures, electrical coupling among ions with different intrinsic mobilities (i.e., different values of ℓ_{ii}) provides a ready explanation of the origin of the off-diagonal coefficients, and hence the coupling of salt fluxes, in the expanded form of Fick's first law (2-10).

(5) Where the phenomenological equations are constructed from independent fluxes and independent forces, and where accurate thermodynamic data are available, experiments have repeatedly shown that the matrix L^a is symmetric. In the case of isothermal, isobaric diffusion, the sometimes controversial criteria established by Onsager (1931a,b) for choosing fluxes and forces do indeed yield symmetrical relations.

(6) Because the matrix L^a and the matrix of thermodynamic factors G are both real, positive-definite and symmetric, it may be shown that the characteristic roots of $[D^V]$ are real and positive and form a complete set even if some roots are repeated.

(7) Traditional methods of measuring $[D^V]$ are burdensome in ternary systems and probably not feasible in systems of four or more components. The difficulties multiply if $[D^V]$ is a function of the activity of oxygen as well as temperature and concentration. A method derived by Gupta and Cooper (1971) promises a substantial reduction in the experimental load; nevertheless, the acquisition of multicomponent diffusion data, especially in solids, presents formidable problems.

(8) Consequently, procedures for estimating multicomponent diffusion coefficients from limiting ionic conductances, tracer diffusion coefficients or ionic conductance coefficients in binary systems are important. Such procedures would allow the determination of the concentration dependence and relative magnitudes of the multicomponent coefficients and, hence, the circumstances in which the fluxes may be uncoupled or the

off-diagonal coefficients safely ignored. Recent tests in binary and ternary aqueous electrolytes suggest that a high degree of accuracy may be attainable.

ACKNOWLEDGMENTS

Much of this paper was written at the Grant Institute of Geology, Edinburgh, and I am grateful to Professor Gordon Craig, FRSE, for his hospitality and kindness during my stay there. I also thank Dr. Robert Freer, Dr. Donald G. Miller, Dr. James Kirkpatrick, David Bieler and, especially, Dr. Donald Graf for various acts of assistance. I am especially indebted to Mrs. Fiona Verth and Mrs. Carol Sanderson for their patience and care with the manuscript. Finally, I would like to thank Dr. Antonio Lasaga for his critical reading of the manuscript.

CHAPTER 6 REFERENCES

Albright, J. G. and Mills, R. (1965) A study of diffusion in the ternary system, labeled Urea-Urea-Water at 25° by measurement of the intradiffusion coefficients of Urea. J. Phys. Chem., 69, 3120-26.

Anderson, D. E. (1976) Diffusion in metamorphic tectonics: lattice-fixed reference frames. Phil. Trans. Royal Soc. Lond., A283, 241-245.

——— and Buckley, G. R. (1974) Modeling of diffusion controlled properties in silicates. In *Geochemical Trnasport and Kinetics,* Hofmann, A. W., Giletti, B. J., Yoder, H. S. Jr., and Yund, R. A., Eds., Carnegie Inst. of Washington Pub. 634, 31-52.

——— and Graf, D. L. (1976) Multicomponent electrolyte diffusion. Ann. Rev. Earth Planet. Sci., 4, 95-121.

——— (1978) Ionic diffusion in naturally-occurring aqueous solutions: use of activity coefficients in transition-state models. Geochim. Cosmochim. Acta, 42, 251-262.

Baluffi, R. W. (1960) On the determination of diffusion coefficients in chemical diffusion. Acta Metall., 8, 871-3.

Brady, J. B. (1975a) Reference frames and diffusion coefficients. Amer. J. Sci., 275, 954-983.

——— (1975b) Chemical components and diffusion. Amer J. Sci., 275, 1073-1088.

Birchenall, C. E., Correa da Silva, L. C. and Mehl, R. F. (1948) Discussion: Diffusion, mobility and their interrelation through free energy in binary metallic system. Trans. AIME, 175, 197-201.

Buening, D.K. and Buseck, P. R. (1973) Fe-Mg lattice diffusion in olivine. J. Geophys. Res., 78, 6852-62.

Burchard, J. K. and Toor, H. L. (1962) Diffusion in an ideal mixture of three completely miscible non-electrolytic liquids-toluene, chlorobenzene, bromobenzine. J. Phys. Chem., 66, 2015, 2022.

Chandok, V. K., Hirth, J. P. and Dulis, E. J. (1962) Effects of cobalt on carbon activity and diffusivity in steel. Trans. AIME, 224, 858-864.

Christensen, N. H. (1977) Ternary diffusion with moving boundaries. J. Amer. Ceram. Soc., 60, 54-58.

Cooper, A. R. (1974) Vector space treatment of multicomponent diffusion. In *Geochemical Transport and Kinetics,* Hofmann, A. W., Giletti, B. J., Yoder, H. S. Jr. and Yund, R. A., Eds., Carnegie Inst. of Washington Pub. 634, 15-30.

——— and Gupta, P. K. (1971) Analysis of diffusion controlled crystal growth. In *Advances in Nucleation and Crystallization of Glasses,* Hency, L. L. and Frieman, S. W., Eds., Amer. Ceram. Soc., p. 131-137.

Correa da Silva, L. C. and Mehl, R. F. (1951) Interface and marker movements in diffusion in solid solutions of metals. Trans. AIMME, 191, 155-173.

Crank, J. (1956) *The Mathematics of Diffusion.* Oxford Univ. Press, London.

Cullinan, H. T. Jr. (1965) Analysis of the flux equations of multicomponent diffusion. I and EC Fund., 4, 133-139.

Darken, L. S. (1948) Diffusion, mobility and their interrelation through free energy in binary metallic systems. Trans. Amer. Inst. Min. Met. Pet. Eng., 175, 189-201.

——— (1949) Diffusion of carbon in austenite with a discontinuity in composition. Trans. Amer. Inst. Min. Metall. Pet. Eng., 180, 430-438.

Dayanda, M. A. and Grace, R. E. (1967) Ternary diffusion in copper-zinc-manganese alloys. Trans. AIME, 233, 1287-1293.

Denbigh, K. (1966) *The Principles of Chemical Equilibrium.* Cambridge Univ. Press, Cambridge.

Duda, J. L. and Vrentas, J. S. (1965) Mathematical analysis of multicomponent free-diffusion experiments. J. Phys. Chem., 69, 3305-3313.

Dunlop, P. J. (1957a) A study of interacting flows in diffusion of the system Raffinose -- $KCl-H_2O$ at 25°C. J. Phys. Chem., 61, 994-1000.

——— (1957b) Interacting flows in diffusion of the system Raffinose-Urea-Water. J. Phys. Chem., 61, 1619-22.

——— and Gosting, L. J. (1955) Interacting flows in liquid diffusion: expressions for the solute concentration curves in free diffusion and their use in interpreting Gouy diffusiometer data for aqueous three-component systems. J. Amer. Chem. Soc., 77, 5238-5249.

——— (1959) Use of diffusion and thermodynamic data to test the Onsager Reciprocal Relation for isothermal diffusion in the system $NaCl-KCl-H_2O$ at 25°C. J. Phys. Chem., 63, 86-93.

________, Steel, B. J. and Lane, J. E. (1972) Experimental methods for studying diffusion in liquids, gases and solids. Techniques of Chemistry, I, 205-349.

Ekman, A., Liukkonen, S. and Kontturi, K. (1978) Diffusion and electric conduction in multicomponent electrolyte systems. Electrochim. Acta, 23, 243-250.

Fisher, G. W. (1977) Nonequilibrium thermodynamics in metamorphism. In *Thermodynamics in Geology,* Fraser, D. G., Ed., Reidel Publishing Co., Holland, p. 381-403.

Freer, R. (1980) Bibliography. Self-diffusion and Impurity diffusion in oxides. J. Mater. Sci., 15, 803-824.

Fujita, H. and Gosting, L. J. (1956) An exact solution of the equations for free diffusion in three-component systems with interacting flows, and its use in evaluation of diffusion coefficients. J. Amer. Chem. Soc., 78, 1099-1106.

________ (1960) A new procedure for calculating the four diffusion coefficients of three-component systems for Gouy diffusiometer data. J. Phys. Chem., 64, 1256-1263.

Gibbs, J. W. (1878) On the equilibrium of heterogeneous substances. In *Scientific Papers of J. Willard Gibbs, I, Thermodynamics,* Dover, N.Y., 55-371.

Groot de, S. R. and Mazur, P. (1962) *Non-equilibrium Thermodynamics,* North Holland Publishing Co., Amsterdam.

Guggenheim, E. A. *Thermodynamics,* North Holland Publishing Co., Amsterdam.

Gupta, P. K. and Cooper, A. R. (1971) The [D] matrix for multicomponent diffusion. Physica, 54, 39-43.

Hall, L. D. (1953) An analytical method of calculating diffusion coefficients. J. Chem. Phys., 21, 87-89.

Hamer, W. J. and Wu, Y. C. (1972) Osmotic coefficients and mean activity coefficients of uni-univalent electrolytes in water at 25°C. J. Phys. Chem. Ref. Data, 1, 1047-99.

Hartley, G. S. (1931) Theory of the velocity of diffusion of strong electrolytes in dilute solution. Phil. Mag., 273, 473-488.

Hildebrand, F. B. (1952) *Methods of Applied Mathematics.* Prentice Hall, Inc., N.Y.

Hollingsworth, C. A. (1967) *Vectors, Matrixes and Group Theory for Scientists and Engineers,* McGraw-Hill Book Co., N.Y.

Hooyman, G. J. (1956) Thermodynamics of diffusion in multicomponent systems. Physica, 22, 751-9.

Howard, R. E., and Lidiard, A. B. (1964) Matter transport in solids. Rept. Prog. Phys., 27, 161-240.

Kim, H., Reinfelds, G. and Gosting, L. J. (1973) Isothermal diffusion studies of water-potassium chloride-hydrogen chloride and water-sodium chloride-hydrogen chloride systems at 25°. J. Phys. Chem., 77, 934-940.

Kirkaldy, J. S. (1957) Diffusion in multicomponent metallic systems. Canad. J. Phys., 35, 435-440.

________ (1970) Isothermal diffusion in multicomponent systems. In *Advances in Materials Res.* 4, 55-100.

________, Weichert, D. and Sia-Ul-Haq (1963) Diffusion in multicomponent metallic systems. VI, Canadian J. Phys., 41, 2166-2173.

________, Lane, J. E. and Mason, G. R. (1963) Diffusion in multicomponent metallic systems VII. Solutions of the multicomponent diffusion equations with variable coefficients. Canadian J. Phys., 41, 2174-2186.

Kirkendall, E. O. (1952) Diffusion of zinc in alpha brass. Trans. AIMME, 147, 104-110.

Kirkwood, J. G., Baldwin, R. L., Dunlop, P. J., Gosting, L. J. and Kegeles, G. (1960) Flow equations and frames of reference for isothermal diffusion in liquids. J. Chem. Phys., 33, 1505-1513.

________, and Oppenheim, I. (1961) *Chemical Thermodynamics.* McGraw-Hill Book Co., N.Y.

Lasaga, A. C. (1979a) The treatment of multicomponent diffusion and ion-pairs in diagenetic fluxes. Amer. J. Sci., 279, 324-346.

________ (1979b) Multicomponent exchange and diffusion in silicates. Geochim. Cosmochim. Acta, 43, 455-469.

________, Richardson, S. M., and Holland, H. D. (1977) The mathematics of cation diffusion and exchange between silicate minerals during retrograde metamorphism. In *Energetics of Geological Processes,* Saxena, S. K. and Bhattacharji, S., Eds., Springer-Verlag New York, p. 353-388.

Li, Y. H. and Gregory, S. (1974) Diffusion of ions in sea water and in deep sea sediments. Geochim. Cosmochim. Acta, 38, 703-714.

Loomis, T. P. (1978a) Multicomponent diffusion in garnet: I. Formulation of isothermal models. Amer. J. Sci., 278, 1099-1118.

——— (1978b) Multicomponent diffusion in garnets: II. Comparison of models with natural data. Amer. J. Sci., 278, 1119-1137.

Manning, J. R. (1968) *Diffusion Kinetics for Atoms in Crystals.* Van Nostrand, Princeton, N.J.

——— (1974) Diffusion kinetics and mechanisms in simple crystals. In *Geochemical Transport and Kinetics,* Hofmann, A. W., Giletti, B. J., Yoder, H. S. Jr. and Yund, B. A., Eds., Carnegie Inst. of Washington Pub. 634, 3-13.

Matano, C. (1933) On the relationship between the diffusion coefficients and concentrations of solid metals. Japan. J. Phys. 8, 109-113.

Miller, D. G. (1959) Ternary isothermal diffusion and the validity of the Onsager Reciprocal Relations. J. Phys. Chem., 63, 570-578.

——— (1965) Definitive test of the Onsager Relations in isothermal ternary diffusion of water-sodium chloride-potassium chloride. J. Phys. Chem., 69, 3374-3376.

——— (1966) Application of irreversible thermodynamics to electrolyte solutions. I. Determination of ionic transport coefficients for isothermal vector transport processes in binary electrolyte systems. J. Phys. Chem., 70, 2639-59.

——— (1967a) Application of irreversible thermodynamics to electrolyte solutions. II. Ionic coefficients l_{ij} for isothermal vector transport processes in ternary systems. J. Phys. Chem., 71, 616-632.

——— (1967b) Application of irreversible thermodynamics to electrolyte solutions. III. Equations for isothermal vector transport processes in n-component systems. J. Phys. Chem., 71, 3588-3592.

——— (1974) The Onsager relations; experimental evidence. In *Foundations of Continuum Thermodynamics.* Domingas, J. J. D., Nina, M. N. R. and Whitelaw, J. H., Eds. John Wiley and Sons, N.Y. p. 185-214.

Moynihan, C. T. (1971) Mass transport in fused salts. In *Ionic Interactions,* Petrucci, S., Ed., Academic Press, N.Y., p. 262-384.

Onsager, L. (1931a) Reciprocal relations in irreversible processes. I. Phys. Rev., 37, 405-26.

——— (1931b) Reciprocal relations in irreversible processes. II. Phys. Rev., 38, 2265-2279.

——— (1945) Theories and problems of liquid diffusion. Ann. N.Y. Acad. Sci., 46, 241-265.

——— and Fuoss, R. M. (1932) Irreversible processes in electrolytes. Diffusion, conductance and viscous flow in arbitrary mixtures of strong electrolytes. J. Phys. Chem., 36, 2689-2778.

Prigogine, I. (1955) *Introduction to Thermodynamics of Irreversible Processes.* Thomas, Springfield, Il.

——— and Defay, R. (1954) *Chemical Thermodynamics.* Longmans, London.

Rard, J. A., Habenschuss, A. and Spedding, F. H. (1977) A review of the osmotic coefficients of acqueous $CaCl_2$ at 25°C. J. Chem. Eng. Data, 22, 180-186.

——— and Miller, D. G. (1979a) The mutual diffusion coefficients of $NaCl-H_2O$ and $CaCl_2-H_2O$ at 25°C from Rayleigh interferometry. J. Soln. Chem., 8, 701-716.

——— (1979b) The mutual diffusion coefficients of $Na_2SO_4-H_2O$ and $MgSO_4-H_2O$ at 25°C from Rayleigh interferometry. J. Soln. Chem., 8, 755-766.

——— (1980a) Mutual diffusion coefficients of $BaCl_2-H_2O$ and $KCl-H_2O$ at 25°C from Rayleigh interferometry. J. Chem. Engin. Data, 25, 211-215.

——— (1981) Isopiestic determination of the osmotic and activity coefficients of aqueous $MgCl_2$ solutions at 25°C. J. Chem. Engin. Data, 26, 38-43.

Revzin, A. (1972) A new procedure for calculating the four diffusion coefficients for ternary systems from Gouy optical data. Application to data for the system $KBr-HBr-H_2O$ at 25°. J. Phys. Chem., 76, 3419-29.

Robinson, R. A. and Stokes, R. H. (1959) *Electrolyte Solutions.* Academic Press, N.Y.

Rush, R. M. (1969) Parameters for the calculations of osmotic and activity coefficients for twenty-two aqueous mixtures of two electrolytes at 25°C. Oak Ridge Nat- Lab. Rept. ORNL-4402.

——— and Robinson, R. A. (1968) A reevaluation of the activity coefficients in the system sodium chloride-potassium chloride-water. J. Tenn. Acad. Sci., 43, 22-25.

Scatchard, G. (1961) Osmotic coefficients and activity coefficients in mixed electrolye solutions. J. Amer. Chem. Soc., 83, 2636-42.

Schuck, F. O. and Toor, H. L. (1963) Diffusion in the three component liquid system methyl alcohol-n propyl alcohol-isobutyl alcohol. J. Phys. Chem., 67, 540-545.

Smigelskas, A. D. and Kirkendall, E. O. (1947) Zinc diffusion in alpha brass. Trans. AIMME, 171, 130-142.

Sucov, E. W. and Gorman, R. R. (1965) Interdiffusion of calcium in soda-lime silica glass at 880° to 1308°C. J. Amer. Ceram. Soc., 48, 426-429.

Sunderlöf, L. O. and Södervi, I. (1963) Free diffusion in multicomponent solution. Arkev Kemi, 21, 143-160.

Swalin, R. A. (1972) *Thermodynamics of Solids.* John Wiley & Sons, N. Y.

Toor, H. L. (1964a) Solution of the linearized equations of multicomponent mass transfer: I. Amer. Inst. Chem. Eng. J., 10, 448-455.

——— (1964b) Solution of the linearized equations of multicomponent mass transfer: II. Matrix methods. Amer. Inst. Chem. Eng. J., 10,460-465.

Tyrrel, H. J. V. (1964) The origin and present status of Fick's diffusion law. J. Chem. Educ. 41, 397-400.

Van Gool, W. (1966) *Principles of Defect Chemistry of Crystalline Solids.* Academic Press, N.Y.

Varshneya, A. K. and Cooper, A. R. (1972) Diffusion in the system K_2O-SrO-SiO_2. III. Interdiffusion coefficients. J. Amer. Ceram. Soc., 55, 312-317.

Vignes, A. and Sabatier, J. P. (1969) Ternary diffusion in Fe-Co-Ni alloys. Trans. AIME, 245, 1795-1802.

Wagner, C. (1969) The evaluation of data obtained with diffusion couples of binary single-phase and multiphase systems. Acta Metall., 17, 99-107.

Wendt, R. P. (1962) Studies of isothermal diffusion at 25°C in the system water-sodium sulfate-sulfuric acid and tests of the Onsager relation. J. Phys. Chem., 66, 1279-1288.

——— (1965) The estimation of diffusion coefficients for ternary systems of strong and weak electrolytes. J. Phys. Chem., 69, 1227-1237.

——— and Shamim, M. (1970) Isothermal diffusion in the system water-magnesium chloride-sodium chloride as studied with the rotating diaphragm cell. J. Phys. Chem., 74, 2770-2783.

Wilkinson, J. H. (1965) *The Algebraic Eigenvalue Problem.* Clarendon Press, Oxford.

Ziebold, T. O. and Ogilvie, R. E. (1967) Ternary diffusion in copper-silver-gold alloys. Trans. AIME, 239, 942-953.

Chapter 7

The ATOMISTIC BASIS of KINETICS: DEFECTS in MINERALS

Antonio C. Lasaga

INTRODUCTION

In the treatment of kinetics in geochemistry a central position must be given to the defect theory of silicates and metal oxides. Defects in crystals provide the machinery whereby many important geochemical phenomena are possible. For example, defects are essential in (a) the homogenization of zoned crystals, (b) the alteration of the composition of minerals, (c) changes in the order-disorder status of tetrahedral cations in aluminosilicates, (d) the resetting of radiometric ages, (e) isotope equilibration, (f) creep in the earth's mantle, (g) formation of exsolution lamellae, and (h) the growth and dissolution of minerals. The first four examples (a-d) are related to the general process of bulk diffusion, which is only possible due to the existence of point defects. The next three (e-g) may be controlled by point defect diffusion or by yet another type of defect, extended defects, *e.g.*, dislocations. The last example (growth and dissolution) is often governed by the number and types of dislocations on the crystal surface, which itself may be thought of as a two-dimensional crystal defect (see Chapter 3).

In this chapter we will summarize some of the salient features of defects in silicates and how they can be applied to geochemical problems. At the outset, the different types of defects mentioned above must be well defined.

Defects in the periodic structure of a crystal caused by the absence of an atom or ion at a lattice site (vacancy), the presence of an additional atom in an interstitial position, or the presence of a foreign, possibly aliovalent, ion at a lattice site (impurity) are called *point defects*. Within this classification, we must distinguish two types of crystals; (a) those with small defect concentrations (<0.1 atomic %) and (b) those with extensive defects. The latter usually have complex defect structures and include defect clusters. In crystals with large degrees of non-stoichiometry such as FeO or FeS, the consideration of the defect concentrations is essential.

Extended defects or thermodynamically irreversible defects can be categorized as follows:

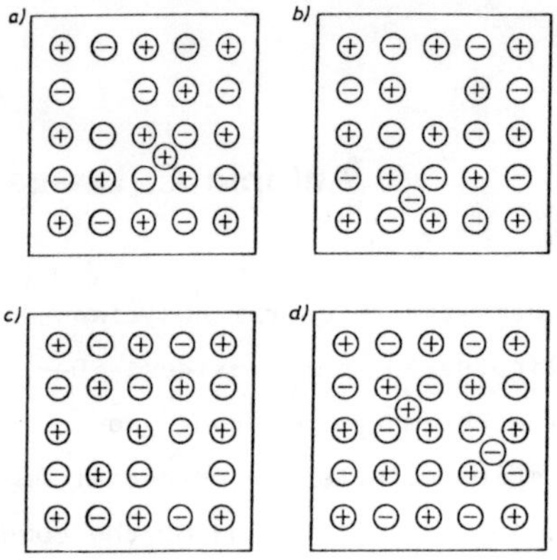

Figure 1. Models of lattice disorder in MX type compounds: (a) the Frenkel type, (b) the anti-Frenkel type, (c) the Schottky type, (d) the anti-Schottky type.

(1) *macroscopic defects* having dimensions exceeding 10^{-3} cm, *e.g.*, cracks, fissures and empty spaces.

(2) *microscopic defects*: These include *dislocations*, also termed *line defects*, which are displacements of lattice planes along a defect boundary line. Also included here are *planar defects:* stacking faults, twin boundaries and grain boundaries.

The understanding of point defects is essential for the treatment of the more complex extended defects. Furthermore, much more data are available on point defects. Therefore, we will discuss point defects first and then turn to the extended defects at the end.

POINT DEFECTS

Most minerals can be classified to a certain extent as ionic crystals. If a crystal contains ions we must further classify the occurrence of the defects, taking into account the restriction of electroneutrality. When an ion is removed from a site in the crystal structure, the crystal acquires a charge imbalance. This imbalance must be corrected if electroneutrality is to be maintained. In simple *ionic* crystals there are two dominant types of *intrinsic* defects which maintain electroneutrality (see Fig. 1):

(1) *Schottky defect:* Stoichiometric proportions of vacancies in each of the two ionic sublattices, *e.g.*, in a crystal MX, equal numbers of cation and anion vacancies.

(2) *Frenkel defect:* Equal numbers of interstitial and vacancies of one ion in the compound. If an anion is inserted into an interstitial site leaving behind an anion vacancy, an anion Frenkel defect (sometimes termed an anti-Frenkel defect) is formed. Similarly, a cation interstitial with a cation vacancy is a cation Frenkel defect.

Table 1. Types of Intrinsic Point Defects Occurring in Ionic Crystals

Substance	Structure	Type of intrinsic disorder
Alkali halides	NaCl	Schottky
MgO	NaCl	Schottky
AgBr, AgCl	NaCl	Cation Frenkel
CsCl, TlCl	CsCl	Schottky
BeO	Wurtzite	Schottky
CaF_2, ThO_2	CaF_2	Anion Frenkel

An *interstitial* site in a mineral refers to *any* normally unoccupied site in the crystal structure. For example, in spinels, which have only one-half of the octahedral sites occupied, the other half of the octahedral sites are interstitial sites. Usually most ionic crystals will have *one* type of intrinsic defect predominant over the others at any given temperature. Which type of defect is important in diffusion depends on the energies required to form and move these defects in the crystal structure. Of course, in minerals with a variety of ion sites and multi-valence impurities, the defects will be more complicated than simple Schottky or Frenkel defects and the defects are often controlled by *extrinsic* factors (such as impurity content). Some well-established types of intrinsic defects for simple ionic crystals are given in Table 1. Obviously, Frenkel defects are important for a small particle in a large interstitial structure. This is the case for carbon diffusion in iron metal, a Frenkel-dominated process of great importance to the steel industry. This is also the case with F^- in CaF_2.

Terminology and Concentration of Point Defects in Crystals

Kröger (1971) and co-workers have made a great contribution to the thermodynamic treatment of defects. Their main idea was to treat point defects as *quasi-chemical* species. For example, vacancies and interstitials may be regarded respectively as solutions of the vacuum and of solute atoms in the host crystal. We therefore assign a chemical potential μ to each defect. The defect chemical potential can be written in the usual manner as $\mu = \mu^\circ + RT\ln(\gamma X)$ where X is the *site fraction* of the defect. Table 2 gives the chemical symbols for each type of defect

Table 2. Kröger Defect Symbols

Symbol	Defect	Effective charge, in units q
e'	Quasifree electron in the conduction band	-1
$h^{\cdot}$	Quasifree hole (lack of electron) in the valence band	+1
V_M^x	M vacancy	0
V_M''	M vacancy	-2
V_X^x	X vacancy	0
M_X''	M at an X site	-2
X_M^x	X at an M site	0
$M_i^{\cdot}$	M at an interstitial site	+1
M_M^x	M at an M site	0
X_X^x	X at an X site	0
V_i^x	Unoccupied interstitial site	0
F_M^x	Foreign atom F at M site	0
$(V_M V_X)^x$	Associate of M and X vacancies at neighboring sites	0
$(F_X X_i)^x$	Associate of F at X site and X at an adjacent interstitial site	0

as introduced by Kröger (1971). In particular, we refer to the pure crystal by the symbol 0. This emphasizes the fact that all defect chemical potentials are relative to the perfect crystal as standard state; it also means that μ(perfect crystal) = 0. Each defect has associated with it an *effective charge* in Table 2. The effective charge is the charge acquired by the crystal as a result of the defect. For example, a Ca^{2+} vacancy in diopside is written as V_{Ca}'' (omitting the *M2* site). The effective charge is -2, since loss of +2 charge at a Ca^{2+} site is the same as inserting a -2 charge at that site. Equivalently, the crystal receives a -2 charge as a result of the defect. Likewise, an $A\ell^{3+}$ interstitial in feldspar is labelled $A\ell_i^{\cdot\cdot\cdot}$ indicating an effective charge of +3. In this case, where the crystal normally has *no* charge at the interstitial site, the insertion of an $A\ell^{3+}$ ion increases the charge of the normal crystal by +3.

Treating the various defects as quasi-chemical species, chemical reactions can be written involving defect species. In formulating chemical reactions involving defects, three rules must be obeyed:

(1) Conservation of atom types.

(2) The sum of *effective* charges on the left and right sides of the reaction must be the same (conservation of charge).

(3) The ratio of the number of sites characteristic of the crystal structure (*e.g.*, tetrahedral versus octahedral) must be maintained.

For example, suppose we are interested in the formation of Mg^{2+} and O^{2-} vacancies in a forsterite crystal. We could then write the defect reaction as

$$Mg_{Mg}^{x} + O_{O}^{x} \rightarrow V_{Mg}'' + V_{O}^{\bullet\bullet} + MgO(\text{surface}). \qquad (1)$$

Equation (1) states that a Mg^{2+} ion on a magnesium octahedral site in forsterite (Mg_{Mg}) and an O^{2-} ion on an oxygen site (O_O) are removed to new sites on the surface of the crystal leaving behind a magnesium site vacancy (V_{Mg}'') and an oxygen site vacancy ($V_O^{\bullet\bullet}$). The Kröger notation, at first a bit cumbersome, is the notation widely used. Note that in forming vacancies the ions are not removed to infinity but rather placed on the surface of the crystal. The surface of the crystal is thereby acting as the source and sink of defects, which proceed to diffuse into the crystal.

Forsterite has two distinct octahedral sites (*M1* and *M2*) as well as three distinct oxygen sites (O1, O2, and O3). Therefore, equation (1) should be more specific. For example, the reaction

$$Mg(M1)_{Mg(M1)}^{x} + O(3)_{O(3)}^{x} \rightarrow V_{Mg(M1)}'' + V_{O(3)}^{\bullet\bullet} + MgO(\text{surface}) \qquad (2)$$

describes the formation of a Schottky defect with the cation vacancy occurring at the *M1* site and the anion vacancy at the O(3) site.

Application of thermodynamics to quasi-chemical reactions involving defects introduces an equilibrium constant for each reaction. For example, the equilibrium constant for reaction (2) would be

$$\frac{X_{Mg(M1),v}\, X_{O(3),v}}{X_{Mg,Mg}\, X_{O,O}} = K_{eq} \qquad (3)$$

where $X_{Mg(M1),v}$ is the fraction of Mg-*M1* sites which are vacant, $X_{Mg,Mg}$ is the fraction of Mg-*M1* sites which are occupied by Mg and similarly for oxygen. Equation (3) assumes that $\gamma = 1$ for all species; this assumption is warranted if the concentration of defect species is small since in this case Henry's law applies to the defect species and Raoult's law to the solvent species. Obviously, it is usually a very good approximation to set $X_{Mg,Mg}$ and $X_{O,O}$ equal to 1. In this case, equation (3) simplifies to

$$X_{Mg(M1),v}\, X_{O(3),v} = K_{eq} = e^{-\frac{\Delta G_f}{RT}} \qquad (4)$$

where ΔG_f is the free energy of formation for the pair of vacancies. Regardless of how many other defects are present, equation (4) is *always* true at equilibrium and shows that the defect concentrations may depend exponentially on temperature. In the particular case that there are no other impurities or defects, or if the particular defects in reaction (2) are the predominant types, charge balance requires that $X_{Mg,v} = X_{O,v}$, and so

$$X_{Mg,v} = X_{O,v} = e^{-\frac{\Delta G_f}{2RT}} \qquad (5)$$

where the *M1* and O(3) labels are omitted in equation (5).

Smyth and Stocker (1975) and Stocker (1978a,b) discuss the various possible defect reactions for pure forsterite as well as for olivine. They suggest that for pure stoichiometric forsterite (intrinsic defects), magnesium Frenkel defects are important:

$$Mg_{Mg}^{x} + V_{i}^{x} \rightarrow Mg_{i}^{\cdot\cdot} + V_{Mg}^{''} .$$

In forsterite a magnesium interstitial would refer to occupation by magnesium of one of the usually *unoccupied* octahedral sites. There are four non-equivalent octahedral sites in Mg_2SiO_4; two (*M1* and *M2*) are occupied and two (*M3* and *M4*) are unoccupied. Morin *et al.* (1979) use recent electrical conductivity data on high-purity forsterite to show that electrical conductivity occurs by magnesium interstitials; whereas, the diffusion data suggest magnesium vacancies are dominant. For a non-stoichiometric forsterite with excess $MgSiO_3$, the defect reaction would be

$$4MgSiO_3 \rightarrow 4Mg_{Mg}^{x} + 2V_{Mg}^{''} + 3Si_{Si}^{x} + Si_{i}^{\cdot\cdot\cdot\cdot} + 12O_{O}^{x}$$

creating both silicon interstitial (since the SiO_2 content of olivine is increased) and magnesium vacancies. The presence of Fe^{2+} in olivine, however, affects the defect structure drastically, as we show in a later section.

The free energy of defect formation in equations (4) and (5) can be written in terms of the enthalpy and entropy of formation:

$$\Delta G_f = \Delta h_f - T\Delta s_f \ . \tag{6}$$

The enthalpy term is, of course, related to the temperature dependence of the defect concentration, which can be experimentally obtained from transport data. Δh_f is the dominant term in equation (6). Table 3 contains some typical values of the defect formation enthalpies for simple ionic crystals. Note that the range of enthalpy of formation is typically 1-3 electron volts or 20-60 Kcal/mole. The size of Δh_f indicates that the defect concentrations can change substantially with changes in temperature. Table 4 gives the variation of defect concentration with temperature in sodium chloride and in MgO. Note the many fewer vacancies in the more refractory MgO at low temperatures. In fact, at low temperatures the defect concentrations are so small that the defect structure of both crystals is definitely impurity controlled (see next section), even for the highest grade synthetic crystals. Close to the melting point of each crystal, however, the intrinsic defect concentration can reach high values in the ppm range.

For cubic metals experiments which measure changes in the length of the lattice parameter of a sample can yield direct information on vacancy concentrations. In fact, the concentration of vacancies, c, is deduced from

$$c = 3\left(\frac{\Delta L}{L} - \frac{\Delta a}{a}\right)$$

where L is the length of the thin sample and a is the lattice parameter as measured by x-ray techniques. The values of c and the energies of formation thus derived are also given in Table 3.

Vacancy Pairs

A cation vacancy and an anion vacancy are oppositely charged defect species (*e.g.*, the effective charge) and it is expected that there would exist a coulombic attraction between them. In an analogous fashion to ion pairs of electrolyte solutions, this attraction will lead to the formation of a *vacancy pair*, if the defect concentration is substantial. In fact, since water has a dielectric constant of 78 while ionic solids have $\varepsilon = 5 - 10$, the crystal should be less effective in keeping charged species apart than aqueous solutions. A vacancy pair is overall electrically neutral and behaves as a *dipole*. The quasi-chemical reaction would be

$$V''_M + V^{\bullet\bullet}_X \rightarrow V^x_{MX} \; . \tag{7}$$

If equilibrium is assumed in equation (7) we can write

$$\Delta\mu = 0 = \mu^o_{MX,v} + kT\ln\frac{X_P}{z} - \mu^o_{M,v} - kT\ln X_{M,v} - \mu^o_{X,v} - kT\ln X_{X,v} \tag{8}$$

where $\mu^o_{MX,v}$ is the free energy of formation of the vacancy pair, z is the number of distinct *orientations* of the pair (*e.g.*, z = 6 in NaCℓ), and X_P is the fraction of cation *or* anion sites occupied by vacancy pairs. However, from the equilibrium reaction

$$0 \rightarrow V''_M + V^{\bullet\bullet}_X \; ,$$

and since $\mu(0) = 0$, it follows that

$$\mu^o_{M,v} + kT\ln X_{M,v} + \mu^o_{X,v} + kT\ln X_{X,v} = 0 \; .$$

Therefore, we can eliminate the last terms of equation (8) and obtain

$$X_P = z e^{-\frac{\mu^o_{MX,v}}{kT}} \; . \tag{9}$$

Note that X_P is *independent* of $X_{M,v}$ of $X_{X,v}$. For a more detailed treatment, see Bollmann (1977).

The values of the enthalpy of formation of vacancy pairs, Δh_p, where

$$\mu^o_{MX,v} = \mu^o_p = \Delta h_p - T\Delta s_p$$

are given in Table 5 for simple crystals. Note that the values of Δh_p are similar in magnitude to those of Δh_s (Schottky) and Δh_F (Frenkel) in Table 3. Figure 2 shows the geometries of known associated defects in alkaline-earth oxides. In the figure the cation vacancy is represented by a square with an effective charge of -2.

The vacancy pair enthalpies in Table 5 were generally obtained from dielectric loss measurements. In an ideal ionic crystal only the electronic polarization and the ion displacement polarization contribute to the dielectric constant (see Lasaga and Cygan, 1981). However, if the crystal contains vacancy-pairs, it is analogous to possessing a permanent dipole moment. Since the vacancy-pair can reorient itself under the influence of an applied alternating field (see Fig. 3), there will

Table 3

Enthalpy of Formation of Defects - ΔH_f

Schottky Defects	(Kcal/mole)		Frenkel Defects		
Compound	ΔH_S	ref	Compound	ΔH_F	ref
LiF	58	a	AgCl	31	a
LiCl	51	a	AgBr	24	a
LiBr	42	a	CaF_2	59	a
NaCl	53	a	SrF_2	53	a
NaBr	40	a	BaF_2	44	a
KCl	53	a			
KBr	56	a			
KI	37	a	Vacancy in Metals		
CsCl	43	a			
CsBr	46	a	Compound	ΔH_f	ref
CsI	44	a			
$PbCl_2$	36	a	Au	22	c
$PbBr_2$	32	a	Ag	25	c
MgO	110	b	Cu	27	c
CaO	100	b	Al	17	c
Al_2O_3	470	b	Na	10	c
			Pb	11	c
			Mg	13	c

a) Barr and Lidiard (1971)
b) Greenwood (1968)
c) Johnson (1973)

*Note 1 ev = 23.056 Kcal/mole

Table 4[a].

Number of Schottky Defects at Thermodynamic Equilibrium

		NaCl		MgO	
t°C	T°K	$X_{Na,v}$	n_s (per cm^3)	$X_{Mg,v}$	n_s (per cm^3)
-273	0	0	0	0	0
25	298	2.5×10^{-19}	5.6×10^{3}	2.1×10^{-37}	1.1×10^{-14}
200	473	1.9×10^{-12}	4.2×10^{10}	7.9×10^{-24}	4.2×10^{-1}
400	673	5.9×10^{-9}	1.3×10^{14}	5.8×10^{-17}	3.1×10^{6}
600	873	4.5×10^{-7}	1.0×10^{16}	3.0×10^{-13}	1.6×10^{10}
800	1073	6.8×10^{-6}	1.5×10^{17}	6.5×10^{-11}	3.5×10^{12}
1000	1273	melts		2.6×10^{-9}	1.4×10^{14}
1500	1773	"		6.9×10^{-7}	3.7×10^{16}
2000	2273	"		1.6×10^{-5}	8.6×10^{17}

a) used ΔG_f = 50.7 Kcal/mole for NaCl

and ΔG_f = 100 Kcal/mole for MgO

Table 5

Enthalpy of Formation of Vacancy Pairs
and Association of Divalent Cations with Vacancies
(Kcal/mole)*

Compound	ΔH_p	Enthalpy of Reorientation	Q_D**	ref
NaCl	30	29	57	a
KCl	31	24	61	a
KBr	33	30	60	a
Mg^{2+} in NaCl	10			b
Ca^{2+} in NaCl	10			b
Ca^{2+} in KCl	12			b
Sr^{2+} in NaCl	12			b

*1 ev = 23.056 Kcal/mole

**Q_D is the activation energy for anion diffusion via pairs

a) Barr and Lidiard (1971)

b) Lidiard (1973)

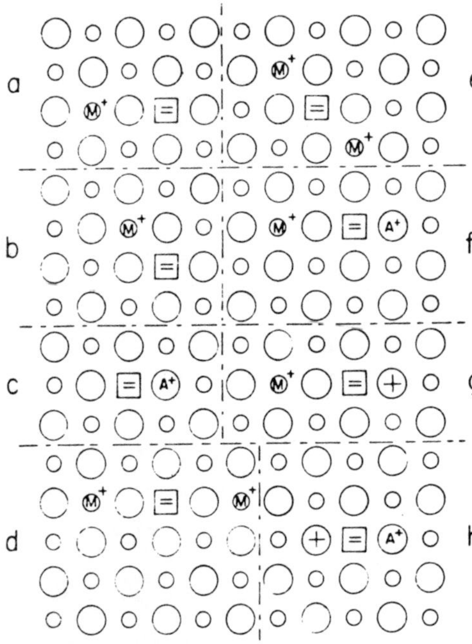

Figure 2. Geometries of known associated-cation-vacancy point defects in alkaline earth oxides. Defect symbols are as follows: M^+ represents a trivalent cation (effective charge = +1); [=] represents a cation vacancy; A^+ represents a monovalent anion (e.g. OH^-); + represents a positive hole i.e. an O^-. (Wertz, 1968.)

be an anomalous dielectric loss when the frequency of the applied electric field matches the appropriate frequency of vacancy pair reorientation. A study of the frequency response as a function of temperature can, with great care, yield the heat of formation of the vacancy pair (see Economau, 1964; for some typical problems in the technique see Bielig and Lilley, 1980). Note that in a similar fashion to the effects of complexing in electrolyte solutions, vacancy pairs do not contribute to the ionic conductivity of a crystal.

Impure Crystals

We can drastically affect the number of defects in a crystal by the introduction of *aliovalent* impurity ions. Consider a divalent cation dissolved in

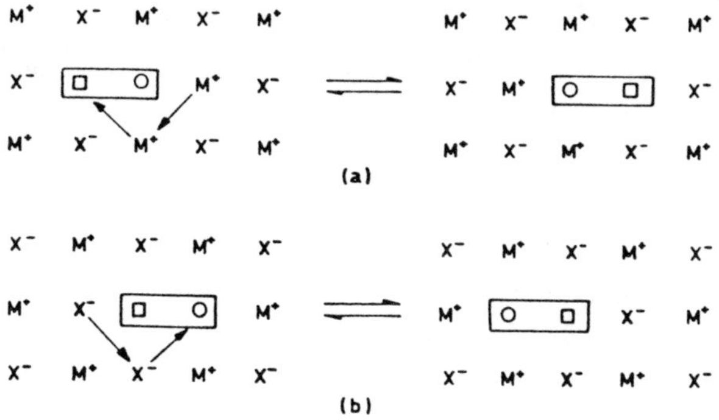

Figure 3. Reorientation of a Schottky vacancy-pair leading to dielectric loss: (a) cation motion, (b) anion motion.

alkali halide, for example, $CaC\ell_2$ in NaCℓ. For each divalent cation added as an impurity there must be one cation vacancy in the cation sites in order to conserve charge. If the mole fraction of impurity cations in cation sites is c, the alkali vacancy mole fraction is X_c, and the halide vacancy mole fraction is X_a, electrical neutrality requires that $X_c = c + X_a$. Furthermore, equilibrium relations such as equation (4) still must be maintained:

$$X_a X_c = e^{-\frac{\Delta G_f}{kT}} \equiv K .$$

Therefore, combining the last two equations, $X_c(X_c - c) = K$, or

$$X_c = \frac{c}{2}\left[1 + \left(1 + \frac{4K}{c^2}\right)^{\frac{1}{2}}\right] \tag{10}$$

$$X_a = \frac{c}{2}\left[\left(1 + \frac{4K}{c^2}\right)^{\frac{1}{2}} - 1\right] \tag{11}$$

Figure 4 shows how the addition of the divalent impurity strongly depresses the anion vacancies, while essentially having $X_c \simeq c$ (especially at low temperatures, where K is small). The scarcity of anion vacancies will *lower* the diffusion coefficient of the anion, as the impurity content increases. This prediction is clearly seen in Figure 5 where the doping of NaCℓ by Sr^{2+} reduces the self-diffusion coefficient of Cl^- until vacancy-pair diffusion becomes important. Figure 4 also shows the concentration of vacancy pairs. Since X_p is independent of the cation or anion vacancy concentration [equation (9)], the vacancy-pair concentration is not influenced by doping.

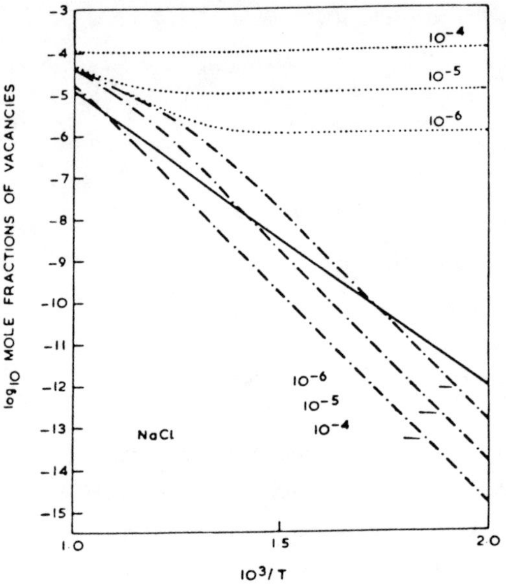

Figure 4. The mole fraction of cation vacancies (........), anion vacancies (-·-·-·-·) and bound anion-cation vacancy pairs (————) in NaCl as functions of $10^3/T$ for various fractions of divalent cations in solid solutions as indicated. The values for anion and cation single vacancies have been calculated from equations (10) and (11) and show the opposite effects which doping has upon the concentrations of the two kinds of defect. Vacancy pairs by contrast are electrically neutral and occur in concentrations uninfluenced by doping. At the lower temperatures and higher impurity concentrations they may thus be "uncovered" and detected in suitable experiments. (The values given here do not allow for pairing between solute ions and cation vacancies and assume that all solute ions are in solution. Pairing of solute ions and precipitation from solid solution will reduce the effective doping and bend the two sets of lines back towards one another. After Lidiard, 1973.)

Most silicates of geochemical interest contain enough impurities that defects are extrinsically controlled and equations (10) and (11) are important. For example, the presence of Fe^{3+} in olivines would determine the number of Fe^{2+} or Mg^{2+} cation vacancies below a certain temperature. The temperature below which the impurities dictate the cation vacancy concentration will depend on the size of ΔG_f for the

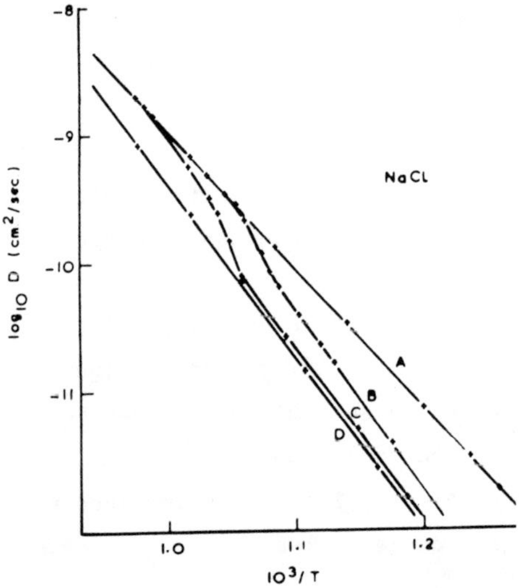

Figure 5. Chlorine self-diffusion in NaCl as a function of T and mole fraction of Sr^{2+} solute ions as

(a) pure (Harshaw crystal),

(b) 2.7×10^{-5}, (c) 6.5×10^{-5},

(d) 3.63×10^{-4}.

The correspondence with Figure 4 indicates that line A is largely diffusion by single anion vacancies while line D for the highly doped crystal is largely diffusion by vacancy pairs; lines B and C show the changeover from one region to the other as the temperature falls corresponding to the crossing of lines shown in Figure 4 (After Lidiard, 1973.)

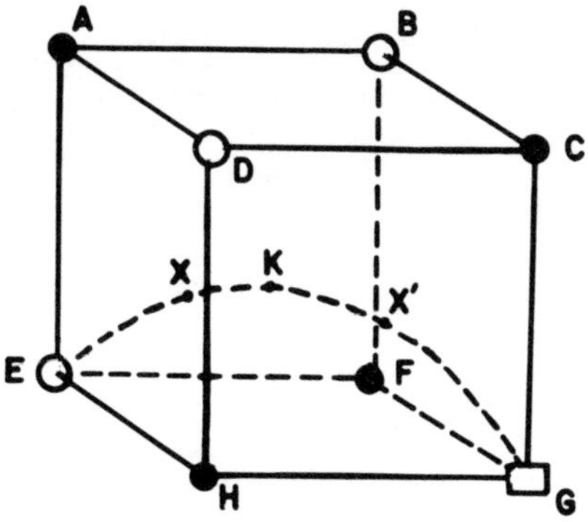

Figure 6. Mg^{2+} migration in MgO. Cation E during its motion E → G passes through the saddle points X and X' (the centers of the equilateral triangles AHF and CHF). Filled circle = oxygen ion; open circle = magnesium ion; square = cation vacancy. (After Varotos and Alexopoulos, 1977.)

intrinsic defect reaction as well as the impurity content. In fact, the temperature at which the intrinsic defect concentration equals the impurity content, c, can be roughly calculated from equation (5)

$$c = e^{-\frac{\Delta G_f}{2RT_c}}$$

or

$$T_c = -\frac{\Delta G_f}{2R\,ln c} . \qquad (12)$$

High impurity content (*i.e.*, low $|ln c|$) or a high ΔG_f will increase T_c. Table 4 gives the intrinsic concentration of Schottky defects in sodium chloride and MgO at different temperatures. T_c would be equal to the temperature in the table where the defect site fraction equals the impurity content. At low temperatures ultra-pure crystals would be needed if intrinsic defects are to be dominant. Furthermore, high melting minerals, like MgO, with higher ΔG_f are more apt to be impurity-controlled than the lower melting alkali halides. The temperature region below T_c, *i.e.*, where impurities control the defect concentrations, is termed the *extrinsic* region. This distinction between intrinsic and extrinsic regions is important in transport phenomena.

Migration Energies

Once vacancies (or interstitials) are present in a crystal, ions can jump into a neighboring vacant site. Since such a motion would push the ion into the repulsive walls of its nearest neighbors, a minimum amount of energy is required to jump from an occupied site to a vacancy. This energy is termed the *migration energy* and is different from the formation energy aforementioned. The migration energy governs the rate of movement of atoms in crystals as well as the movement of the vacancies (by complementarity). A major task of current solid state theories of minerals is to calculate these migration energies. Careful diffusion and/or conductivity experimental data (see next section) can also yield the migration energy. Table 6 gives the energies for

Table 6

Enthalpies of Migration

Vacancy Migration (Kcal/mole)			
	Cation	Anion	
Compound	Δh_m	Δh_m	ref
LiF	16	15	a
LiCl	9	---	a
NaCl	16.4	23	a
NaBr	18	27	a
KCl	16.3	23	a
KBr	15	21	a
KI	28	---	a
CsCl	14	8	a
CsBr	13	6	a
$PbCl_2$	---	8	a
$PbBr_2$	---	7	a
CaF_2		12.9	a
BaF_2		13	a
MgO	78	62	a,b
CaO	81	---	a
PbO	67	---	a
BeO	56	---	a

Anion Interstitial (Kcal/mole)		
Compound	Δh_m	ref
CaF_2	38	a
SrF_2	23	a
BaF_2	18	a

Cation Interstitial (Kcal/mole)		
Compound	Δh_m	ref
AgCl	3	a
AgBr	4	a
$PbCl_2$	36	b
ZrO_2	25	b
UO_2	28	b

Metals (Kcal/mole)		
Compound	Δh_m	ref
Au	20	c
Ag	19	c
Cu	20	c
Al	15	c
Ni	32	c
Pt	32	c

a) Barr and Lidiard (1971)
b) Greenwood (1968)
c) Johnson (1973)

Note 1 ev = 23.056 Kcal/mole

the migration of vacancies or Frenkel defects in simple crystals. Figure 6 shows the path that a Mg^{2+} ion would follow in moving into an adjacent cation vacancy in MgO. Obviously, this motion would also move the vacancy (hence the term "vacancy migration"). In going from point E to point X, the Mg^{2+} ion would have to impinge on the nearby oxygen ions, and thus it requires energy (due mainly to ion repulsion) to pass from E to G in Figure 6. We shall discuss below some of the methods available to obtain these defect migration energies.

POINT DEFECTS AND DIFFUSION

The presence and mobility of point defects enables diffusion to take place in a crystal. The diffusion coefficient of an atomic species will depend on the type and the energetics of certain point defects. If the diffusion occurs via vacancy motion, then the number of atoms jumping from their sites into adjacent vacancies will depend on (1) the fraction of sites which are vacant and (2) the fraction of atoms, which being next to a vacancy, have sufficient thermal energy to go over the migration activation barrier and into the vacancy. The diffusion coefficient is then given by an equation of the form

$$D = \frac{d^2\nu}{3} X_v \, e^{-\frac{\Delta h_m}{RT}} \tag{13}$$

(Jost, 1960, p. 137) which incorporates both terms mentioned above. In equation (13) d is the interatomic distance traveled in a jump, X_v is the site fraction of vacancies, and Δh_m is the energy barrier to go from a filled site to a neighboring vacant site. ν is a vibrational frequency which sometimes is set to the Einstein frequency of a crystal or some fraction of the Debye frequency (see Flynn, 1972). Of course, the pre-exponential factor in equation (13) is quite approximate and only gives a rough idea of the magnitude of D. More detailed treatments involve a sophisticated account of the atomic dynamics in a crystal (Flynn, 1972; see also Chapter 4). A similar equation would follow if the diffusion mechanism involved interstitial sites (Jost, 1960). Both X_v and the exponential term are very sensitive to temperature variations (the fundamental reason behind their important use as geospeedometers!) and X_v is also possibly sensitive to the impurity content and/or oxidation state of the crystal, as we have seen. These terms are therefore critical in the evaluation of diffusion coefficients.

In elucidating the variation with temperature of the term X_v in equation (13) (or similarly the term X_i, if an interstitial mechanism is involved), we must make the important distinction between an extrinsic diffusion mechanism and an intrinsic diffusion mechanism. At high temperatures (intrinsic region) the size of X_v is determined from the energy needed to form the vacancy, Δh_f, *i.e.*, from equations (5) and (6)

$$X_v \propto e^{-\frac{\Delta h_f}{2RT}} \,. \tag{14}$$

In this case the exponential dependence on temperature in equation (14) will contribute to the overall exponential dependence of D in equation (13). We can write the temperature variation of the diffusion coefficient as

$$D = D_o e^{-Q_D/RT} \,.$$

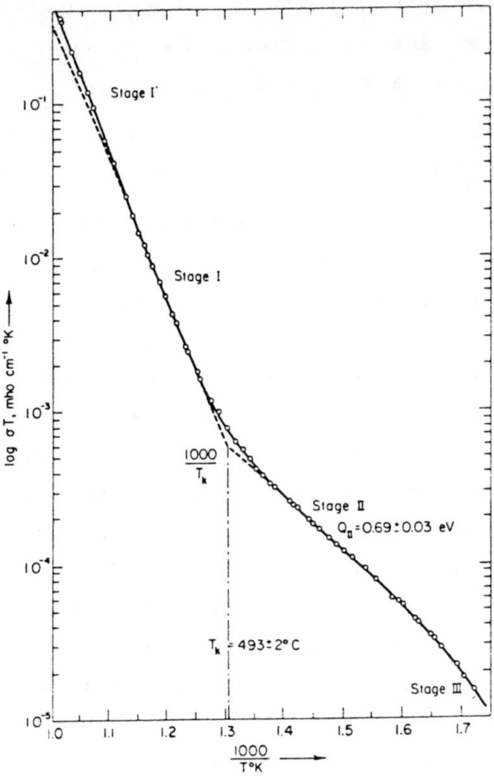

Figure 7. The ionic conductivity of "pure" NaCl as a function of temperature. Intrinsic conduction occurs in stages I and I', while stage II corresponds to conduction by cation vacancies present as a result of the divalent cation impurities in solution. In stage III these vacancies are becoming associated with the impurities to form neutral bound pairs. (After Barr and Lidiard, 1971.)

The overall activation energy for diffusion in the *intrinsic* region is then given by

$$Q_D = \frac{\Delta h_f}{2} + \Delta h_m \ . \quad (15a)$$

However, as the temperature is dropped, we reach the extrinsic region and the number of vacancies induced by the impurities (*e.g.*, aliovalent ions) is much bigger than the number of vacancies generated intrinsically (*i.e.*, by thermal equilibrium) as expressed in equation (12). Below this temperature, X_v is then fixed by the concentration of impurities [equation (10)] and is therefore *independent* of temperature. As a result, an exponential term from equation (13) is lost and the activation energy for diffusion decreases drastically. In the extrinsic region we have

$$Q_D = \Delta h_m \ . \quad (15b)$$

If Q_D is measured in *both* regions, it follows that both Δh_m and Δh_f can be evaluated. In this manner, the experimental numbers in Tables 3 and 6 were obtained. A problem with high melting oxides, however, is that the intrinsic region occurs at rather high temperatures (see Table 4), where diffusion measurements are hard to carry out. In these latter cases, only Δh_m can be extracted from Q_D. The variation of the activation energy with temperature predicted by (15a) and (15b) manifests itself as a *kink* in the *ln* D versus 1/T curve. A well-documented example of this behavior is seen in the accurate experimental data for ionic conductivity in NaCl (Fig. 7) or in the self-diffusion of Cl^- in NaCl doped with $SrCl_2$ (Fig. 5). Stage I in Figure 7 represents intrinsic ionic conduction and hence has a higher activation energy (increased slope). Stage II represents extrinsic ionic conduction, and this has a lower activation energy. Note the extensive data available in this case, which clearly show the change in diffusion or conduction mechanism. Stage III refers to the association of vacancies with impurities; this process leads to neutral defects and so does not contribute to the conduction. Therefore, the conductivity curve bends downwards in this region. In the case of Cl^- diffusion (Fig. 5), the anion vacancy concentration [equation (11)] is given by

$$X_{Cl^-,v} = \frac{K}{c} = \frac{e^{-\frac{\Delta G_f}{RT}}}{c} \propto e^{-\frac{\Delta h_f}{RT}}$$

in the extrinsic region (low T), where $X_{Na^+,v} \simeq c$. In this region, the diffusion activation energy from equation (13) is given by

$$Q_D = \Delta h_f + \Delta h_m \ . \tag{15c}$$

On the other hand, in the intrinsic region the activation energy is given by equation (15a). Hence, the Cl^- diffusion coefficient curve *increases* its slope as the temperature drops from the high T (intrinsic) to the low T (extrinsic) region in Figure 5. This is the *reverse* of the change in slopes for cation diffusion. Note that the kink occurs at a higher temperature as the impurity content increases in accord with equation (12). Figure 5 also shows another kink occurring at even lower temperatures with a decrease in slope. The latter region refers to vacancy-pair controlled diffusion, which according to equation (9) is independent of the impurity content. Clearly, a prediction of where these temperature regions lie is necessarily dependent on the energetics of the point defects, *e.g.*, equation (12). It is also important to understand where these changes in slope occur, since they will seriously affect the validity of any extrapolation of experimental data.

The electrical conductivity of forsterite single crystals, which seems to occur via cation motion (Pluschkell and Engel, 1968), also exhibits this change in slope. Figure 8 is a plot of the logarithm of the electrical conductivity of forsterite and MgO versus the reciprocal of the absolute temperature from Shankland (1968). The conductivity plot

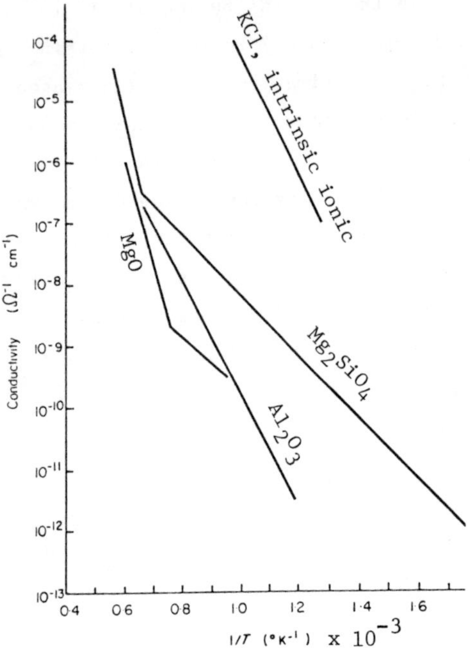

Figure 8. Electrical conductivities of several single-crystal insulators. (From Shankland, 1968.)

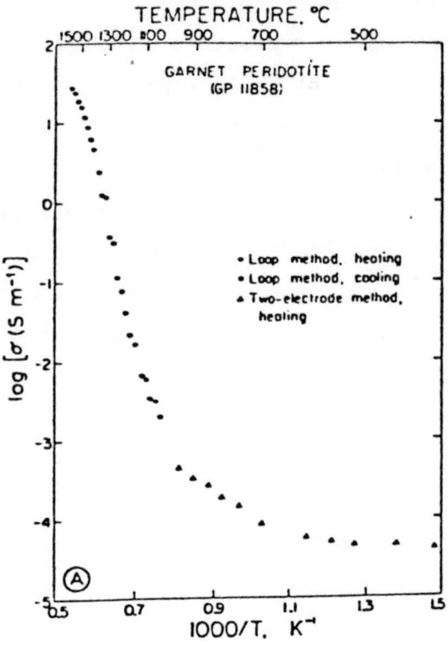

Figure 9. log σ versus 1/T for a garnet peridotite (From Rai and Manghnani, 1978.)

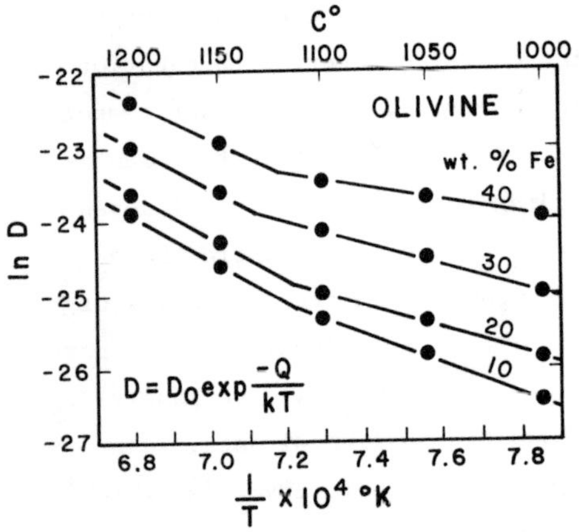

Figure 10. Arrhenius plot of *ln*D vs 1/T (*c*-axis, $pO_2 = 10^{-12}$atm). D = D_o *exp*(Q/kT).

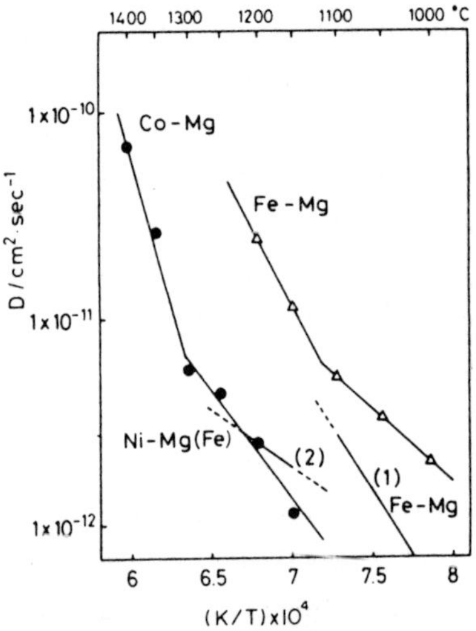

Figure 11. D vs 1/T for the *c*-axis. D is the extrapolated value for pure forsterite. Solid circles = Co-Mg, open triangles = Fe-Mg (Buening and Buseck, 1973). Solid line (1) = Fe-Mg (Misener, 1974). Solid line (2) = Ni-Mg (Fe) interdiffusion in olivine containing 93.7% forsterite (Clark and Long, 1971). (From Morioka, 1980.)

for forsterite shows two lines with different slopes meeting at a temperature of 1150°C. Furthermore, the slope of the line for high temperatures (low 1/T) is higher than that for low temperatures in accord with equations (15a) and (15b). We expect, therefore, the low-temperature region to be an extrinsic region and the high-temperature region to be an intrinsic region. Results obtained by Kobayashi and Maruyama (1971) show another change in slope at 800°K. Their activation energies for conduction in the temperature range 800-1200°K lie close to 18 Kcal/mole. This may reflect yet another mode of electron transport. At higher temperatures values of 53 Kcal/mole have been reported (Duba, 1972). Electrical conductivity plots similar to that of olivine have also been measured in ultramafic rocks of interest in the earth's mantle. For example, Figure 9 (Rai and Manghnani, 1978) contain data for a garnet peridotite.

Diffusion data on olivines exhibit patterns similar to the conductivity data, yielding an activation energy for cation diffusion along the *c*-axis of 30 Kcal/mole at $T < 1125°C$ and 62 Kcal/mole at $T > 1125°C$ when extrapolated to pure forsterite (see Fig. 10 after Buening and Buseck, 1973). It is interesting that Buening and Buseck also obtain a change in slope in the log D versus 1/T plot at a temperature of 1125°C (Fig. 10), which they suggest may be due to a change from an intrinsic to an extrinsic diffusion mechanism. Misener (1974) found an Fe-Mg interdiffusion activation energy of 58 Kcal/mole for the *c* direction in forsterite in the temperature range 900-1100°C. Data on Ni^{2+} diffusion along the *c* axis in olivine with 93% forsterite (Clark and Long, 1971) yield an activation energy of 46 Kcal/mole between 1149-1234°C. These numbers are similar in the high-temperature region. However, Misener's data yield a high activation energy to lower temperatures than Buening and Buseck's data. The Co-Mg cation diffusion data of Morioka (1980) are shown in Figure 11. Note that the change in slope occurs at a higher temperature (∿1300°C) in this case. This difference may be partly due to a difference in oxidation of Co and Fe in olivine; greater amounts of Co^{3+} or impurities would increase the temperature needed for intrinsic diffusion, T_c [see equation (12)]. However, a higher ΔG_s (the activation energy for diffusion of Co-Mg at $T > 1300°C$ is ∿126 Kcal/mole (Morioka, 1980)) will also increase T_c for Co-Mg diffusion according to equation (12).

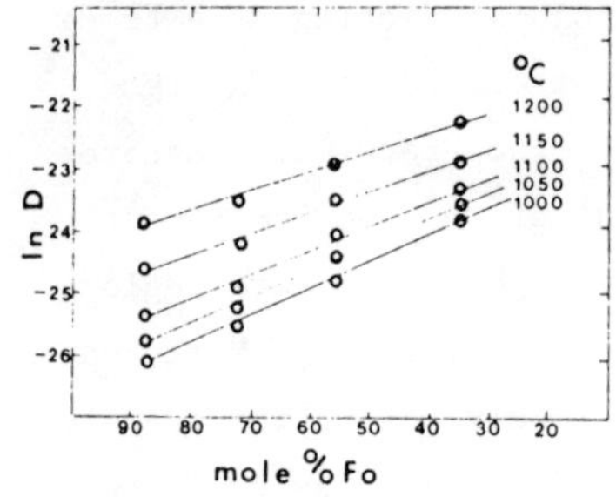

Figure 12. *lnD* versus mole percent forsterite along the *c*-axis at $pO_2 = 10^{-12}$ atm (Buening and Buseck, 1973.)

All the diffusion data presented above find that cation diffusion along the *c* axis of olivine is faster than that along the *a* or *b* axis. However, the oxygen diffusion data of Reddy *et al.* (1980) suggests that the *b*-axis is the fastest direction for anion transport. The cation diffusion anisotropy is in accord with recent crystal structure calculations (Lasaga, 1980) to be discussed below.

Figure 12 shows the variation in the Fe-Mg interdiffusion coefficient of olivines as a function of Fe content at constant f_{O_2} and temperature. At constant f_{O_2} the ratio of Fe^{3+}/Fe^{2+} in olivines is fixed. Therefore, increasing the Fe^{2+} content will increase Fe^{3+} and so will also increase the number of cation vacancies. This increase should then increase the diffusion coefficient, as is found in Figure 12. Of course, it should be stressed that since the data in Figure 12 refer to *interdiffusion* coefficients, only *part* of the variation in D can be related to the change in Fe^{3+} content; the rest is related to changes in the actual lattice itself (see Chapter 6 by Anderson). The latter point is clear from the reduced but non-trivial variation of D with Fe^{2+} content in the high-T intrinsic region (see also Fig. 10).

For general purposes, some of the available diffusion data on silicates, including the data on olivines mentioned above, is summarized in Table 7. Most of the low-temperature data is expected to be in the extrinsic region and therefore should relate to the migration of defects.

Finally, it should be mentioned that recent work on the conductivity of forsterite (Morin *et al.*, 1977, 1979) suggests that intrinsic electrical conduction occurs at higher temperatures than those of Shankland or Pluschkell and Engel. They claim that the earlier experiments may have been influenced by thermionic emission from the contacts. More work is needed to establish firmly the dominant defects of transport in olivines and their temperature dependence.

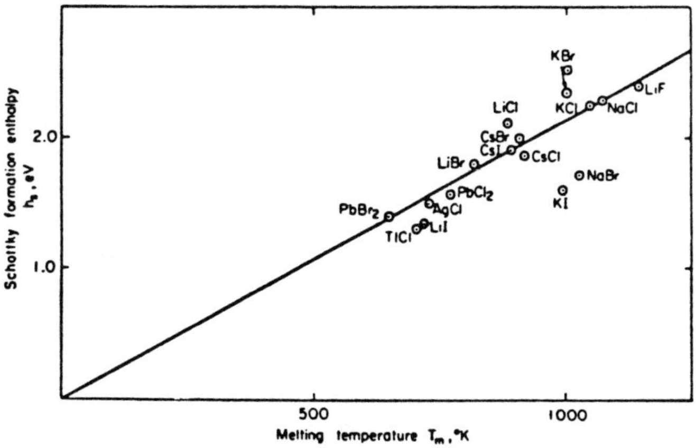

Figure 13. Correlation of the enthalpy of formation of Schottky defects with melting temperature T_m of various halides (Barr and Lidiard, 1971.)

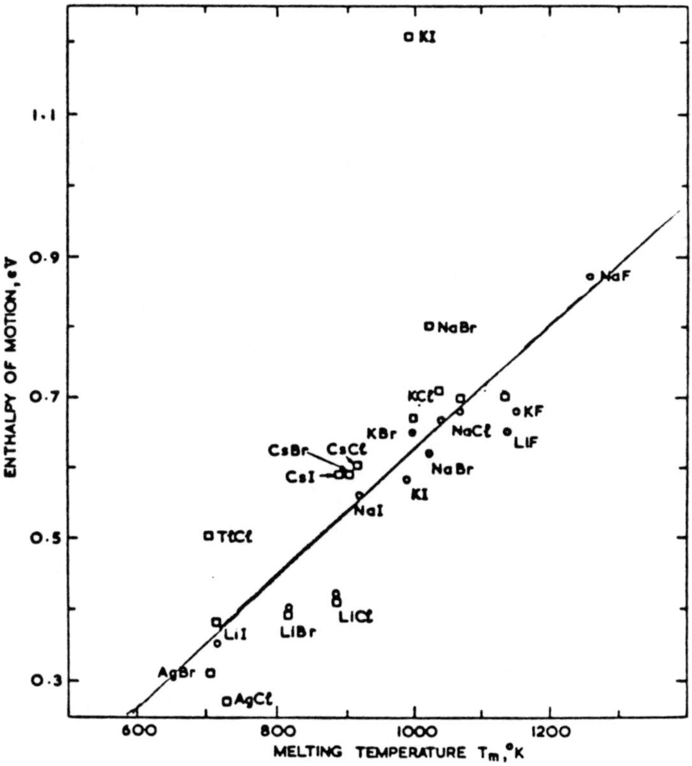

Figure 14. Correlation of the enthalpy of motion of vacancies with the melting temperature T_m of various halides. Circles: bound vacancy, Squares: free vacancy. (After Barr and Lidiard, 1971.)

Table 7

Diffusion in Silicates

Mineral	Species	T-range	D_o (cm^2/sec)	Q_D (Kcal/mole)	reference
Mg_2SiO_4	c-axis Fe-tracer	900-1100°C	4.1×10^{-3}	58.9	Misener (1974)
	c-axis Ni-tracer	1149-1234°C	1.1×10^{-5}	46.0	Clark and Long (1973)
	c-axis Fe-tracer	1125-1200°C $f_{O_2} = 10^{-1/2}$	3.1×10^{-2}	61.1	Buening and Buseck (1973)
	c-axis Fe-tracer	1000-1125°C	8×10^{-7}	31.7	ibid
	a-axis Fe-tracer	1125-1200°C	1.3×10^{-1}	67.6	ibid
	c-axis Co-tracer	1300-1400°C	1.9×10^{6}	126	Morioka (1980)
	c-axis Co-tracer	1150-1300°C	2×10^{-5}	47	ibid
	a-axis b-axis	1275-1625°C	7.3×10^{-5}	77	Reddy et al. (1980)
	O-self diffusion $f_{O_2} = 1$	1275-1625°C	3.5×10^{-3}	89	ibid
	c-axis O-self diffusion	1275-1625°C	1.3×10^{-2}	95	ibid
Fe_2SiO_4	c-axis Mg-tracer	900-1100°C	1.5×10^{-2}	49.8	Misener (1974
	c-axis Mg-tracer	1125-1200°C $f_{O_2} = 10^{-12}$	2.9×10^{-2}	58.8	Buening and Buseck (1973)
	c-axis Mg-tracer	1000-1125°C $f_{O_2} = 10^{-12}$	7.6×10^{-7}	29.5	ibid
Co_2SiO_4	Si^{30}-tracer	1200-1300°C	8.9×10^{-3}	79.4	Schmalzried (1978)
$KAlSi_3O_8$	O self-diffusion	350-700°C 1kb hydrothermal	4.5×10^{-8}	25.6	Giletti et al. (1978)
Adularia	O self-diffusion	400-700°C 2kb hydrothermal	5.3×10^{-7}	29.6	Yund and Anderson (1974)

Table 7 continued

Orthoclase	Na-tracer	500-800°C 2kb hydrothermal	8.9	52.7	Foland (1974)
	K-self-diffusion	500-800°C 2kb hydrothermal	16.1	68.2	ibid
	Rb-tracer	500-800°C 2kb hydrothermal	38	73	ibid
	Sr^{90}-tracer	800-870°C	6×10^{-4}	41.2	Misra and Venkatasubramanian (1977)
Microcline	O-self-diffusion	400-700°C 2kb hydrothermal	2.8×10^{-6}	29.6	Yund and Anderson (1974)
	K^{40}-self-diffusion	600-800°C 0.5-2kb hydrothermal	133.8	70.0	Lin and Yund (1972)
	Sr^{90}-tracer	800-870°C	5×10^{-4}	38.6	Misra and Venkatasubramanian (1977)
$NaAlSi_3O_8$	O-self-diffusion	350-800°C 1kb hydrothermal	2.3×10^{-9}	21.3	Giletti et al. (1978)
		600-800°C 2kb hydrothermal	2.5×10^{-5}	37	Anderson and Kasper (1975)
	Na-self-diffusion	600-940°C	5×10^{-4}	35	Bailey (1971)
		200-600°C	2.3×10^{-6}	19	Lin and Yund (1972)
Biotite $K(Mg,Fe)_3AlSi_3O_{10}(OH)_2$	c-axis K-self-diffusion	550-700°C 2kb hydrothermal	2.7×10^{-10}	21	Hoffmann and Giletti (1970)
	K^{40}-self-diffusion	$D_a \sim D_b > 10^{2-4} D_c$			Hoffmann et al. (1974)
Anorthite $CaAl_2Si_2O_8$	O-self-diffusion	350-800°C 1kb hydrothermal	1.4×10^{-7}	26.2	Giletti et al. (1978)
Garnet $Fe_{2.4}Mg_{.6}Al_2Si_3O_{12}$	Mg-Fe interdiffusion	500-700°C	3.7×10^{-4}	60	Lasaga et al. (1977)
SiO_2	^3H_2O-tracer		2.8×8.5	22-26	Shaffer et al. (1974)

ESTIMATION OF Δh_m, Δh_s, Q_D, AND D_o

It would be very useful if a simple scheme existed that allowed the calculation of the point defect energetics and diffusion constants in minerals. Currently, there are available methods, which are well founded and give reasonable results for simple ionic crystals (Lasaga, 1980). We will comment on them in a later section. Here we will concentrate on simple *empirical* relationships, which have been used in the past, and on their applicability to silicates.

The simplest relation established has been between defect energies and the *melting temperature* of solids. Figures 13 and 14 show the *linear* correlation found between the enthalpy of vacancy formation, Δh_s, or the enthalpy of vacancy migration, Δh_m, and the melting temperatures of alkali halides. The lines fit the following equations:

$$\Delta h_s = 2.14 \times 10^{-3}\ T_m (\text{ev}) \qquad (T_m \text{ in } {}^\circ\text{K}), \tag{16a}$$

$$\Delta h_m = 6 \times 10^{-4}\ T_m (\text{ev}) \qquad (T_m \text{ in } {}^\circ\text{K}), \tag{16b}$$

or

$$\Delta h_s = 0.0493\ T_m \qquad (\text{Kcal/mole}),$$

$$\Delta h_m = 0.014\ T_m \qquad (\text{Kcal/mole}).$$

Although there is scatter in the figures, the relationship holds rather well. Some papers have tried to justify the linear relation between Δh_s, Δh_m and T_m (see Varotsos and Alexopoulos, 1977; Glyde, 1967; Mukherjee, 1965; Couchman and Ryan, 1978). However, the theoretical explanation has not been settled in a satisfactory manner as yet. Note that if both Δh_s and Δh_m are linearly related to the melting temperature, it must also follow that they are linearly related to each other, *i.e.*, $\Delta h_m = A\ \Delta h_s$ for some constant A. [$A = 0.28$ for alkali halides.] While the theoretical explanation for the *linear* relationship is not clear, it is not surprising to relate Δh_s, Δh_m to T_m. RT_m is related to the energy needed to destroy the crystal structure, at least on a long range level, and hence is related to the creation and motion of defects at large.

Figures 15 and 16 give similar data for vacancies in metals. For metals the linear relation becomes (Johnson, 1973; Trivari *et al.*, 1976):

$$\Delta h_s = 7.5 \times 10^{-4}\ T_m\ \text{ev} \qquad (T_m \text{ in } {}^\circ\text{K}) \tag{17}$$

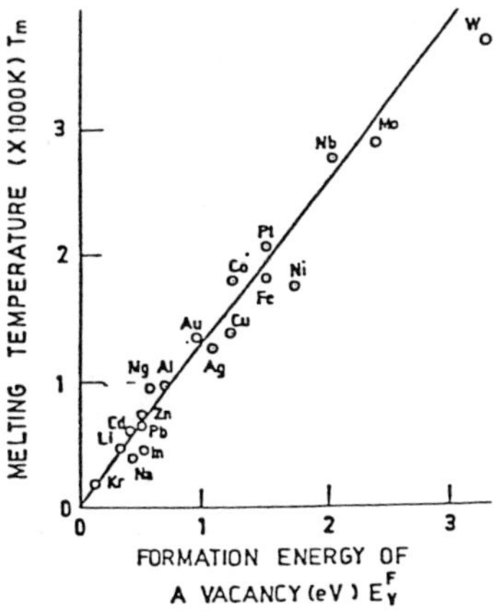

Figure 15. The relation between the for-tormation energy of a vacancy E_V^F and mel-ting temperature T_m.

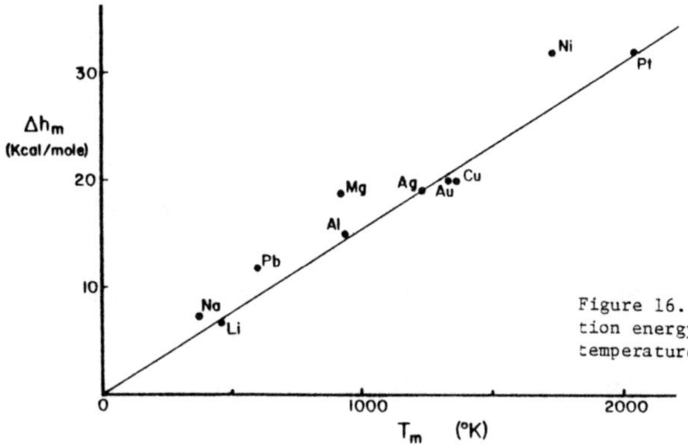

Figure 16. The relation between the migra-tion energy of a vacancy, Δh_m and melting temperature.

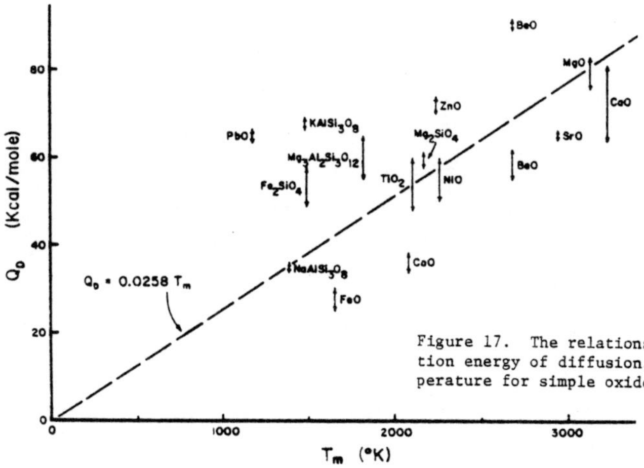

Figure 17. The relationship between the activa-tion energy of diffusion, Q_D, and melting tem-perature for simple oxides and silicates.

or $= 0.0173\ T_m$ (Kcal/mole),

$$\Delta h_m = 6.0 \times 10^{-4}\ T_m\ \text{ev} \qquad (T_m \text{ in } °K) \tag{18}$$

or $= 0.0138\ T_m$ (Kcal/mole).

Note that the size of Δh_s in metals is generally lower than that in ionic crystals, reflecting the looser bonds in metals.

If both Δh_s and Δh_m are proportional to T_m then the diffusion activation energy, Q_D, should also be proportional to T_m by virtue of equations (15a) and (15b). The data must pertain to *self-diffusion* coefficients (see Chapter 6) since the relationship refers to the intrinsic properties of the crystal structure itself (*e.g.*, T_m). Barr and Lidiard (1971) used equation (16a) to estimate Δh_s for simple oxides, where experimental data is lacking. We have plotted some of the known diffusion activation energies for a variety of oxides including some silicates in Figure 17. Note that in some cases there are still large uncertainties in Q_D as shown by the lines in the plot. The scatter in the plot is greater than that for alkali halides or metals. However, the increased scatter in part reflects the bigger variety in structural types represented in Figure 17. Another reason for increased scatter in Figure 17 stems from the possible dependence on oxygen fugacity, the possible inclusion of both intrinsic and extrinsic diffusion data and diffusion anisotropy. The dashed line in the figure obeys the equation

$$Q_D = 0.0258\ T_m \text{ Kcal/mole} \qquad (T_m \text{ in } °K). \tag{19}$$

The proportionality in equation (19) is intermediate between that in equations (16a) and (16b) for halides.

In a recent paper (Lasaga, 1979a) the point was made that the activation energies for tracer diffusion in NaCl and MgO were similar among the different cations and thus were a function only of crystal structure, if the charge on the ions was the same. This observation would be in accord with a general relation such as that between Q_D and T_m. A refined explanation of this result, however, is given in a later section.

It should be noted that if Q_D is proportional to T_m, then the *pressure* dependence of the diffusion activation energy could be

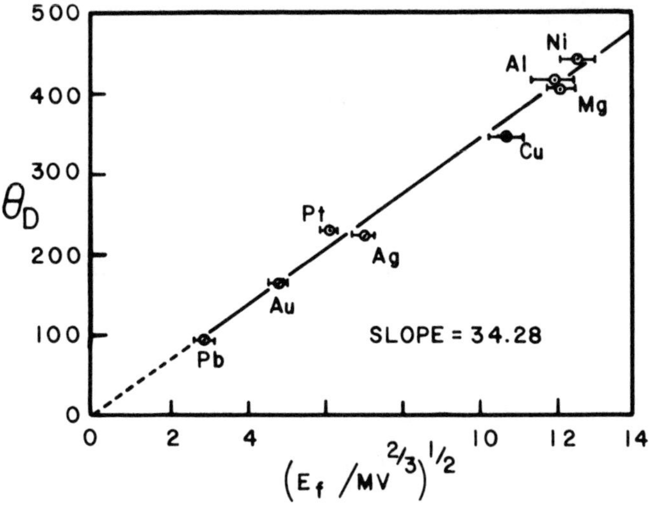

Figure 18. Debye characteristic temperature Θ_D versus $(E_f/MV^{2/3})^{1/2}$ for close-packed metals. E_f is the formation energy of vacancies in cal/gram-atom and M and V are the mass and volume respectively of a gram-atom (Mukherjee, 1965.)

be obtained from the pressure dependence of the melting point, *i.e.*, from the Clausius-Clapeyron equation.

Other papers (*e.g.*, Trivari *et al.*, 1976; Glyde, 1967; Mukherjee, 1965) have also related Δh_s and Δh_m to the square of the Debye temperature, the heat of fusion and the cohesive energy of the lattice. Figure 18 is a plot of Θ_D, the Debye temperature, versus the square root of the vacancy formation energy, Δh_f, for metals from Mukherjee (1965). It is found that Θ_D satisfies the following equation

$$\Theta_D = 34.3(\Delta h_f/MV^{\frac{2}{3}})^{\frac{1}{2}} \qquad (\Delta h_f \text{ in cal/mole}), \tag{20}$$

where M and V are the atomic weight and molar volume of the metal. Again a reasonably straight line is obtained. Glyde (1967) has given a theoretical account of the relation between Δh_m or Δh_s and Θ_D^2. Finally, to close out the loop, Couchman and Ryan (1978) and Couchman (1979) discuss the basis and consequences for the so-called *Lindemann relation* which relates T_m and Θ_D:

$$T_m \propto \Theta_D^2 \ . \tag{21}$$

Equation (21) links the T_m and Θ_D^2 relations found in the literature [equations (17) and (20].

Grimes (1972) has made an interesting application of relation (20) to compounds with the spinel structure. Since both Δh_s and Δh_m

are proportional to Θ_D^2, the activation energy for diffusion, Q_D is expected also to be proportional to Θ_D^2. Grimes rewrites the relation obtained by Glyde (1967) for face-centered cubic metals as

$$Q_D = 2.21 \times 10^{-2} \; (2\pi)^2 (\frac{k}{h})^2 m \; \Theta_D^2 (\frac{a}{n})^2 \tag{22}$$

where m is the mass of the diffusing ion, a is the lattice constant, and n is a pure number such that a/n is the size of the diffusion jump (*e.g.*, for fcc metals n = $\sqrt{2}$). Grimes assumes that, since the packing of oxygen atoms in spinels approximates cubic closest packing (∿fcc), equation (22) is also applicable to spinels. An important problem in using equation (22) is the determination of the Debye temperature, Θ_D, which characterizes the vibrational spectrum of the crystal structure. Normally, Θ_D can be derived from the measurements of the heat capacity of spinels close to 0°K (see Kittel, 1956). Grimes chose to compute Θ_D from the root mean square of the temperatures determined from the positions of the four infra-red absorption bands of spinels:

$$\Theta_i \equiv \frac{hc\nu_i}{k} \tag{23}$$

where ν_i is one of the four infra-red frequencies.

If Q is known from diffusion data, the lattice constant a from x-ray data and Θ_D computed from the infra-red spectrum, n in equation (22) can be calculated from

$$n = 2 \times 10^{-4} \; a \; \Theta_D \sqrt{\frac{m}{Q_D}} \tag{24}$$

(a in Å, Θ_D in °K, m in atomic mass units and Q_D in ev). Table 8 gives the results obtained for nine spinels. The calculations were applied to both cation and anion (O^{2-}) diffusion. The n values obtained range from 2 to 6 and give a certain validity to equation (22).

In fact, Grimes uses the n values to predict the crystallographic defect diffusion path. For example (see Fig. 19) a cation in an octahedral site has 12 nearest octahedral sites (six actually occupied) at a distance of $a\sqrt{2}/4$. In this case a jump from octahedral to octahedral sites would give n = $4/\sqrt{2}$ = 2.83. Similarly, an alternating octahedral-tetrahedral jump traverses a distance of $a\sqrt{3}/8$ and would yield n = $8/\sqrt{3}$ = 4.62. Finally, a tetrahedral-tetrahedral jump has n = 4. Therefore, the data in Table 8 suggest that the migration of divalent ions, *e.g.*,

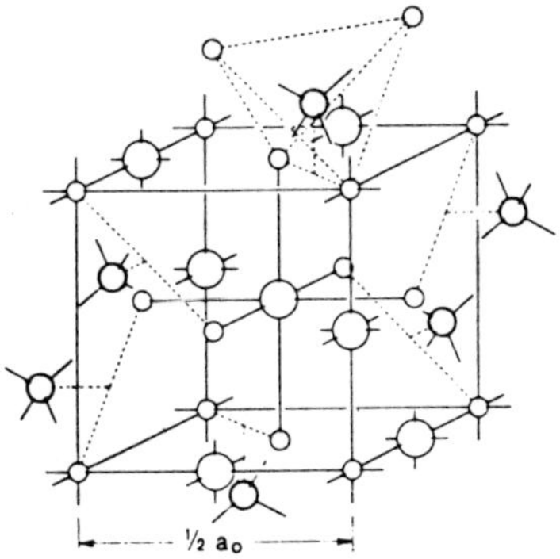

Figure 19. The structure of $MgAl_2O_4$, (spinel.) ne eighth of the unit cell is shown.

Small circles = oxygen.
Intermediate circles = Mg.
Large circles = Al.

Table 8. Data for Derivation of Parameter n

Compound	Lattice constant a in Å	Oxygen u parameter	Θ_D °K	Diffusing species	Crystal Site occupied at room temperature	Q ev	n
$MgAl_2O_4$	8.080 (a)	0.387 (a)	739	$^{28}Mg^{2+}$	Tetrahedral	3.74	3.27
$NiAl_2O_4$	8.046 (b)	0.381 (b)	720	$^{63}Ni^{2+}$	Tet. + Oct.	2.31	6.03
				$^{65}Ni^{2+}$	Tet. + Oct.	2.39	6.04
$ZnAl_2O_4$	8.086 (c)	0.389 (c)	752	$^{65}Zn^{2+}$	Tetrahedral	3.39	5.32
$CoCr_2O_4$	8.332 (d)		660	$^{60}Co^{2+}$	Tetrahedral	2.22	5.72
				$^{51}Cr^{3+}$	Octahedral	3.04	4.50
$NiCr_2O_4$	8.317 (d)	0.387 (e)	645	$^{63}Ni^{2+}$	Tetrahedral	2.66	5.22
				$^{65}Ni^{2+}$	Tetrahedral	3.24	4.80
				$^{51}Cr^{3+}$	Octahedral	3.15	4.31
				$^{18}O^{2-}$	Oxygen	2.84	2.70
$ZnCr_2O_4$	8.327 (d)	0.390 (f)	670	$^{65}Zn^{3+}$	Tetrahedral	3.72	4.66
				$^{51}Cr^{3+}$	Octahedral	3.52	4.25
$FeFe_2O_4$	8.394 (g)	0.379 (g)	548	^{55}Fe	Tet. + Oct.	2.39	4.42
$NiFe_2O_4$	8.325 (h)	0.382 (h)	585	$^{55}Fe^{3+}$	Tet. + Oct.	3.56	3.83
				$^{18}O^{2-}$	Oxygen	2.65	2.54
$ZnFe_2O_4$	8.443 (i)	0.387 (i)	555	$^{65}Zn^{2+}$	Tetrahedral	3.74	3.91
				$^{59}Fe^{3+}$	Octahedral	3.56	3.82

(from Grimes, 1972)

Co^{2+}, Ni^{2+}, Zn^{2+}, occurs by alternating jumps between octahedral and tetrahedral sites.

In closing this section it should be noted that these relations are *empirical* and therefore should be used with caution. However, the relations are useful as "estimates" of defect energetics and diffusion energies and also as guidelines for theoretical research. Of course, for complex silicates the variety of coordination number, structures and oxidation state require a greater refinement of these relations.

Compensation Law

Another interesting empirical observation, named the *compensation law*, relates the pre-exponential factor and the activation energy for diffusion (Winchell, 1969; Winchell and Norman, 1969). The temperature dependence of the diffusion coefficient can be written as

$$D = D_o \, e^{-\frac{Q_D}{RT}} \qquad (25)$$

in the usual manner. An increase in Q_D would decrease the size of D (at fixed T) through the exponential term in equation (25). However, an increase in D_o can offset or compensate the effect of a rise in Q_D. Therefore, if D_o and Q_D both increase or decrease simultaneously, one can speak of a compensation law. Most people who have perused diffusion data probably have noticed that usually, if the activation energy, Q_D, is decreased, so is the pre-exponential term, D_o, and vice versa (see Table 7). In most cases, the compensation law is much more explicit and relates log D_o and Q_D in a linear fashion (see also Chapter 4):

$$Q_D = a + b \log_{10} D_o \quad . \qquad (26)$$

Most of the Winchell's data (1969) were obtained from diffusion in silicate glasses and are shown in Figure 20. If materials with similar structure are chosen then the scatter can be reduced (Fig. 21). Winchell found that equation (26) when applied to *all* ion species in a variety of structures (Fig. 20) yielded the constants a, b as (D_o in cm^2/sec)

$$Q_D = 41 + 6.9 \log_{10} D_o \qquad \text{(Kcal/mole)} \qquad (27)$$

with a correlation coefficient of 0.87.

The previous section noted the relationship between Q_D and Θ_D^2 and theory relates D_o with Θ_D [equation (13)], it is not surprising therefore to find that Q_D and D_o are related. However, this relation is not

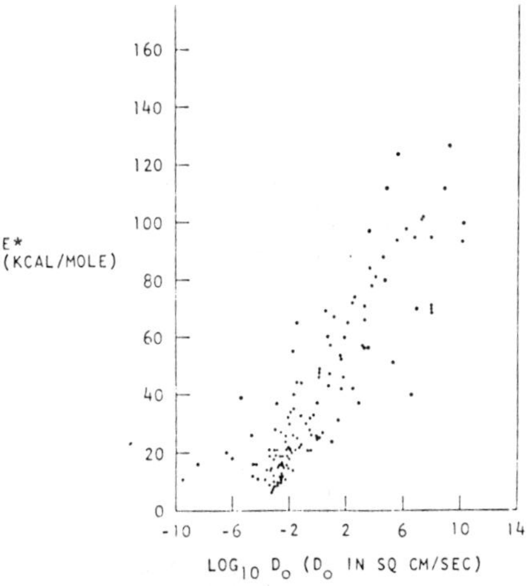

Figure 20. Compensation law correlation for diffusion in silicates. E* is the activation energy for diffusion. D_o is the pre-exponential factor. (From Winchell, 1969.)

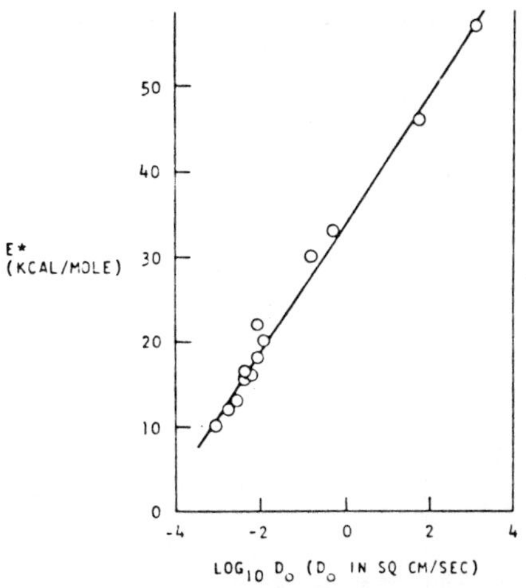

Figure 21. Example of compensated diffusion in silicates for the transport of Na, K, and Ca in $Na_2O - K_2O - R_xO - SiO_2$ glasses (R = Li, Na, K, Mg, Sr, Ca, Ba). (From Winchell, 1969.)

expected to be universal for *all* minerals; nor is the relationship expected apriori to be linear as in equation (27).

One interesting consequence of the compensation law in equation (26) is the prediction that *all* the diffusion coefficients become equal at a universal temperature! To show this, insert equation (26) into equation (25) to obtain

$$D = D_o\, e^{-\frac{a + b \log_{10} D_o}{RT}} \quad \text{or} \quad \log_{10} D = \log_{10} D_o - \frac{a + b \log_{10} D_o}{2.303\ RT}$$

It can be checked that if T is such that

$$\log_{10} D_o - \frac{b \log_{10} D_o}{2.303\ RT} = 0 \ ,$$

then
$$\log_{10} D = - \frac{a}{2.303\ RT} \ .$$

Therefore we conclude that at the unique temperature

$$T^* = \frac{b}{2.303\ R} \ , \qquad (28)$$

D has the value:
$$D = 10^{-a/b} \qquad (29)$$

for *all* species in *all* crystals! T* is equal to 1508°K from equation (28) and the data in (27). Therefore, at a temperature of 1508°K, *all* the diffusion coefficients should equal $10^{-a/b}$ or $1.1 \times 10^{-6}\ cm^2/sec$.

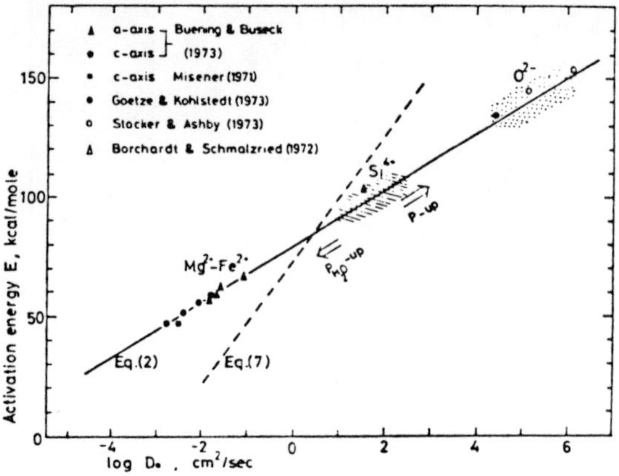

Figure 22. Compensation correlation between activation energy E and logarithm of pre-exponential factor D_o from the lattice diffusion coefficients in olivine. The solid line is drawn on the basis of the experimental data of Mg-Fe interdiffusion from Misener (1971) and Buening and Buseck (1973). The hatched area and dotted area show the estimated domains of the lattice self-diffusion coefficients of Si and O in olivine (Fo_{90}) to be found. (From Tsukahara, 1976.)

The compensation law has been recently used by Tsukahara (1976) in predicting diffusion creep in olivine. Tsukahara plotted the olivine diffusion data from Misener (1971) and Buening and Buseck (1973) for Fe-Mg interdiffusion and the estimated values for oxygen diffusion in olivine from Stocker and Ashby (1973) in a Q_D versus $\log_{10}D_o$ plot (Fig. 22). The line in Figure 22 obeys the relation

$$Q_D = 80 + 12 \log D_o \quad \text{(Kcal/mole)} \tag{30}$$

which is *different* from Winchell's relation [equation (27)]. Now the universal temperature is much higher, *i.e.*, $T^* = 2622°K$; also the common D is equal to $2 \times 10^{-7} cm^2/sec$ at T^*. Tsukahara then estimates Q_D for Si and uses the compensation law to obtain D_o and therefore D. Unfortunately, recent diffusion data obtained by proton activation analysis (Reddy *et al.*, 1980) for oxygen self-diffusion in forsterite plots substantially off the compensation line in Figure 22 (see Table 7).

The use of the compensation law in silicates will have to be modified to recognize the variations in structure and in the types of defects which are possible in minerals. This is obvious in comparing equations (27) and (30). For example, the data in Table 7 for cation and oxygen diffusion along the *c*-axis of forsterite yield

$$Q_D = 83 + 6.4 \log_{10}D_o \quad (r = 0.83 \text{ for 7 points}).$$

Likewise, the data for oxygen self-diffusion in feldspar yield

$$Q_D = 54 + 3.9 \log_{10}D_o \quad (r = 0.99 \text{ for 4 points}),$$

and the data on hydrothermal cation diffusion in orthoclase yield

$$Q_D = 26 + 31 \log_{10} D_o \qquad (r = 0.92 \text{ for 3 points}).$$

While there are significant variations in the a and b coefficients, the high correlation factors (r) suggest that, if the law is applied within a certain class of minerals (*e.g.*, feldspars or olivines), it could be a useful empirical tool.

NON-STOICHIOMETRY, DEFECTS AND DIFFUSION

We have already discussed the profound effect that changing the iron content has on diffusion and electrical conduction in olivines. While olivines are generally considered "stoichiometric" minerals, it is obvious that impurities such as Fe^{3+} are not negligible. Minerals such as $Fe_{1-x}O$ or $Fe_{1-x}S$ can have large deviations from stoichiometry. We will relate the non-stoichiometry of these minerals and the size of diffusion coefficients to the defect structure and to such external variables as the oxygen fugacity. In fact, it was not until scientists realized the importance of f_{O_2} on conductivity in olivine that the wide disagreement between different measurements was reconciled. In many compounds containing atoms with multiple valencies, which includes most silicates, the oxygen fugacity can be an important factor in the defect structure. We will illustrate the relation between non-stoichiometry and diffusion with a series of geochemically important iron-containing oxides and silicates. A classic example is wüstite, $Fe_{1-x}O$. The Fe/O ratio of $Fe_{1-x}O$ will clearly depend on f_{O_2}. For example, we can write thermodynamically

$$(1-x)FeO + \frac{x}{2} O_2 \rightleftarrows Fe_{1-x}O \ . \qquad (31)$$

We can, therefore, relate x to f_{O_2}. A similar relation is also found between x in $Fe_{1-x}S$ and f_{S_2}, a relation often used by economic geochemists. However, equation (31) is not sufficient to characterize the *defect* structure in $Fe_{1-x}O$. There are at least two ways of obtaining an excess of oxygen atoms in $Fe_{1-x}O$: oxygen interstitials or iron vacancies. Equation (31) does not predict which defect predominates. Nonetheless, there is a direct relationship between the defect structure and the diffusion behavior of a mineral as already seen in equation (13). If f_{O_2} influences the non-stoichiometry of $Fe_{1-x}O$ and the defects influence the diffusion rates, the diffusion coefficient, D, will be dependent on f_{O_2}. In fact, the f_{O_2} dependence of D can be predicted, *if* we know the diffusion mechanism. Furthermore, this relation between D and f_{O_2} will enable a very important check on the nature of the defects playing a dominant role in diffusion.

For example, suppose that Fe diffusion in $Fe_{1-x}O$ occurred via cation vacancies. We can write a defect reaction for the formation of vacancies in $Fe_{1-x}O$ as follows

$$2Fe^{2+} + 1/2\ O_2 \rightarrow 2Fe^{3+} + V_{Fe^{2+}} + O^{2-}\ , \tag{32a}$$

or using the notation of Kröger

$$2Fe^{x}_{Fe} + 1/2\ O_2 \rightarrow 2Fe^{\cdot}_{Fe} + V''_{Fe} + O^{x}_{O}\ . \tag{32b}$$

Note that equation (32b) conserves charge on both sides of the equation (*i.e.*, $2^{\cdot}$ and 2'). The equilibrium constant for the defect reaction is

$$K = \frac{X^2_{Fe^{3+}}\, X_{v,Fe}\, X_{O^{2-}}}{X^2_{Fe^{2+}}\, f^{1/2}_{O_2}} \tag{33}$$

where $X_{Fe^{3+}}$, $X_{Fe^{2+}}$, $X_{v,Fe}$ are the mole fraction of cation sites occupied by Fe^{3+}, Fe^{2+} and a vacancy, respectively, and $X_{O^{2-}}$ is the fraction of oxygen sites occupied by oxygen. It will be a very good approximation to set $X_{Fe^{2+}} \sim 1$ and $X_{O^{2-}} \sim 1$. Moreover, if electroneutrality is maintained in the structure (or equivalently from the stoichiometry of reaction (32))

$$2X_{v,Fe} = X_{Fe^{3+}}\ . \tag{34}$$

As a result,

$$K = \frac{X_{v,Fe}(2X_{v,Fe})^2}{f^{1/2}_{O_2}}$$

or

$$X_{v,Fe} = \left(\frac{K}{4}\right)^{1/3} f_{O_2}^{\ 1/6}\ . \tag{35}$$

In a vacancy-dominated diffusion mechanism equations (13) and (35) imply that

$$D_{Fe} \propto X_{v,Fe} \propto f^{1/6}_{O_2}\ . \tag{36}$$

Therefore, the self-diffusion of Fe in $Fe_{1-x}O$ should *increase* with f_{O_2}

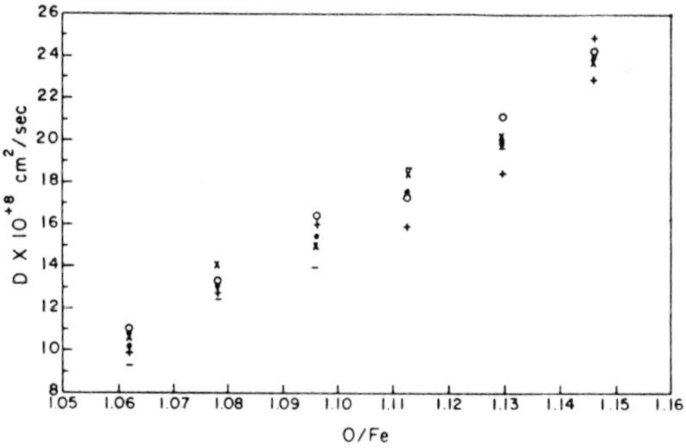

Figure 23. The tracer diffusion coefficient of iron in wüstite at 1100° C vs the atomic ratio of oxygen to iron. (.) Average of four tracer coefficient determinations. (°) and (-) represent the determination of the tracer coefficient from the same specimen at one composition at the end of 30 min (°) and 60 min (-). (+) and (x) represent the determination of tracer coefficients from the same specimen at one composition at the end of 30 min (+) and 60 min (x). (From Hembree and Wagner, 1969.)

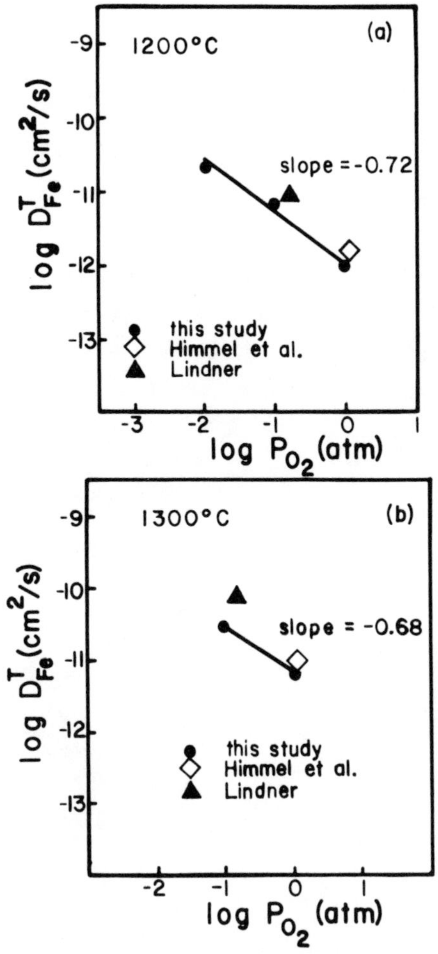

Figure 24. Iron tracer diffusion coefficient of hematite as a function of oxygen partial pressure at (a) 1200°C and (b) 1300°C (Chang and Wagner, 1972.)

and vary proportional to $f_{O_2}^{1/6}$. Since the number of defects is extensive in $Fe_{1-x}O$, it is expected that equation (36) holds only approximately due to the pairing or clustering of defects (*e.g.*, a $V''_{Fe} \cdot Fe^{\bullet}_{Fe}$ complex). Nonetheless, experimental data from Hembree and Wagner (1969) support equation (36). Hembree and Wagner measured the iron tracer diffusion coefficient of wüstite at 1100°C as a function of composition. They fixed the oxygen fugacity by using a predetermined CO_2-CO mixture. Figure 23 shows the tracer diffusion coefficient as a function of the O/Fe ratio in wüstite and the P_{CO_2}/P_{CO} ratio of the experiment. [O/Fe was obtained from O/Fe = 0.0841 $\log(P_{CO_2}/P_{CO})$ + 1.0869.] The straight line indicates that

$$D_{Fe} \sim A\left(\frac{P_{CO_2}}{P_{CO}}\right)^{0.36\pm0.04}$$

Since $f_{O_2}^{1/2} \propto P_{CO2}/P_{CO}$, the experimental data yield $D_{Fe} \propto f_{O_2}^{0.18\pm0.02}$. Therefore, the experimental data agree within experimental uncertainty with

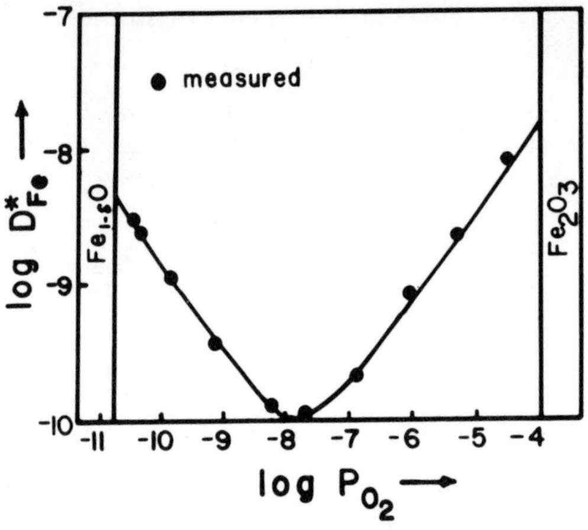

Figure 25. Iron tracer diffusion coefficients in $Fe_{3-\delta}O_4$ as a function of oxygen partial pressure (in atmosphere) at 1100°C. (From Dieckmann and Schmalzried, 1975.)

the 1/6 power law (0.17) theoretically predicted in equation (36). In turn, the agreement between theory and experiment supports the proposed vacancy mechanism for iron diffusion in wüstite and the defect structure implied by reaction (32).

In the case of hematite, diffusion seems to occur via an *interstitial* mechanism. Chang and Wagner (1972) carried out iron tracer (Fe^{55}) diffusion experiments as a function of f_{O_2} at 1200°C and 1300°C. The experimental data, while not extensive, yield

$$D_{Fe} \propto f_{O_2}^{-0.72} \qquad [T = 1200°C] \tag{37a}$$

and

$$D_{Fe} \propto f_{O_2}^{-0.68} \qquad [T = 1300°C] \tag{37b}$$

(see Fig. 24). If the diffusion is of the interstitial type, the relevant defect reaction becomes

$$Fe_{Fe}^{x} + 3/2\ O_O^{x} \rightarrow Fe_i^{x} + 3/4\ O_2\ . \tag{38}$$

The equilibrium constant expression is

$$K = \frac{X_{i,Fe}\ f_{O_2}^{3/4}}{X_{Fe^{3+}}\ X_{O^{2-}}}\ . \tag{39}$$

Since $X_{Fe^{3+}}$, $X_{O^{2-}} \sim 1$ as usual, we obtain immediately

$$X_{i,Fe^{3+}} \propto f_{O_2}^{-3/4}\ . \tag{40}$$

Equation (40) is in close agreement with the experiment.

Magnetite shows unusual behavior. Figure 25 exhibits the iron tracer diffusion coefficients in $Fe_{3-\delta}O_4$ as a function of f_{O_2} obtained at 1100°C by Dieckmann and Schmalzried (1975). It is obvious that two different diffusion mechanisms are operative with different f_{O_2} dependencies. The solid curve in Figure 25 is obtained from

$$D = D_I f_{O_2}^{-2/3} + D_v f_{O_2}^{2/3} \tag{41}$$

with $D_I = 3.3 \times 10^{-16}$ cm^2/sec and $D_v = 7.5 \times 10^{-6}$ cm^2/sec. The f_{O_2} dependence of the D_v term arises from a vacancy mechanism. In this case the defect reaction is

$$3Fe^{2+}_{oct} + 2/3\ O_2 \rightarrow 2Fe^{3+}_{oct} + V_{Fe^{2+},oct} + 1/3\ Fe_3O_4\ \text{(surface)}, \tag{42a}$$

or using the Kröger notation

$$3Fe^{oct^x}_{Fe^{oct}} + 2/3\ O_2 \rightarrow 2Fe^{oct^{\cdot}}_{Fe^{oct}} + V''_{Fe^{oct}} + 1/3\ Fe_3O_4\ \text{(surface)}, \tag{42b}$$

depicting the creation of an Fe^{2+} vacancy in the octahedral sites and the oxidation of two Fe^{2+} sites to Fe^{3+}. At equilibrium

$$K = \frac{X^2_{Fe^{3+}}\ X_{v,Fe^{2+}}}{X^3_{Fe^{2+}}\ f^{2/3}_{O_2}}\ .$$

Since the fraction of Fe^{3+} ions in octahedral sites, $X_{Fe^{3+}}$, and the fraction of the Fe^{2+} ions in octahedral sites are both constant and very nearly equal to 1/2, we obtain

$$X_{v,Fe^{2+}} \propto f^{2/3}_{O_2}\ . \tag{43}$$

Equation (43) justifies the second term in (41).

The f_{O_2} dependence of the D_I term arises from an interstitial mechanism. In this case, the defect reaction is

$$Fe^{oct^x}_{Fe^{oct}} + V^x_i \rightarrow V''_{Fe^{oct}} + Fe^{\cdot\cdot}_i\ , \tag{44}$$

indicating the creation of a Fe^{2+} interstitial and the formation of an Fe^{2+} vacancy at the octahedral Fe^{2+} site. At equilibrium

$$K = \frac{X_{v,Fe^{oct}}\ X_{i,Fe^{2+}}}{X_{Fe^{oct}}\ X_{v,i}}\ .$$

Since the fraction of interstitials which are vacant, $X_{v,i}$, is close to unity and $X_{Fe^{oct}} \sim 1/2$, we have

$$X_{i,Fe^{2+}} \propto \frac{1}{X_{v,Fe^{oct}}}\ . \tag{45}$$

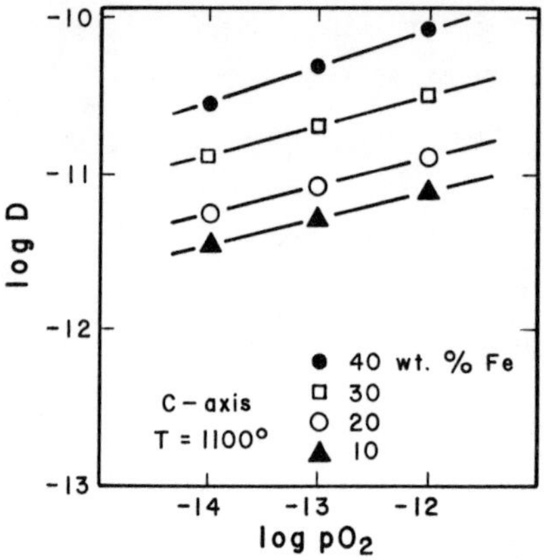

Figure 26. Log D versus log pO_2 for olivine with 10 (triangles), 20 (open circles), 30 (squares), and 40 (solid circles) wt % Fe (*c* axis; T = 1100°C). (From Buening and Buseck, 1973.)

Combining equations (43) and (45) we obtain

$$X_{i,Fe^{2+}} \propto f_{O_2}^{-2/3} ,$$

which is the dependence observed.

We close these examples with data on olivine. Figure 26 shows the f_{O_2} dependence of the olivine Fe-Mg interdiffusion coefficient measured by Buening and Buseck (1973). The hexagonal close packing of oxygen ions in olivine suggests a vacancy diffusion mechanism rather than an interstitial one for divalent cations (however, see below). As in the earlier examples, a vacancy-forming defect reaction can be written for iron in olivine:

$$3Fe^{2+} + 1/2\ O_2 \rightarrow 2Fe^{3+} + V_{Fe^{2+}} + FeO\ (surface); \qquad (46a)$$

$$3Fe^{x}_{Fe} + 1/2\ O_2 \rightarrow 2Fe^{\bullet}_{Fe} + V''_{Fe} + FeO\ (surface). \qquad (46b)$$

Note that it is the presence of vacancies arising from the oxidation of Fe^{2+} which influences the diffusion in olivines. Pure forsterite (with no Fe) would not show this behavior. At equilibrium, equation (46) yields

$$K = \frac{X^2_{Fe^{3+}} X_{v,Fe}}{X^3_{Fe^{2+}} f_{O_2}^{1/2}} .$$

Once more, from the stoichiometry of equation (46) we must have $2X_{v,Fe} = X_{Fe^{3+}}$. Therefore,

$$X_{v,Fe} = (\frac{K}{4})^{1/3} f_{O_2}^{1/6} . \qquad (47)$$

The data from Buening and Buseck yields

$$D_{Fe} \propto f_{O_2}^{\,0.172 \pm 0.022} \quad . \tag{48}$$

A similar result has been found for cobalt diffusion in Co_2SiO_4 (see Schmalzried, 1978).

ELECTRONIC DEFECTS

Some of the defects discussed in the previous section involving changes in valency fall under the category of electronic defects. The literature on electronic defects, especially those produced by various kinds of radiation damage, is extensive (*e.g.*, see Klick, 1972; Hughes and Henderson, 1972; Henderson and Wertz, 1977). We will comment on these only briefly. Electronic defects represent either an "excess" of electrons (*particles*) or a lack of electrons (*holes*) in a region of the crystal structure. The oxidation of an ion creates an electron hole. Thus, Fe^{3+} in FeO is an electron hole or O^- in MgO is also an electron hole. In both cases the effective charge is positive. On the other hand, Fe^+ in olivine is a particle (excess electron). Figure 27 shows the so-called *F-center* in alkali halides. In this case, an anion vacancy traps an extra electron, which is held there by the surrounding positive ions. Likewise, an F' center (effective charge = -1) contains *two* electrons in an anion vacancy. An F' center can contribute to electrical conduction, while an F-center, with no effective charge, cannot. In the case of simple oxides, where an anion vacancy has an effective charge of +2, trapping one electron (net charge +1) forms an F^+ center. Similar meaning is attached to F, F^- centers (two and three electrons trapped). The interaction of an electron with the ionic lattice induces a large polarization (see next section), which can be treated by the "polaron" model (a "polaron" is an electron such as in Figure 27 and the surrounding polarized lattice; it is not fully localized and can move in the crystal).

The analogous terminology for cation vacancies uses the letter V. Therefore, a V^- center in a simple oxide indicates a cation vacancy, adjacent to which is a single positive hole (*i.e.*, an O^- in a near-neighbor oxygen ion). The net effective charge of the defect is -1

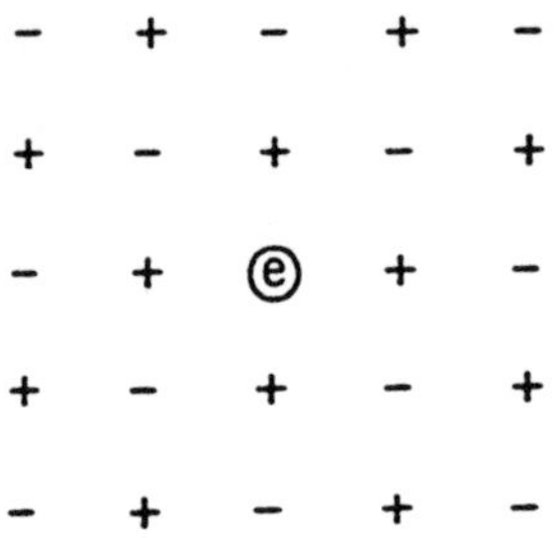

Figure 27. The F center in alkali halides. Positive alkali ions are shown as +; negative halide ions as −. The electron e is held in the halide ion vacancy by the electric field of the surrounding ions.

(+1 for a hole and −2 for the cation vacancy). Figure 28 shows some typical electronic defects in alkaline earth oxides. Many ionic crystals are transparent to photons with wavelengths in the visible. Electronic defects can introduce energy levels into the band gap of these crystals and thereby cause selective absorption of light in the visible. The crystals then become colored and this has given rise to the term *color center* for these electronic defects.

An important method of studying electronic defects is by electron spin resonance spectroscopy (ESR). ESR measures the response of a magnetic dipole (*e.g.*, an unpaired electron) to a magnetic field. The response is a function of the local environment surrounding the unpaired electron and therefore can be used to characterize the nature of the electronic defect (see Henderson and Wertz, 1977).

One of the dominant types of electronic defects in oxides involves the oxidation of a cation, for example, the formation of Fe^{3+} in a crystal containing Fe^{2+}. These electron *holes* can migrate without the motion of the entire ion, since an Fe^{2+} can transfer an electron to a

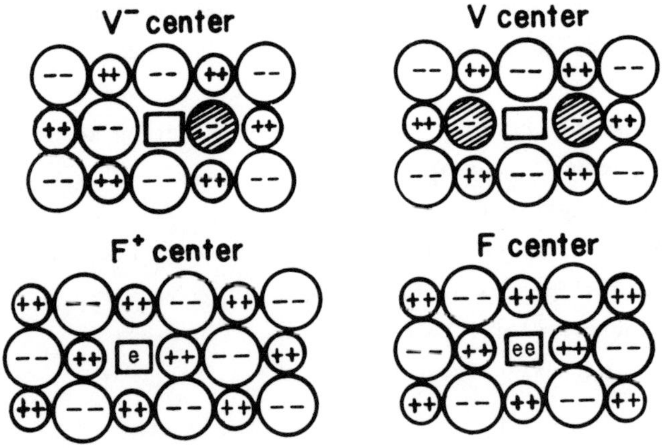

Figure 28. Structure of single anion and cation vacancy centers in the alkaline earth oxides: e represents a trapped electron; ee represents two trapped electrons (Henderson and Wertz, 1977.)

neighboring Fe^{3+} ion. Hence, the electron holes can contribute an additional component to the electrical conductivity (but not to the diffusion).

Forsterite, a good insulator, has a large band gap >7.5 ev (Shankland, 1968). As a consequence, electronic transport is most likely to be extrinsic. Some of the electronic defects are discussed by Morin *et al.* (1977). An Mg^{2+} vacancy in forsterite has an effective charge of -2. The negative charge on the vacancy may be compensated by the presence of a hole, *e.g.* Fe^{3+}, on a magnesium site next to the vacancy. This association produces a $V_{Mg}^{-}(Fe^{3+})$ defect. Another electronic defect discussed by Morin *et al.* (1977) in forsterite is the ion Fe^{4+}. However, they claim that the electrical conductivity is controlled by electron transfer using Fe^{3+} impurities.

Electronic defects may play an important part in recent treatments of multicomponent diffusion in silicates. Lasaga (1979a) recently proposed a model to estimate the multicomponent diffusion coefficients, D_{ij}, in silicates. Cooper (1965) had earlier proposed a similar model. Both papers assumed that the *ionic* motions in the lattice would have to ensure electroneutrality. This constraint is well known in electrolyte theories (Lasaga, 1979b). However, if electron holes or particles in electronic defects are of sufficient abundance and mobility, these defects may be able to ensure electroneutrality by the motion of electrons. In this latter case, the coupling of the motion of the diffusion ions is reduced to only volume effects (Darken, 1948) or defect association phenomena. It is therefore of great importance to sort out the electronic defects of silicates, especially those containing multiple valence ions.

DEFECT CALCULATIONS IN MINERALS

Up to now, our attention has focused on the various possible defect reactions in minerals and the concomitant equilibrium relations between defect concentrations. The free energies needed to evaluate the equilibrium constants have been obtained either by careful experiments or estimated by empirical correlations. However, the most general and fundamental approach to defects (and diffusion) is based on calculations using an atomistic solid state theory. The treatment of the

Table 9

Calculations of Enthalpies of Formation for Point Defects

Substance	Cation Frenkel	ΔH_f^a (Kcal/mole) Anion Frenkel	Schottky	ΔH_f Experimental
NaCl	66	106	41-54	50-58[b]
NaBr	59	112	38-51	40[b]
KCl	80	86	44-51	52-60[b]
KBr	73	96	42-49	53-58[b]
CaF_2		60-62		51-65[c]
SrF_2		51-55		39-53[c]
BaF_2		37-44		44[c]

a - calculated values of ΔH_f

b - experimental ΔH_f for Schottky defects

c - experimental ΔH_f for Anion Frenkel defects

(From Lidiard, 1971)

energetics of point defects in ionic crystals from an atomistic point of view dates back to the classic paper by Mott and Littleton (1938). In this theory the energies of formation and migration of the defects are related to the electronic, optical and elastic properties, as well as the crystal structure of each mineral (see Barr and Lidiard, 1971; Lasaga, 1980). Table 9 compares calculated enthalpies of formation of Schottky and anion Frenkel defects of simple crystals with experimental data. The results are encouraging. These defect calculations open up new avenues of research as well as a deeper understanding of the details of the transport processes and the interatomic forces in minerals.

Most current theories still employ the same approach as Mott and Littleton, although treatment of the various terms in the original theory has been refined. The energy needed to move ions in an ionic structure or to create vacancies and interstitials can be attributed essentially to three terms: (1) the Madelung energy; (2) the short-range energy (mainly repulsion) due to ionic overlap; and (3) the polarization energy. The physical property which distinguishes ionic crystals and creates problems in theoretical calculations is the presence of a large polarization energy.

The evaluation of the Madelung energy is perhaps the most well-established and oldest of the procedures. The Coulomb energy at a site with position vector $\vec{r}_i$ of an ion with charge z_i is given by the series:

$$V_{coul} = \sum_j \frac{z_i\, z_j\, e^2}{|\vec{r}_j - \vec{r}_i|} \tag{49}$$

where the sum is over all the ions in the lattice. The direct summation of the Coulomb potential, equation (49), converges very slowly and Fourier techniques must be used to speed the convergence (see Ohashi, 1980; Lasaga, 1980; Ewald, 1921). While the exact Ewald formula requires a summation over the reciprocal lattice vectors, the very simple sum

$$V_{coul} = \sum_j \frac{z_j}{r_j} \operatorname{erfc}(\sqrt{\eta}\, r_j) - 2z_i(\eta/\pi)^{1/2} \tag{50}$$

has been found to be completely adequate if η is less than about 0.025 Å^{-2} (see Lasaga, 1980). In equation (50) the sum is over all the ions (j) in the crystal other than the *i*th ion and r_j is the distance between the *j*th ion and the reference *i*th ion; erfc(x) is the complement to the error function (Crank, 1975) and η is an *arbitrary* parameter. The expression for V_{coul} in equation (50) should be *independent* of the value of η chosen (if η is small, otherwise the full Ewald expression [equation (7) in Lasaga, 1980] must be used). Therefore, the Ewald method essentially multiplies the direct series in equation (49) by an error function term, erfc, which makes the series converge very fast. Of course, as $\eta \to 0$ equation (50) reduces to the straight summation (49). Most applications of the Ewald formula seek the Coulomb potential *at* a particular lattice site in an ionic crystal. To compute this potential, V_{coul}, either the full Ewald formula [equation (7) in Lasaga, 1980] or the equally valid equation (50) (if η is less than $\sim$0.025 Å^{-2}) can be used. If the Coulomb potential at some *arbitrary* position (not an occupied site) in the crystal is desired, however, the second term in equation (50) (or the same term in the full formula) must be removed. This second term derives from a substraction of the self-energy contribution of the ion *at* the site to the Coulomb potential, a contribution which is not relevant at an arbitrary site. Therefore for an arbitrary position (not a lattice site)

$$V_{coul} = \sum_j \frac{z_j}{r_j} \operatorname{erfc}(\sqrt{\eta}\, r_j) \qquad (51)$$

where the sum is over *all* the ions in the crystal and r_j is the distance from the *j*th ion to the arbitrary point in the crystal. Of course, equation (51) is *independent* of the value of η chosen (to within six significant figures as long as η is any value less than 0.025 $Å^{-2}$). For recent papers using equation (50) in silicates see Lasaga (1980) and references therein.

Short-range Potential

While the method of evaluation of the long-range summations needed to obtain the Madelung potential is well developed, the description of the repulsive forces between ions is not completely understood. This results, in part, from the necessity to include quantum mechanical effects and the possibility of many-body forces in evaluating the ion-ion short-range potentials. By far the most common treatment uses the empirical Born-Mayer repulsive potential

$$V_{Born} = A \exp(-r/\rho) \qquad (52)$$

which is a pairwise potential between two ions at a distance r (A and ρ are constants). For simple cubic crystals, the parameters A and ρ in equation (52) are usually obtained not from quantum mechanical calculations but from the equilibrium cation-anion nearest bond distance and the compressibility of the crystal.

A recent paper by Lasaga (1980) uses the so-called electron gas model of Gordon and Kim (1972). The basic assumption of the model is to treat ions as spherically symmetric charge distributions of known electron density (analogous to the densities that x-ray crystallographers use) and then evaluate the interionic interaction by using the well-known electron gas theory, which relates the energy of an electron gas to its electron density. The electron density at any one point is given by the sum of the electron densities of each ion separately at that point. This model works very well in predicting the positions (r_{eq}) and depths of potential minima for the intermolecular potentials of the rare gases (which are isoelectronic with many of the common ions) and of closed shell ions (Gordon and Kim, 1972). In particular, it predicts very well the important potential wall for distances closer

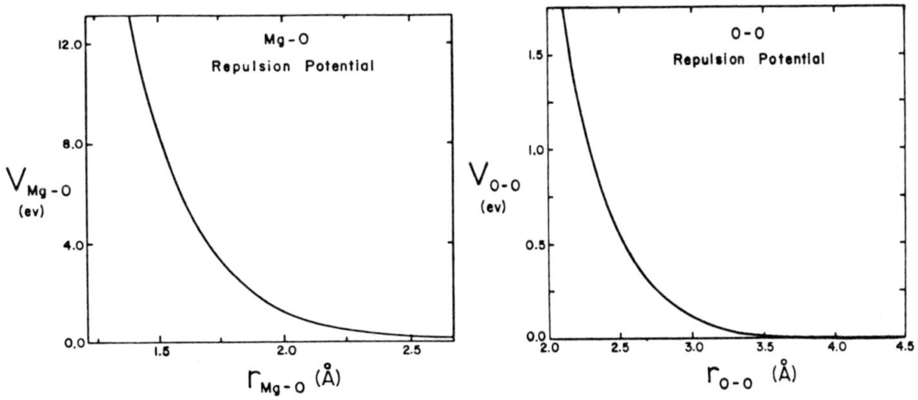

Figure 29. Repulsion potentials calculated from the electron gas model using the Yamashita-Asano oxygen wavefunction: (a) the Mg-O repulsion potential, (b) the O-O repulsion potential.

than the internuclear equilibrium distance, r_{eq}. Figure 29 shows the repulsion potentials calculated for Mg-O and O-O interactions. Note that for large distances there is a small attraction in the O-O potential.

The need for accurate short-range potentials for atoms in silicates is still with us. This is particularly the case with the partly covalent Al-O and Si-O bonds. Current work using quantum mechanics (Tossell and Gibbs, 1978; Lasaga, 1981) will enable a calculation of these potentials in the near future. An alternative is to use semi-empirical potentials and constrain the parameters from spectroscopic data on crystals. This latter approach was used by Lasaga (1980) in obtaining the Si-O potential.

Polarization Energy

A very important term in the energetics of point defects is the polarization energy. The sign of the polarization energy is the opposite of that of the Coulomb energy term. Therefore, the polarization term reduces significantly the energy needed to remove an ion from a lattice. It is the importance of this term, its long-range nature, and the difficulty of evaluating it that makes the treatment of defects in ionic lattices much harder than that needed in metals.

When an ion is placed in an electric field $\vec{E}$, a dipole moment, $\vec{\mu}_i$, is induced on the ion by the field. The two vectors are related to each other by the polarizability of the ion, α_i:

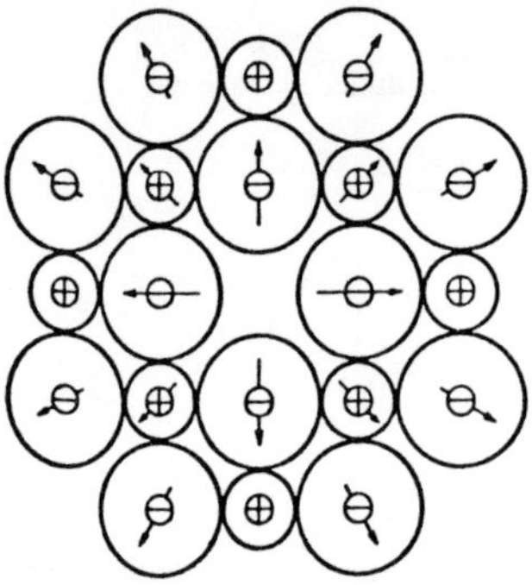

Figure 30. Polarization effects of a vacancy in an ionic crystal. Induced dipoles are shown by an arrow.

$$\vec{\mu}_i = \vec{\alpha}_i \vec{E}_i \ . \qquad (53)$$

Normally in an electrically neutral crystal the ions are situated at sites which by symmetry have no net or at most a very small local electric field present. However, if a vacancy is formed in a crystal, that is equivalent to the introduction of an excess charge in the lattice (see Table 2). The electric field associated with this excess charge will now induce dipole moments on all the ions in the crystal (see Figure 30). Of course, since the magnitude of the electric field diminishes as $1/r^2$, the dipole moment induced on ions far away from the vacancy will also diminish rapidly. Each of these induced dipoles will interact with the effective charge of the vacancy (excess charge) to produce the polarization energy. The potential energy of one dipole $\vec{\mu}_i$ interacting with a charge q at a distance $\vec{r}_i$ is given by

$$V_{dipole} = - \frac{\vec{\mu}_i \cdot \vec{r}_i}{r_i^3} q \ . \qquad (54)$$

To obtain the polarization energy we sum equation (54) over all the induced dipole moments. Further details are given in Lasaga (1980).

A simple treatment due to Jost (1960) readily shows the importance of the polarization energy. The polarization energy of a charged sphere of radius r and charge q embedded in a dielectric continuum with dielectric constant ε_0 is given by

$$V_{polar.} = -1/2 \frac{q^2}{r} (1 - \frac{1}{\varepsilon_0}) \qquad (55)$$

(see Jost, 1960, p. 105; or Jackson, 1962, p. 115). In this case r would be the size of the cation or anion hole and is comparable to an ionic radius. Compare equation (55) to the Coulomb potential in an NaCl lattice at a lattice site:

$$V_{coul} = 1/2 \frac{1.746 \ q^2}{a} \qquad (56)$$

where a is the cation-anion distance and 1.746 is the usual Madelung constant. Because the hole size, r, in equation (55) is nearly equal

Table 10 Defect Energies in Forsterite

Site	V_{coul} (ev)	V_{pol} (ev)	V_{rep} (ev)	V_{tot} [b]
Mg-M1	45.8049	33.5103	4.5872	7.7074
Mg-M2	49.3143	32.3256	4.1799	12.8088
O-1	54.9676	25.9265	26.2177	2.8234
O-2	54.7608	26.3449	25.5061	2.9098
O-3	52.5915	25.7443	25.6634	1.1838

[a] $\alpha_{Mg} = 0.91$, $\alpha_{Si} = 0.20$, $\alpha_0 = 2.39$, $\varepsilon_0 = 6.44$. Yamashita-Asano O^{2-} used in the calculations.

[b] The actual energy needed to remove the ion is half this amount as explained in the text.

to a/2 and ε_0 is approximately 5, the magnitudes of the terms in equations (55) and (56) are similar.

The polarization term has the opposite sign to the Coulomb term; it cannot, therefore, be ignored in evaluating the overall energetics of the defects. Table 10 shows the results obtained by Lasaga (1980) for forsterite. Table 10 points out explicitly some important points. First, the repulsion energy for the magnesium ions is only 10% of the Coulomb energy. This is a typical result. On the other hand, the polarization energy is a significant fraction of the Coulomb energy. An important point is that the energy differences between the sites

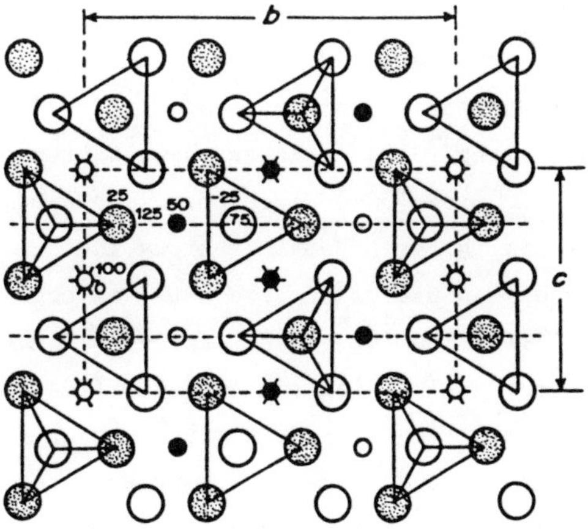

Figure 31. Olivine structure parallel to (100) plane. Small open circles are Mg atoms at x = zero; small solid circles are Mg atoms at $x = 0.5$. Radial lines on small circles designate the *Ml* octahedral sites. Big circles are oxygens. Si atoms, at centers of tetrahedra, are not shown. (Redrawn from Deer *et al.*, 1962.)

(*e.g.*, *M1* and *M2* sites) depend on differences in all three energy contributions. The repulsion energy for oxygen is a much bigger fraction of the Coulomb energy than in the case of magnesium. The reason is that the Si-O bond is partly covalent and was modeled by fitting the Born-Mayer potential to the bond properties. In fact, it is surprising and encouraging that the absolute size of the energies needed to form oxygen vacancies is so close to what is expected!

In the previous section on diffusion, we noted that in forsterite cation diffusion was fastest along the *c*-axis and oxygen diffusion was fastest along the *b*-axis. The results in Table 10 predict that *M1* cation vacancies are easiest to form and hence should be most abundant. This should be true even in the extrinsic region, since the Fe^{3+} content will affect the total cation vacancy concentration and not their partitioning among *M1* or *M2* sites. Since the *M1* sites line up in a row along the *c* direction (see Fig. 31), *M1* vacancies should facilitate diffusion along the *c*-axis. On the other hand, for an Mg atom to move along the *a* or *b* direction, it would have to hop alternately from an *M1* to an *M2* site. Since there are much fewer *M2* vacancies, diffusion along the *a* or *b* direction should be harder. Therefore, calculations are in accord with the observed anisotropy of cation diffusion. The results in Table 10 also predict that the O3 oxygen vacancies should be the most abundant vacancies. Continuous O3 jumps can be performed along either the *a* or *c* direction of forsterite but not along the *b* direction (see Fig. 31). Therefore, oxygen diffusion should be slowest along the *b* axis. In contrast, the data from Reddy *et al.* (1980) indicates that the oxygen diffusion along [010] is faster by a factor of ~ 2 than the diffusion along the [100] or [001] directions. However, the activation energy along the *a*-axis is *less* than that along the *b*-axis. The results in Table 10 and the oxygen diffusion data would be consistent with dislocation controlled diffusion (see next section). Since very few oxygen vacancies are expected, dislocation controlled diffusion is a reasonable assumption.

Work is being carried out currently to predict the migration energies of ions in silicates. The basic theory follows along similar lines as discussed in this section. Since diffusion is often extrinsically controlled, these migration energies will enable a calculation of diffusion rates in silicates.

LARGE DEVIATIONS FROM STOICHIOMETRY

As the number of defects in a crystal increases beyond roughly 0.1 atomic percent, interactions between the defects cannot be ignored. In particular, defects cannot be treated as a nearly random homogeneous array. For example, for minerals such as $Fe_{1-x}O$ or $Fe_{1-x}S$, the deviations from stoichiometry can be as high as x = 0.10. In these cases, neutron diffraction experiments have shown that the defects (vacancies) are *ordered*. This ordering can lead to isolated microdomains whose local symmetry approximate that of another crystal structure. For example, Fe^{3+} ions in FeO begin to enter the tetrahedral sites in such an ordered way as to produce an inverse spinel of Fe_3O_4 composition within a microdomain (Roth, 1960). An Fe^{3+} ion in a tetrahedral site along with two adjacent vacancies in FeO is termed a *Roth complex.*[1]

EXTENDED DEFECTS: DISLOCATIONS AND PLANAR DEFECTS

The theory of extended defects is more difficult than that of point defects. Part of the problem is the "irreversible" or non-equilibrium nature of the defects. For example, extended defects can occur in crystals as a result of imperfect crystal growth or from previous deformation. Recent work using high resolution transmission electron microscopy suggests that even the most carefully grown "pure" crystals contain a significant number of extended defects. One of the most common extended defects is the dislocation or line defect (see Figs. 32-36). In this defect the disturbance of the periodic structure of the crystal occurs along a *line* of atoms or ions. There are two limiting types of dislocations: the *edge dislocation* and the *screw dislocation*. An edge dislocation occurs as a result of a slip of one plane relative to a neighboring plane by a distance of only a few atoms (see Figs. 32 and 33). The result is an extra vertical half-plane of atoms, which ends at the line imperfection, as in the upper half of Figure 32, which also shows that the structure is everywhere normal and unstrained except close to the dislocation. The length of a dislocation can be many tens of thousands of unit cells. Dislocations are important because only small stresses are required to move them; as a plane of atoms moves, the breaking of old bonds is accompanied simultaneously by the formation of new bonds (see Fig. 34).

- - - - - - - - - - - - -

[1] See Greenwood (1968) for more details.

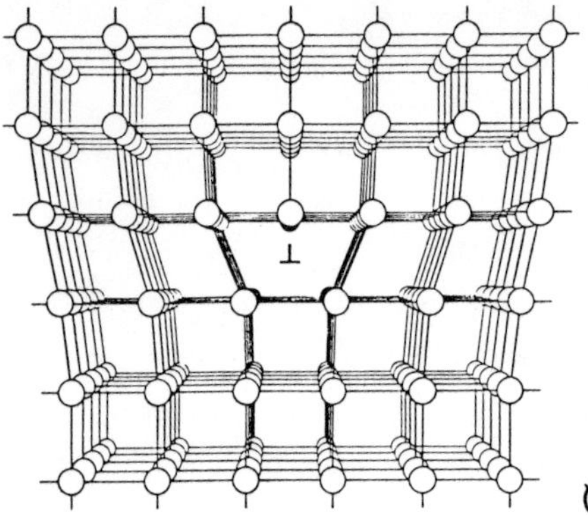

Figure 32. Three-dimensional scheme of a lattice with an edge dislocation (⊥ shows the dislocation line).

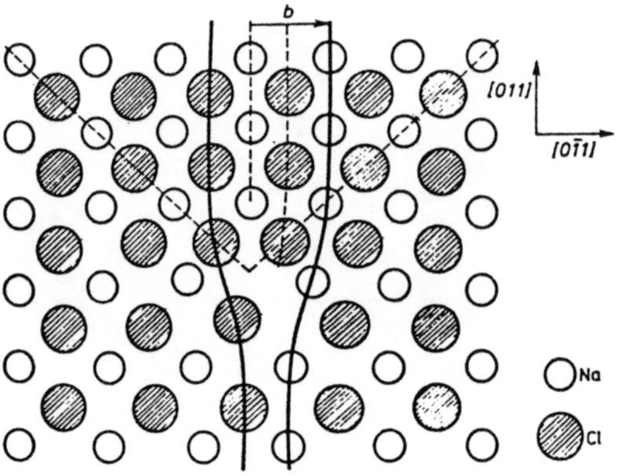

Figure 33. An edge dislocation in the structure of sodium chloride (From Mrowec, 1980.)

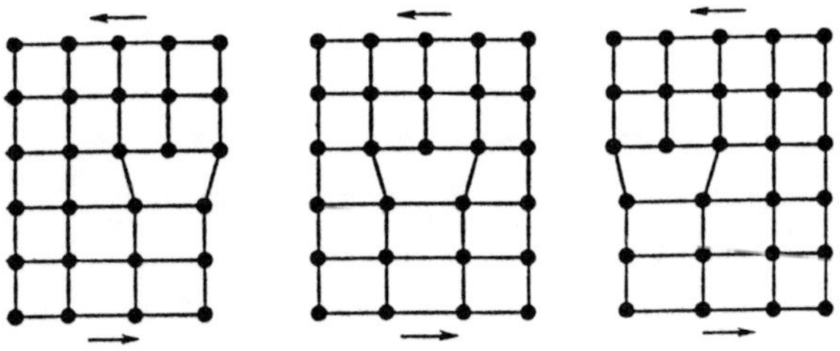

Figure 34. Motion of an edge dislocation (glide) under a shear stress tending to move the upper surface of the specimen to the left.

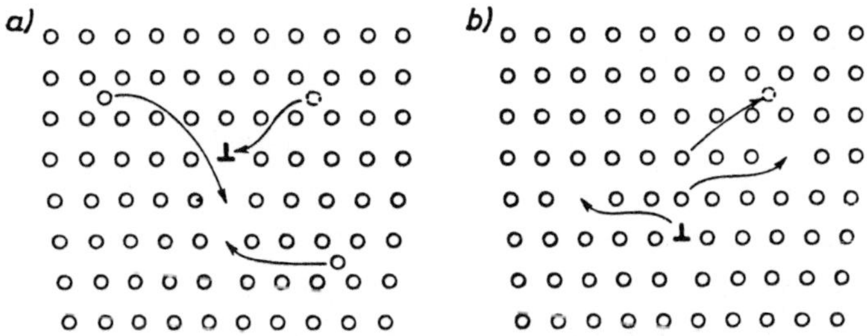

Figure 35. The scheme of climbing of edge dislocations caused by: (a) addition of atoms (annihilation of interstitial defects); (b) release of atoms (generation of interstitial defects or annihilation of vacancies). (From Mrowec, 1980.)

Therefore, dislocations increase the elasticity of crystals by several orders of magnitude. In fact, it was the discrepancy between theoretical calculations of the limits of elasticity of perfect crystals and the measured elastic properties of real crystals that led to the postulation of extensive line dislocations. The type of motion depicted in Figure 34 is termed a dislocation "*glide*" (or slip). Glides are important in the plasticity of crystals, and they represent responses to applied shear stresses.

Another important type of motion of edge dislocations is termed "*climb*," and it is shown in Figure 35. This motion depends on the generation or absorption of point defects and therefore, like diffusion, is an *activated process*. This ability of dislocations to change the local defect concentration makes them very important in maintaining the thermal equilibrium concentration of point defects. Otherwise, the defect generation or annihilation would have to take place at the surface of the crystal. A screw dislocation marks another boundary between a slipped and an unslipped plane in a crystal. But in this case the boundary is parallel to the slip direction rather than perpendicular to it as in the edge dislocation (see Fig. 36). Another method of characterizing line defects is by means of a closed loop surrounding the dislocation line. This loop is called the *Burgers circuit* and contains integral multiples of lattice translations passing through the undisturbed region of the crystal. The loop has equal numbers of unit translations in both the + or - directions and is situated in the plane perpendicular to the dislocation line. With a dislocation present, the circuit is not closed and a quantity called a *Burgers vector*, $\vec{b}$, is missing (see Fig. 36). For an edge dislocation $\vec{b}$ is perpendicular to the dislocation line; whereas, for a screw dislocation $\vec{b}$ is parallel to the dislocation line. Figure 37 shows that "glide" movement of a screw dislocation is also relatively easy. However, simple climbing is *not* possible for a screw dislocation.

The density of dislocations in crystals is determined from the number of dislocation lines that intersect a unit area of the crystal. The density ranges from 100 dislocations/cm^2 for nearly perfect crystals to 10^{11}-10^{12} dislocations/cm^2 for strongly defective crystals (Kittel, 1956). One of the most powerful modern tools available to the mineralogist for studying extended defects is the transmission

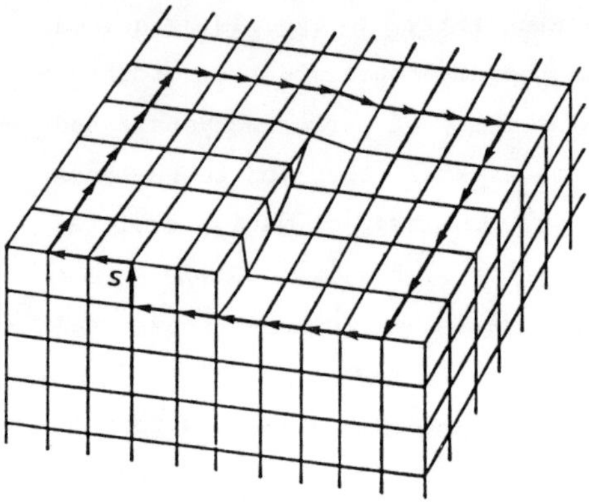

Figure 36. A screw dislocation. The scheme of Burgers circuit for screw dislocation is also shown.

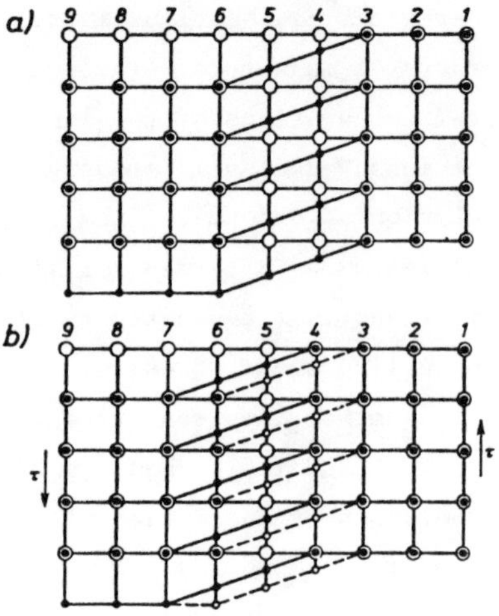

Figure 37. The scheme of shifts of atoms during the shift of screw dislocation by one lattice parameter (From Mrowec, 1980.)

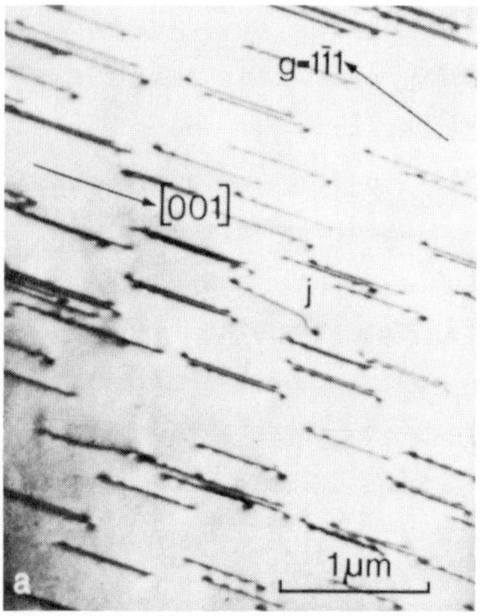

Figure 38. Dislocation structures in experimentally deformed olivine (Phakey *et al.*, 1972) (a) Crystal deformed at 800°C, containing generally straight screw dislocations parallel to [001]; note the jog in the dislocation at j.

electron microscope. While x-ray diffraction data yield the Fourier transform of electron densities averaged over many thousands of unit cells, high resolution transmission electron microscopy, HRTEM, can resolve structural detail down to a scale of several Ångströms. The TEM has been widely used in studying dislocations and submicroscopic twins and exsolution lamellae. For example, Figure 38 shows screw dislocations in a TEM image of deformed olivine crystals. Note that the dislocations align themselves preferentially along the [001] direction. Other recent work on olivine includes a study by Boland (1976) and Sande and Kohlstedt (1976). An interesting use of HRTEM in analyzing defects is illustrated in Buseck and Veblen (1978).

Dislocations are important in the following kinetic processes:

(1) Providing paths for accelerated diffusion (termed *short circuit diffusion*;

(2) Heterogeneous nucleation of new phases in the original crystal (*exsolution*);

(3) Crystal growth;

(4) Diffusion creep in the mantle.

We will comment only briefly on some of the salient features here.

Short circuit diffusion has been mentioned already in the discussion of oxygen diffusion in olivines. If a crystal, such as olivine, contains sufficient dislocations from previous deformations or from its original growth, the diffusion of certain ions may be controlled by the dislocations. For example, the atoms in the extra half-plane of Figure 35b (close to the dislocation boundary) are highly strained. It is much easier, therefore, to move these atoms into nearby vacancies, as indicated. This process ("climb") reduces the migration energy and hence increases the diffusion rate along the dislocation line. Diffusion along dislocations will be dominant if the density of dislocations is high and if both the impurity-induced and the intrinsic concentration of point defects is very small, so that point defect diffusion is restricted. This requirement may be met by the oxygen ion vacancies in forsterite, where the presence of Fe^{3+} does not increase the number of vacancies, which otherwise is expected to be very small. However, extrinsic point defect diffusion is important for cations in olivines, because cation vacancies are enhanced by impurities. Hence, oxygen diffusion may be dislocation-controlled; whereas, cation diffusion may be point defect-controlled. In dislocation controlled diffusion, the fastest diffusion direction will be determined by the direction with the largest density of dislocations.

Dislocations can play a major role in the nucleation of exsolution lamellae. Figure 39 shows pigeonite exsolution lamellae which have nucleated on dislocations in the host augite crystal. Nord *et al.* (1976) point out that the habits which nucleate are those in which the principal misfit of the precipitate is approximately parallel to the Burgers vector of the dislocation. This is roughly obeyed by (001) pigeonite lamellae, whose principle misfit direction, [001]*, lies only 15° from a possible Burgers vector of [001]. Furthermore, the TEM photo in Figure 39 shows both partial and unit dislocations. Unit locations, which have a "zig-zag" appearance, are complete edge dislocations as described earlier. Partial dislocations are straight in the figure

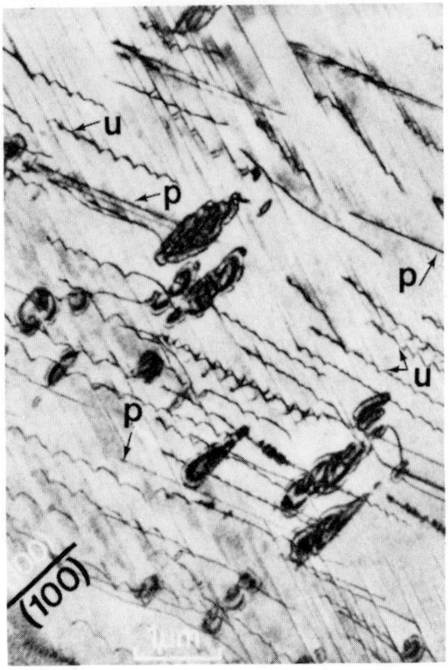

Figure 39. (001) lamellae of pigeonite heterogeneously nucleated on isolated unit u and partial p dislocations in augite. Micrographs are bright-field at 1000kV.

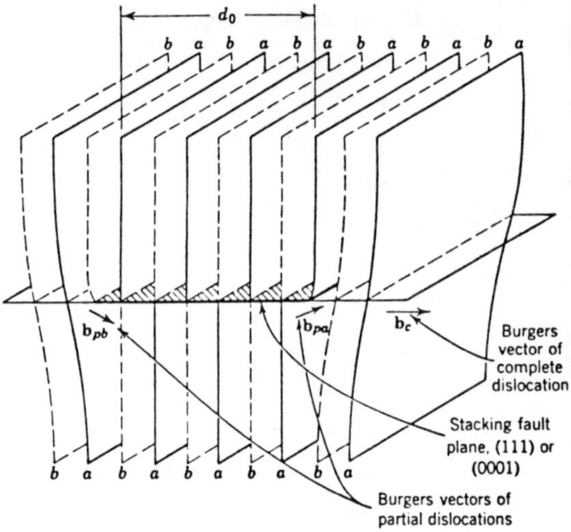

Figure 40. Partial dislocation in hexagonal close-packed structure.

and have fewer pigeonite lamellae associated with them than the unit dislocations. Partial dislocations occur if the extra planes of a complete edge dislocation (containing several extra planes) are separated from each other. Figure 40 illustrates a partial dislocation in a hexagonal close-packed structure. The partial dislocations have a reduced strain energy and this may be the reason for their inefficiency as nucleating agents.

The nucleation and growth of crystals from solution or melt at low supersaturation would be very small indeed if there were no extended defects in crystal surfaces. For example, screw dislocation growth has been well documented in crystal growth theory (see Chapter 8). This type of growth represents non-equilibrium growth even near saturation and hence the kinetics follows non-linear rate laws (see Chapter 4). Another role of surface dislocations is to provide energetic sites (see discussion on adsorption in Chapter 1) for preferential dissolution of the mineral. This effect manifests itself in etch pits, which develop during dissolution (see Chapter 3).

Finally, dislocations aid the important diffusion creep of minerals. The steady-state creep rate, $\dot{\varepsilon}$, has been found to depend on the diffusivity of the slowest moving species and the differential stress, σ, *i.e.*,

$$\dot{\varepsilon} \propto Df(\sigma)$$

where f is an empirical function of σ (see Vander Sande and Kohlstedt, 1976). This function, f, depends on the spacing between partial dislocations in olivine; the smaller the spacing the easier it is to deform the olivine by either cross-slip or climb of dislocations. Vander Sande and Kohlstedt found dislocations with total Burgers vectors $\vec{b}$ = [100] and [001] and with a spacing between partial dislocations of 37 ± 5 Å. Recall that a complete dislocation can "dissociate" into partial dislocations. The Burgers vectors of the partial dislocations add up to the complete Burgers vector (which is one of the translation vectors of the specific lattice). For example, a dislocation with $\vec{b}$ = [100] in the (010) plane can dissociate into

$$[100] = [1/6,\ 1/36,\ 1/4] + [2/3,\ 0,\ 0] + [1/6,\ \bar{1}/36],\ \bar{1}/4]$$

(Vander Sande and Kohlstedt, 1976). The strain of a dislocation is proportional to $|\vec{b}|^2$ (Lardner, 1974). Therefore, slip in a mineral is expected to be caused predominantly by dislocations with small Burgers vectors. Furthermore, the glide usually occurs on the planes of closest packing in the crystal (Lardner, 1974). For more details on dislocations and elastic deformations see Green (1976) and Lardner (1974).

SUMMARY

An understanding of defects in minerals is essential for an atomistic description of kinetics. This chapter has discussed the methods and concepts available for a quantitative treatment of defects. Most of the basic theory deals with point defects, since these defects are the best understood and also are central in the kinetics of transport (conduction, diffusion) in minerals. The variety and energetics of point defects in a mineral depend on intrinsic properties such as structure, size of the ions, dielectric constant, polarizability, Debye frequency and melting point. On the other hand, extrinsic factors, such as impurities in the crystal and oxidation state, can significantly affect the defect behavior of a mineral. The interplay of intrinsic and extrinsic defects gives rise to changes in the mechanism of diffusion or electrical conduction. Because transport phenomena control many important geochemical processes, these changes in mechanism are of great interest. Ideally, one would like to predict the point defects and the transport properties of minerals if the pressure, temperature, and composition are known.

The treatment of extended defects is much more complex. At present there are abundant data and theoretical studies of dislocations in metals. However, the theoretical description of dislocations in ionic crystals is still being developed. These non-equilibrium defects are important in a variety of geochemical processes also, and so more work should be carried out on silicates.

ACKNOWLEDGMENTS

The author thanks Greg E. Muncill and Randy Cygan for assistance in preparing the manuscript and Deb Detwiler for help with the typing. Support from National Science Foundation Grant EAR-7801785 is also gratefully acknowledged.

Anderson, T. F. and Kasper, R. B. (1975) Oxygen self-diffusion in albite under hydrothermal conditions. Trans. Am. Geophys. Union, 56, 459.

Bailey, A. (1971) Comparison of low-temperature with high-temperature diffusion of sodium in albite. Geochim. Cosmochim. Acta, 35, 1073-1081.

Barr, L. W. and Lidiard, A. B. (1971) Defects in ionic crystals. In *Physical Chemistry, An Advanced Treatise,* Vol. 10, Jost, W. (ed.), p. 152-228.

Bielig, G. A. and Lilley, E. (1980) The dielectric loss of crystals of NaCl; Cd^{2+} following plastic deformation. Phil. Mag. A., 41, 745-760.

Boland, J. T. (1976) An electron microscope study of dislocation image contrast in olivine (Mg, $Fe)_2SiO_4$. Phys. Stat. Sol. (A), 34, 361-367.

Bollman, W. (1977) Extrema of the temperature-dependent degree of association of point defects in ionic crystals. Phys. Stat. Sol. (A), 43, 175-184.

Boyd, F. R. and England, J. L. (1959) Pyrope. Carnegie Institution Yearbook, 58, 82-89.

Buening, D. K. and Buseck, P. R. (1973) Fe-Mg lattice diffusion in olivine. J. Geophys. Res., 78, 6852-6862.

Buseck, P. R. and Veblen, D. R. (1978) Trace elements, crystal defects, and high resolution electron microscopy. Geochim. Cosmochim. Acta, 42, 669-678.

Chang, R. H. and Wagner, J. B., Jr. (19/2) Direct-current conductivity and iron tracer diffusion in hematite at high temperatures. J. Amer. Ceram. Soc., 55, 211-213.

Christie, J. T. and Ardell, A. J. (1976) Deformation structures in minerals. In *Electron Microscopy in Mineralogy,* Wenk, H. R. (ed.), Springer-Verlag, New York, p. 374-403.

Clark, A. M. and Long, J. V. P. (1971) The anisotropic diffusion of nickel in olivine. In *Diffusion Processes,* Vol. 2, Sherwood, J. N., *et al.* (eds.), p. 511-521.

Cohen, A. J. and Gordon, R. G. (1975) Theory of the lattice energy, equilibrium structure, elastic constants, and pressure-induced phase transitions in alkali-halide crystals. Phys. Rev. B., 12, 3228-3241.

——— and Gordon, R. G. (1976) Modified electron-gas study of the stability elastic properties and high-pressure behavior of MgO and CaO crystals. Phys. Rev. B., 14, 4593-4605.

Cooper, A. R. (1965) Model for multicomponent diffusion. Phys. Chem. Glasses, 6, 55-61.

Couchman, P. R. (1979) The Lindemann hypothesis and the size dependence of melting temperature. Phil. Mag. A, 37, 369-373.

Crank, J. (1975) *The Mathematics of Diffusion.* Clarendon Press, Oxford, p. 410.

Darken, L. S. (1948) Diffusion mobility and their interrelation through free energy in binary metallic systems. Trans. AIME, 175, 184-201.

Dieckmann, R. and Schmalzried, H. (1975) Point defects and cation diffusion in magnetite. Zeit. Phys. Chem., Nene Folge, 96, 331-333.

Duba, A. (1972) Electrical conductivity of olivine. J. Geophys. Res., 77, 2483.

Economou, N. A. (1964) Dielectric loss of potassium chloride. Phys. Rev., A, 135, 1020-1022.

Ewald, P. P. (1921) Die Berechnung Optischer und elektrostatischer Gitterpotentiale. Ann. Physik, 64, 253-287.

Feit, M. D. (1971) Some formal aspects of a dynamical theory of diffusion. Phys. Rev. B, 3, 1223-1229.

——— (1972) Dynamical theory of diffusion II, comparison with rate theory and the impurity isotope effect. Phys. Rev. B., 5, 2145-2153.

Flynn, C. P. (1972) *Point Defects and Diffusion,* Clarendon Press, Oxford, p. 826.

Foland, K. A. (1974) Alkali diffusion in orthoclase. In *Geochemical Transport and Kinetics,* Hoffman, Giletti, Yoder, Yund, (eds.), Carnegie Institution of Washington, p. 77-98.

Gjostein, N. A. (1973) Short circuit diffusion. In *Diffusion,* Aaronson, H. I. (ed.), American Society Metals, p. 241-274.

Glyde, H. R. (1967) Relation of vacancy formation and migration energies to the Debye temperature in solids. J. Phys. Chem. Solids, 29, 2061-2065.

Gordon, R. G. and Kim, Y. S. (1972) Theory for the forces between closed-shell atoms and molecules. J. Chem. Phys., 56, 3122-3133.

Green, H. W. II (1976) Plasticity of olivine in peridotites. In *Electron Microscopy in Mineralogy,* Wenk, H. R. (ed.), p. 443-464.

Greenwood, N. N. (1968) *Ionic Crystals, Lattice Defects and Nonstoichiometry.* Butterworths, London, p. 194.

Grimes, N. W. (1972) Self-diffusion in compounds with spinel structure. Phil. Mag. 25, 67-76.

Hembree, P. and Wagner, J. B., Jr. (1969) The diffusion of Fe^{55} in wüstite as a function of composition at 1100°C. Trans. Met. Soc. AIME, 245, 1547-1552.

Henderson, B. and Wertz, J. E. (1977) *Defects in the Alkaline Earth Oxides*. Halstead Press, New York, p. 159.

Hofman, A. W. and Giletti, B. J. (1970) Diffusion of geochronologically important nuclides in minerals under hydrothermal conditions. Eclogae geol. Helv, 63, 141-150.

________, Giletti, B. J., Hinthorne, J. R., Anderson, C. A. and Comaford, D. (1974) Ion microscope analysis of a potassium self-diffusion experiment in biotite. Earth Planet. Sci. Lett., 24, 48-52.

Hughes, A. E. and Henderson, B. (1972) Color centers in simple oxides. In *Point Defects in Solids*, Vol. 1, Crawford and Slifkin (eds.), p. 381-384.

Johnson, R. A. (1973) Comparison of calculated and measured values of ΔH_m for vacancies in pure metals. In *Diffusion*, Aaronson, H. I. (ed.), p. 25-46.

Jost, W. (1960) *Diffusion in Solid State Physics*. Academic Press, New York, p. 558.

Kittel, C. (1956) *Introduction to Solid State Physics*. 2nd ed. Wiley, New York, p. 617.

Klick, C. C. (1972) Properties of electron centers. In *Point Defects in Solids*, Vol. 1, Crawford and Slifkin (eds.), p. 291-323.

Kobayashi, Y. and Maruyama, H. (1971) Electrical conductivity of olivine single crystals at high temperature. E.P.S.L., 11, 415-419.

Kroger, F. A. (1971) The chemistry of compound semiconductors. In *Physical Chemistry--An Advanced Treatise*, Vol. 10, Jost, W. (ed.), p. 229-259.

________ (1974) *The Chemistry of Imperfect Crystals*, Vol. 2: *Imperfection Chemistry of Crystalline Solids*. American Elsevier, New York, p. 988.

Lardner, R. W. (1974) *Mathematical Theory of Dislocations and Fracture*. University of Toronto Press, p. 363.

Lasaga, A. C. (1979a) Multicomponent exchange and diffusion in silicates. Geochim. Cosmochim. Acta, 43, 455-469.

________ (1979b) The treatment of multicomponent diffusion and ion pairs in diagenetic fluxes. Amer. Jour. Sci., 279, 324-346.

________ (1980) Defect calculations in silicates: olivine. Amer. Mineral., 65, 1237-1248.

________ and Cygan, R. (1981) Polarizabilities of ions in silicates. (Submitted to Amer. Mineral.)

________, Richardson, S. W., and Holland, H. D. (1977) The mathematics of cation diffusion and exchange under metamorphic conditions. In *Energetics of Geological Processes*, Saxena & Bhattacharji (eds.), Springer-Verlag, New York, p. 353-388.

Lidiard, A. B. (1973) Atomic transport in strongly ionic crystals. In *Diffusion*, Aaronson, H. I. (ed.), Am. Soc. Metals, p. 275-308.

Lin, T. H. and Yund, R. A. (1972) Potassium and sodium self-diffusion in alkali feldspar. Contrib. Mineral. Petrol., 34, 177-184.

Misener, D. J. (1974) Cationic diffusion in olivine to 1400°C and 35 kbar. In *Geochemical Kinetics*, Yoder, H. S., Jr. (ed.), Carnegie Institution of Washington, p. 117-129.

Misra, N. K. and Venkatasubramanian, V. S. (1977) Strontium diffusion in feldspars--a laboratory study. Geochim. Cosmochim. Acta, 41, 837-838.

Moffatt, W. G., Pearsall, G. W. and Wulff, J. (1964) *The Structure and Properties of Materials*, Vol. 1, *Structure*. John Wiley and Sons, New York, p. 236.

Morin, F. J., Oliver, J. R. and Housley, R. M. (1972) Electrical properties of forsterite, Mg_2SiO_4. Phys. Rev. B., 16, 4434-4445.

________, Oliver, J. R. and Housley, R. M. (1979) Electrical properties of forsterite, Mg_2SiO_4, II. Phys. Rev. B., 19, 2886-2894.

Morioka, M. (1980) Cation diffusion in olivine - I. Cobalt and magnesium. Geochim. Cosmochim. Acta. 44, 759-762.

Mott, N. F. and Littleton, M. J. (1938) Conduction in polar crystals I. Electrolytic conduction in solid salts. Trans. Farad. Soc., 34, 485-499.

Mrowec, S. (1980) *Defects and Diffusion in Solids: An Introduction*. Materials Science Monographs, 5, Elsevier, New York, p. 466.

Mukherjee, K. (1965) Monovacancy formation energy and Debye temperature of close-packed metals. Phil. Mag. 12, 915-918.

Nord, G. L., Heuer, A. H. and Lally, J. S. (1976) Pigeonite exsolution from augite. In *Electron Microscopy in Mineralogy*, Wenk, H. R. (ed.), Springer-Verlag, New York, p. 220-227.

Ohashi, Y. (1980) Extension of Fourier techniques to more generalized potential calculations in crystals. Geol. Soc. Amer. Abstr. with Prog., 12, 494.

Pluschkell, W. and Engell, H. G. (1968) Ionen- und Elektronenleitung im Magnesiumortho-silikat. Ber. Deutsch. Keram. Ges., 45, 388.

Rai, C. S. and Manghnani, M. H. (1978) Electrical conductivity of ultramafic rocks to 1820° Kelvin. Phys. Earth Planet. Int., 17, 6-13.

Reddy, K. P. R., Oh, S. M., Major, L. D. and Cooper, A. R. (1980) Oxygen diffusion in forsterite. J. Geophys. Res., 85, 322-326.

Roth, W. L. (1960) Defects in the crystal and magnetic structures of ferrous oxide. Acta Crystallogr., 13, 140-149.

Schmalzried, H. (1978) Reactivity and point defects of double oxides with emphasis on simple silicates. Phys. Chem. Minerals, 2, 279-294.

Shaffer, E. W., Shi-Lan Sang, J., Cooper, A. R. and Heuer, A. H. (1974) Diffusion of tritated water in β-quartz. In *Geochemical Transport and Kinetics,* Hofmann, Giletti, Yoder, Yund (eds.), Carnegie Institution of Washington, p. 131-138.

Shankland, T. J. (1968a) Band gap of forsterite. Science, 161, 51.

———— (1968b) Transport properties of olivines. In *The Application of Modern Physics to the Earth and Planetary Interior,* Runcorn, F. K. (ed.), Wiley Interscience, New York, p. 175-190.

Smyth, D. M. and Stocker, R. L. (1975) Point defects and nonstoichiometry in forsterite. Phys. Earth Planet. Int., 10, 183-192.

Stocker, R. L. (1978a) Point-defect formation parameters in olivine. Phys. Earth Planet. Int., 17, 108-117.

———— (1978b) Influence of oxygen pressure on defect concentrations in olivine with a fixed cationic ratio. Phys. Earth Planet. Int., 17, 118-129.

Tossell, J. A. and Gibbs, G. V. (1976) A molecular orbital study of shared-edge distortions in linked polyhedra. Amer. Mineral., 61, 287-294.

Trivari, G. P. and Patil, R. V. (1976) Correlation of diffusion data with valence bond strength. Trans. Japanese Inst. Met., 16, 476-480.

Tsukahara, H. (1976) Diffusion and diffusion creep in olivine and ultrabasic rocks. J. Phys. Earth, 24, 89-103.

Vander-Sande, J. B. and Kohlstedt, D. L. (1976) Observation of dissociated dislocations in deformed olivine. Phil. Mag., 34, 653-658.

Varotsos, P. and Alexopoulos, K. (1977) Comment on a correlation between the migration enthalpy of a cation vacancy in alkali halides with NaCl structure and their melting points. Phys. Review B., 15, 5994-5994.

Wertz, J. E. (1968) Characterization of point defects in oxides. In *Mass Transport in Oxides,* Wachtman, J. B. and Franklin, A. D., (eds.), p. 11-23.

Winchell, P. (1969) The compensation law for diffusion in silicates. High Temp. Sci., 1, 200-215.

————, and Norman, J. H. (1969) A study of the diffusion of radioactive nuclides in molten silicates at high temperatures. In *Third International Symp. on High-Temperature Technology Proc.*, Butterworth, London, p. 479-492.

Yund, R. A. and Anderson, T. F. (1974) Oxygen isotope exchange between potassium feldspar and KCl solution. In *Geochemical Transport and Kinetics,* Hofmann, Giletti, Yoder, Yund (eds.), Carnegie Institution of Washington, p. 99-105.

Chapter 8

KINETICS of CRYSTALLIZATION of IGNEOUS ROCKS

R. James Kirkpatrick

INTRODUCTION

Petrologists have always been interested in the processes occurring in igneous rocks. Within the last decade, however, there has been an increasing awareness that a theoretical and experimental knowledge of the rates and mechanisms of the crystallization process is important in understanding these rocks and in interpreting observations in igneous rocks. Most of the applications of kinetic ideas to igneous rocks have been done with volcanic rocks both because their thermal and mechanical histories are more easily understood than those of plutonic rocks and because their deviation from equilibrium is larger (although see Donaldson, 1977 and Maale, 1976). At present, the theoretical, experimental, and observational understanding of the kinetics of the processes operating in igneous bodies is growing rapidly, and it seems likely that the kinetics now being applied to volcanic rocks will soon find extensive application to plutonic rocks and even to magma generation and migration problems.

The primary objectives of this chapter are to bring together the theory underlying our present understanding of the kinetics of crystallization of igneous rocks and to illustrate the use of this theory in understanding experimental and observational data. A complete summary of all applications of kinetics to igneous rocks would be prohibitively long, but hopefully this chapter will serve as an introduction to current thinking about the kinetics of igneous rocks and provide a basis for understanding other work.

The viewpoint of this chapter will be that *the rate of any chemical reaction, including the crystallization of igneous rocks, is zero at equilibrium and proceeds at a finite rate only at a finite deviation from equilibrium.* Thus, an understanding of the processes operating in igneous rocks requires an understanding of how deviation from equilibrium affects the rates and mechanisms of the processes occurring during crystallization.

The crystallization of a silicate melt, or any other phase transformation involving large compositional or structural changes, requires two steps: (1) initiation of a small volume of the product phase (nucleation) and (2) growth of the nuclei of the product phase at the expense of the reactant phase. In at least some igneous rocks it appears that crystallization occurs by simultaneous nucleation and growth (Kirkpatrick, 1976). A comprehensive theory must, then, be able to treat both processes simultaneously. We will discuss separately the kinetics of nucleation and growth and then discuss the overall rate of bulk crystallization. For each of these we will first discuss the theory and then illustrate its use in understanding experimental or observational data.

The rate equations will all be cast in terms of the *undercooling*, ΔT, defined as the difference between the liquidus temperature of the existant liquid and the actual temperature (i.e., $\Delta T = T_L - T$). This liquidus temperature is generally not the liquidus temperature of the bulk composition. After crystallization of an incongruently melting rock composition begins, the liquid composition will change and the liquidus temperature of the existant liquid will decrease. Care must be taken in the application of these expressions to use only the liquidus temperature of the liquid actually present and not that of the original bulk composition. The expressions could equally well have been cast in terms of the *supersaturation*, S, defined as the difference between the actual concentration of some component and the equilibrium concentration. In some cases this formulation may be of some advantage (Lofgren *et al.*, 1975). It has the disadvantage of being undefined in terms of stable equilibria at temperatures below the solidus temperature. In many cases crystallization may continue at temperatures well below the equilibrium solidus, and in those cases supersaturation may be a difficult parameter to use. The undercooling is always defined. The formulation used in a given situation is probably a matter of convenience.

NUCLEATION

Theory

Steady State Nucleation. Nucleation is the submicroscopic process by which atoms of a reactant phase rearrange themselves into a bit of the product phase large enough to be thermodynamically stable. Because it is submicroscopic, nucleation is difficult to study experimentally and to describe theoretically. The classical theory of nucleation is based on visualizing the clustering of atoms in a very simple way and assigning macroscopic properties to the clusters of atoms. We will see that the theory describes the qualitative time and temperature variations of the nucleation rate, but for the most thoroughly investigated cases it does not work quantitatively. More sophisticated statistical mechanical treatments are of limited usefulness because the expressions developed usually cannot be evaluated. These will not be discussed.

The classical theory is due mainly to Volmer and Weber (1926), Becker and Döring (1935), and Turnbull and Fisher (1949), although there have been many other significant contributions. Useful reviews are those of Fine (1964) at an introductory level and Zettlemoyer (1969) and Christian (1975) at a more advanced level.

Nucleation theory is based on the concept of heterophase fluctuations, first discussed by Volmer and Weber (1926). A statistical treatment of materials leads to the idea that in any single phase small spontaneous variations in density, composition, defect concentration, and other characteristics, called *homophase fluctuations*, are always forming and disappearing. If these transient variations are large enough, the clusters of atoms may take on the characteristics of another phase. These fluctuations are called *heterophase fluctuations*. Nucleation theory assumes that these heterophase clusters grow by the additions of individual atoms. Because fluctuations of any kind cause local changes in the free energy, the equilibrium distribution (number per unit volume) of clusters is given by the *Boltzmann distribution*,

$$N_i = N_v \exp(-\Delta G_i/RT) \tag{1}$$

where N_i is the number of clusters per unit volume containing i atoms, N_v is the number of atoms per unit volume of reactant phase, ΔG_i is the free energy of a cluster containing i atoms, R is the gas constant, and T is absolute temperature. Equation (1) holds closely as long as i is less than the critical size to be discussed below. If the reactant phase is within its stability field, the number of clusters decreases rapidly with increasing size, and no heterophase fluctuation is thermodynamically stable. This distribution is shown in Figure 1. If a reactant phase exists metastably outside of its stability field, the

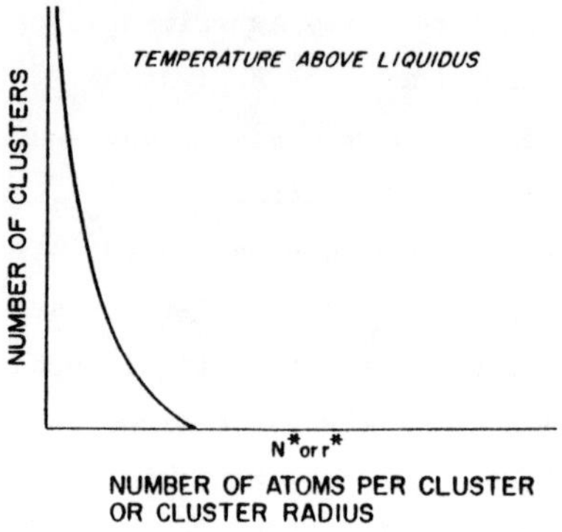

Figure 1. Steady state size distribution of heterophase clusters at temperatures above the liquidus.

chemical free energy of a heterophase cluster of another phase may be lower. Because many of the atoms in the heterophase cluster will be in relatively high energy sites near the surface of the cluster, however, the total free energy of the cluster will not begin to decrease until after it has reached a critical size at which the chemical free energy begins to dominate the surface free energy. Nucleation theory treats this problem by assigning all the extra energy to the surface of the cluster and calling it the *surface energy*. In some cases the surface energy calculated from nucleation rate data may be close to the macroscopically determined surface energy (Matusita and Tashiro, 1973), but there is no theoretical reason that it need be.

In this theory, the free energy change for the formation of a cluster from its constituent atoms, ΔG_{tot}, can be written as the sum of a chemical free energy and a surface free energy term,

$$\Delta G_{tot} = i\Delta g + i^{\frac{2}{3}}\,\varepsilon\sigma \qquad (2)$$

where Δg is the free energy change per atom, ε is a shape factor, i is the number of atoms in the cluster, and σ is the surface energy per unit area. For spherical clusters of radius r, ΔG_{tot} can be written

$$\Delta G_{tot} = \frac{4}{3}\pi r^3\,\Delta G_v + 4\pi r^2\sigma \qquad (3)$$

where ΔG_v is the chemical free energy change per unit volume transformed. In this expression the surface energy term dominates at small radii and the volume free energy term at large radii. Thus, at temperatures below the liquidus the total free energy of the cluster first increases with increasing radius, passes through a maximum, and then decreases. This free energy versus radius relationship is shown in Figure 2. The radius r*, with the maximum free energy is called the critical radius, the maximum free energy, ΔG^*_{tot}, the critical free energy, and the cluster with a radius r* the critical nucleus. r* can be obtained by noting that $d\Delta G_{tot}/dr = 0$ at r*, differentiating equation (3) with respect to radius, and rearranging. For a spherical cluster

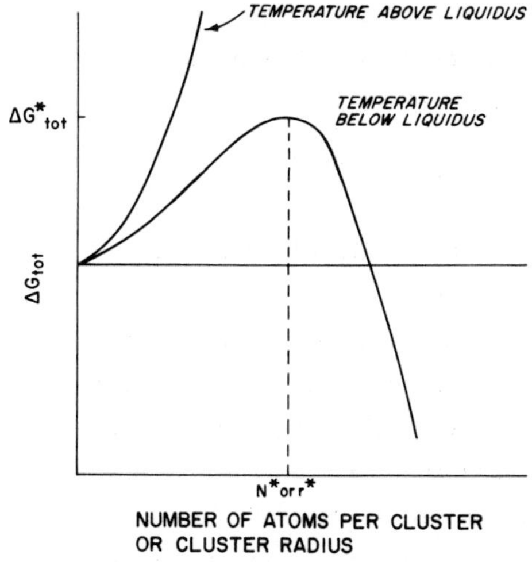

Figure 2. Free energy relationships for heterophase clusters above and below the liquidus.

$$r^* = -2\sigma/\Delta G_v \ . \qquad (4)$$

Substituting this back into equation (3) we find

$$\Delta G^*_{tot} = \frac{16\pi}{3} \frac{\sigma^3}{\Delta G_v^2} \ . \qquad (5)$$

The equilibrium concentration of critical nuclei is

$$N_{r^*} = N_v \exp(-\Delta G^*_{tot}/RT) \ . \qquad (6)$$

In their initial work Volmer and Weber assumed that once a cluster reaches the critical radius it grows rapidly and never decreases in size. The nucleation rate is then the product of the concentration of critical nuclei and the rate at which atoms attach to the critical clusters.

This distribution is shown in Figure 3. Their original derivation was for liquid clusters in a vapor, but Turnbull and Fisher (1949) extended their analysis to condensed systems by using transition state theory (see Chapter 4 by Lasaga). In the Turnbull and Fisher analysis there is an activation energy barrier, ΔG_a, that must be overcome every time an atom attaches to a cluster. Within transition state theory, the probability, P, of an atom having this energy is

$$P = \exp(-\Delta G_a/RT) \ . \qquad (7)$$

The frequency of attachment of an atom to a cluster is then the product of the number of atoms in the reactant phase next to the cluster, n, the frequency with which an atom tries to overcome the barrier, ν ($\nu = kT/h$ in transition state theory; see Chapter 4), and the probability of an atom having the sufficient energy to overcome the barrier. This can be written

$$\text{attachment frequency} = n^*\nu \exp(-\Delta G_a/RT) \qquad (8)$$

where n* is the number of atoms in the reactant phase next to the critical nucleus (note that $n^* \sim 10$-100 atoms). The nucleation rate,

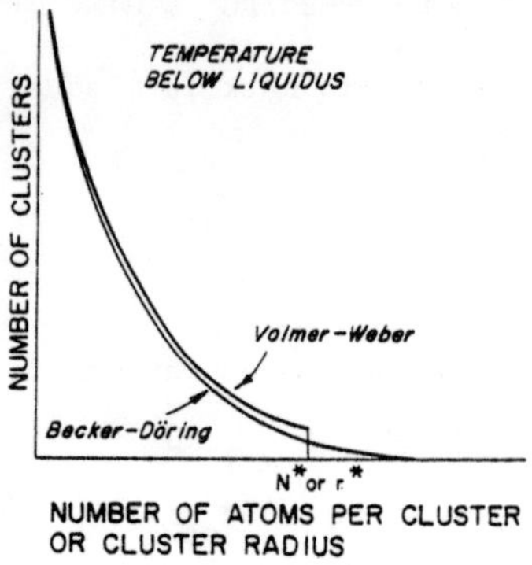

Figure 3. Steady state size distribution of heterophase clusters at temperatures below the liquidus for both the Volmer-Weber and Becker-Döring models. Both show a finite concentration of critical size nuclei.

I, is the product of this and the concentration of critical nuclei [equation (6)],

$$I = n^{*}\nu\, N_v \exp(-\Delta G_a/RT)\exp(-\Delta G^{*}/RT) \qquad (9)$$

where I has units of nuclei/cm^3 of reactant phase/sec.

Becker and Döring (1935) extended the original work of Volmer and Weber by allowing clusters larger than the critical radius to have a finite probability of shrinking back to subcritical size and disappearing. The number of clusters versus radius for this situation is also shown in Figure 3. In condensed systems the nucleation rate for this situation is given by

$$I = \frac{n^{*}}{i^{*}}\, \nu\, N_v \left(\frac{\Delta G^{*}}{3\pi RT}\right)^{\frac{1}{2}} \exp(-\Delta G_a/RT)\exp(-\Delta G^{*}/RT) \qquad (10)$$

where i* is the number of atoms in the critical cluster. The only difference between this and equation (9) is in the pre-exponential term. Turnbull and Fisher (1949) argue that $\frac{n^{*}}{i^{*}} \left(\frac{\Delta G^{*}}{3\pi RT}\right)^{\frac{1}{2}}$ is usually close to 1, and that the pre-exponential term can be approximated by νN_v. Although the justification for this approximation is not entirely clear in all cases, most workers use it. We will see that predictions of nucleation rates using this formulation are often many orders of magnitude in error, although the temperature dependence is qualitatively correct.

To make equation (10) predictive it is necessary to make some approximations and substitutions. In almost all cases it is necessary to find some way of approximating the activation energy and the surface energy. In many cases complete thermochemical data are not available, and it is also necessary to approximate the chemical free energy. The approximation methods discussed here are due to Uhlmann (1969) and references cited in that paper.

The activation energy term (ΔG_a), and the frequency, ν, may be approximated by a term involving a diffusion coefficient ($cm^2 sec^{-1}$).

This diffusion coefficient should not be confused with a bulk diffusion coefficient. The diffusion coefficient, D, can be written

$$D = a_o^2 \nu \exp(-\Delta G_a/RT) \ , \tag{11}$$

where a_o is an interatomic distance (see Chapter 6).

This can be substituted into equation (10) to give

$$I = \frac{N_v}{a_o^2} D \exp(-\Delta G^*/RT) \ , \tag{12}$$

assuming the pre-exponential term in equation (10) is simply νN_v. Diffusion coefficients in fluids are often approximated by the *Stokes-Einstein relationship*, $D = \frac{kT}{3\pi a_o \eta}$ where η is the viscosity and k is the Boltzmann constant. Substituting this and the expression for ΔG^* given by equation (5) into equation (12) gives,

$$I = \frac{N_v kT}{3\pi a_o^3 \eta} \exp[-\frac{16\pi\sigma^3}{3\Delta G_v^2}/RT] \ , \tag{13}$$

which also assumes spherical clusters. The assumption implicit in this approximation is that the atomic processes occurring during viscous flow are the same as those occurring at the surface of the cluster when atoms attach to it. In some cases this substitution seems to be satisfactory (James, 1974), but in others less so (Kalinia *et al.*, 1980; Hofmann and Magaritz, 1977). It is certainly not generally true. Since viscosity is the most easily measured thermally activated process in melts, it seems likely that this substitution will continue to be used.

If the viscosities, surface energies, and thermochemical data are known, equation (13) can be used to calculate the nucleation rates. If the viscosities, thermochemical data, and nucleation rates are known, equation (13) can also be used to calculate the surface energies from plots of log ηI versus $1/\Delta G_v^2 T$, the slope being $(16\pi/3)\sigma$ (Matusita and Tashiro, 1973).

If thermochemical data are not known, ΔG_v can be approximated by noting that $\Delta G_v = 0$ at the liquidus temperature and that $\partial \Delta G_v/\partial T = \Delta S_f$ (note sign) to yield

$$\Delta G_v \cong -\Delta H_f \frac{\Delta T}{T_L} = -\Delta S_f \Delta T \tag{14}$$

where ΔH_f is the latent heat of fusion, ΔT is the undercooling ($\Delta T = T_L - T$), and T_L is the liquidus temperature. The expression

$$\Delta G_v \cong -\Delta H_f \frac{\Delta T}{T_L} \frac{T}{T_L} \quad (15)$$

may be more accurate at large undercoolings (Uhlmann, 1969). However, if ΔC_p is known, it should be used to obtain a more accurate expression for ΔG_v. Substituting equation (15) into equation (13) yields

$$I = \frac{N_v kT}{3\pi a_o^3 \eta} \exp\left(-\frac{16\pi}{3} \frac{\sigma^3}{RT_L \ \Delta H_f^2 \ \Delta T_r^2 \ T_r^3}\right) \quad (16)$$

where ΔT_r is the reduced undercooling ($\Delta T/T_L$) and T_r is the reduced temperature (T/T_L).

Liquid-crystal surface energies can be approximated by the relationship

$$\sigma = \beta \Delta H_f / (A^{\frac{1}{3}} V^{\frac{2}{3}}) \quad (17)$$

where β is an experimentally determined parameter, A is Avogadro's number, and V is the molar volume. β is about 0.33 for non-metals and about 0.5 for metals (Neilson and Weiberg, 1979). Matusita and Tashiro find β to be about 0.45 for lithium disilicate. Substituting this into equation (16) yields

$$I = \frac{N_v kT}{3\pi a_0^3 \eta} \exp\left(-\frac{16\pi}{3} \frac{\beta^3 \Delta H_f}{AV^2 \ RT_L \ \Delta T_r^2 \ T_r^3}\right) . \quad (18)$$

This expression can be used to calculate approximate nucleation rates if the liquidus temperatures, viscosities, and latent heats of fusion are known.

Either equation (16) or (18) can be used to explain the shape of the nucleation curve, when nucleation rate is plotted versus the undercooling, ΔT. If ΔT (or ΔT_r) is very small, the exponential term in (16) or (18) controls the shape of the curve and so I will increase as ΔT increases. This increase will continue with additional increases in ΔT (i.e., lower temperatures) until the viscosity, η, in the denominator of (16) or (18) begins to increase at a rate faster than the exponential term. At this temperature, I will have a maximum and for lower temperatures (higher ΔT), the rapid increase in η will cause a decrease in the nucleation rate. The resulting curve, therefore, is bell-shaped, although the curve is not symmetrical about the maximum. Experimental verification of this nucleation curve will be given in the examples to follow.

Transient Homogeneous Nucleation. An assumption made in the theory discussed above is that there is always a constant distribution of clusters available to form nuclei. When the physical conditions imposed on a reactant phase change, however, the distribution of clusters must also change, and it takes a finite time for the new distribution to become established. Because the nucleation rate depends on the cluster distribution, it too will change during this time. This transient period has been observed in many experiments.

If a phase undergoes a phase transformation at temperature T_e, the distribution of heterophase clusters above T_e is shown in Figure 1. If the temperature is lowered to below T_e, the steady state distribution will change to a relationship similar to those shown in Figure 3. Since the nucleation rate depends on N_{r^*}, there will be no nucleation until some clusters reach r*, and the nucleation rate will be lower than the steady state rate under the steady state value of N_{r^*} is reached. The earliest treatments of this problem by Zeldovich (1943) and Frenkel (1946) gave the approximate relationship

$$I = I_o \exp(-\tau'/t) \ , \tag{19}$$

where I_o is the steady state nucleation rate, τ' is a time constant, and t is time after the physical conditions change. A more exact treatment by Kaschiev (1969) that appears to agree fairly well with results in silicate compositions yields the relationship

$$I = I_o[1 + 2 \sum_{n=1}^{\infty} (-1)^n \exp(-n^2 t/\tau)] \ , \tag{20}$$

where n is an integer and τ is another time constant. Integrating this to give the total number of new particles formed as a function of time, M_t, gives (James, 1974)

$$\frac{M_t}{I_o \tau} = \frac{t}{\tau} - \frac{\pi^2}{6} - 2 \sum_{n=1}^{\infty} \frac{(-1)^n}{n^2} \exp(-n^2 t/\tau) \ . \tag{21}$$

For $t \geq 5\tau$

$$M_t = I_o(t - \frac{\pi^2 \tau}{6}) \ . \tag{22}$$

The $M_t - t$ and $I_r - t$ relationships are shown in Figure 4 (James, 1974). For both equation (19) and equation (22) the nucleation rate is low at first, increases rapidly, and then approaches the steady state value more slowly. The Kashchiev expression approaches I_o more rapidly.

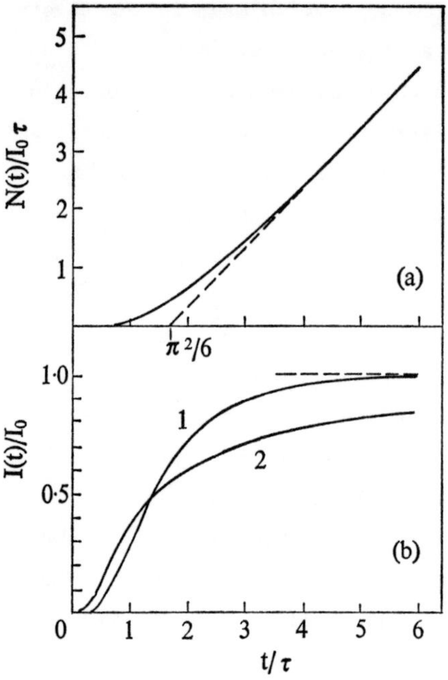

Figure 4. (a) Relationship between total number of crystals nucleated and time for transient nucleation assuming no initial clusters. (b) Nucleation rate versus time during the transient period. Curve 1 is from equation (20); Curve 2 is from equation (19) (after James, 1974).

In Kashchiev's analysis the time constant, τ, is given by

$$\tau = \frac{8kT}{\pi^2 \gamma^* Z \alpha^*} \tag{23}$$

where $\alpha^* = -(\partial^2 \Delta G^*/\partial i^2)_{i=i^*}$, γ^* is the surface area of the critical nucleus, and Z is the rate of addition of atoms. τ will in general decrease exponentially like other thermally activated processes. So that equation (23) can be written

$$\tau = K \exp(\Delta G^*/RT) \tag{23a}$$

where K is not a strong function of temperature.

Heterogeneous Nucleation. It has long been known that the presence of foreign particles can greatly affect the rate of nucleation in a gas or liquid. In bulk samples, in fact, it is difficult to purify the sample completely enough to prevent this heterogeneous nucleation from occurring. Classical nucleation theory treats this problem by considering the reduction in surface energy of a cluster if a heterogeneous surface is available. If the surface energy between the cluster and the surface, $\sigma_{\beta s}$, is less than the surface energy between the reactant phase and the surface, $\sigma_{\alpha s}$, there will be some surface energy reduction if the cluster forms on the surface. Since the surface energy of the cluster is lower, the critical radius will be smaller, the number of critical nuclei larger, and the nucleation rate larger.

For a spherical cap nucleating on an undeformable substrate, Figure 5, the equation for mechanical equilibrium (i.e., balance forces at the three-phase contact) is

$$\sigma_{\alpha s} = \sigma_{\beta s} + \sigma_{\alpha\beta} \cos\theta \qquad (0 \leq \theta \leq \pi) \ . \tag{24}$$

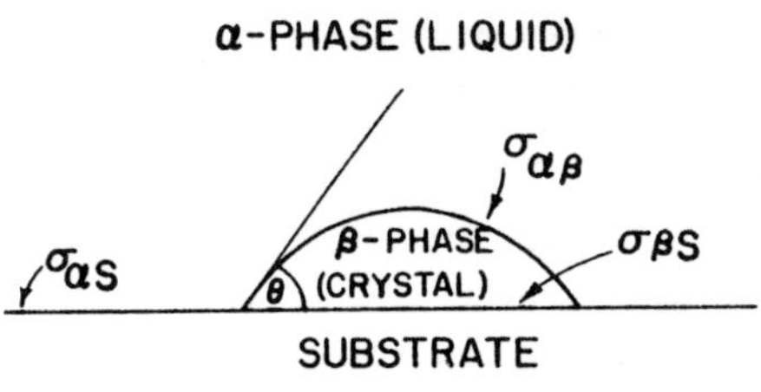

Figure 5. Cross section through a heterophase cluster heterogeneously nucleated on a substrate. θ is the contact angle between the surface of the cluster and the substrate, and the σ's are the three different surface energies.

In the limit of $\sigma_{\beta s} = 0$, $\cos\theta$ and θ approach zero, and the product phase wets the surface. In this case the only barrier to nucleation is the line energy associated with the edge of the cap. If $\sigma_{\beta s} = \sigma_{\alpha s}$ or is larger, θ equals 180° or larger, and there is no surface energy advantage for nucleation on the surface.

Calculating the area of a spherical cap and using equation (24), the energy barrier to nucleation, ΔG_h^*, can be written

$$\Delta G_h^* = \frac{4\pi}{3}\frac{\sigma_{\alpha\beta}^3}{\Delta G_v^2}(2 - 3\cos\theta + \cos^3\theta) \ . \qquad (25)$$

This is the same as equation (5) for homogeneous nucleation except for the term $\frac{1}{4}(2 - 3\cos\theta + \cos^3\theta)$, which reduces the barrier if the surface energy relationships are satisfactory.

To calculate the nucleation rate we also need an expression for the size distribution of the clusters on the surface. Assuming that the number of atoms in the reactant phase from which the clusters form is equal to the number of atoms in the reactant phase next to the surface, N_s, the equilibrium size distribution of clusters with i atoms, N_{ih}, is

$$N_{ih} = N_s \exp(-\Delta G_{ih}/RT) \ . \qquad (26)$$

The rate of nucleation in the Becker-Döring formulation (as modified by Turnbull and Fisher, 1949) is, then,

$$I = \frac{n^*}{i^*}\nu N_s \frac{\Delta G_h^*}{3\pi RT}\exp(-\Delta G_a/RT)\exp(-\Delta G_h^*/RT) \ . \qquad (27)$$

In most real cases the substrate on which heterogeneous nucleation is occurring is unlikely to be flat. Most surfaces have some scratches or pits. If this is ture, nucleation will be greatly enhanced, and a nucleus of a phase may be stable out of its normal stability field if it is in a hole or scratch. Christian (1975) gives a derivation for the effect of a simple cylindrical cavity that illustrates most of the principles.

Solid State Nucleation. The theory discussed above is, in principle, applicable to all states of matter. For a solid nucleating in a solid, however, significant volume change may occur during a phase transformation which cannot be readily occupied by flow of the reactant phase to fill the space. This results in significant strain energy, which must be added to the energy barrier to nucleation. We will be dealing mostly with nucleation from liquids and will not discuss the problem further. Christian (1975) gives a review.

Experimental Nucleation Data in Simple Silicate Compositions

In recent years the number of experimental studies of the rates of nucleation in silicate compositions has increased greatly. This has allowed a more comprehensive comparison of experiment and theory than has previously been possible. Most of these experiments have been done at large undercoolings for compositions of interest to the ceramics industry. They do, however, illustrate the experimental methods and the principles used to interpret the data. In addition, there are a few studies of nucleation rates for composition of geologic interest.

The System Li_2O-SiO_2. This system is the most thoroughly studied silicate system in terms of its nucleation behavior. We will discuss primarily the data of James (1974), because it illustrates the experimental methods, shows both steady state and transient nucleation, and because it has been the basis for comparison of theory and experiment. We will also briefly discuss the data of Matusita and Tashiro (1973). References to other work in this system are listed in those papers.

The experimental method used by James (1974) is to make a batch of melt of the desired composition, quench it to a glass, reheat pieces of the glass at the desired nucleation temperatures (undercoolings of hundreds of degrees) for known times, heat the same samples at a higher temperature to grow the crystals to visible size, and count the number of crystals. This is the method used by most workers, although in some cases the samples may be brought to the nucleation temperature directly from above the liquidus. The higher temperature growth treatment after the nucleation is necessary because the temperature range at which nucleation rates can be most easily measured is usually a range of very low growth rates, and the higher temperature growth runs allow the crystals to be more easily visible. The nucleation rate is the slope of

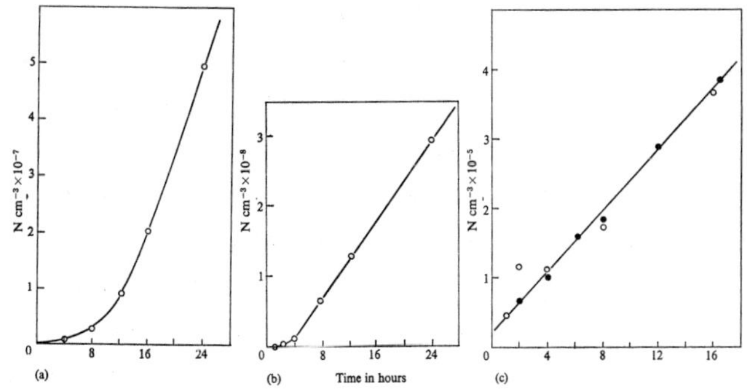

Figure 6. Number of crystals versus time nucleated internally in lithium disilicate glass. (a) 440°, (b) 454°, (c) 527°. Open points for a growth temperature of 560°C; filled points for a growth temperature of 620°C (From James, 1974).

the plot of number of crystals per unit volume versus time for all runs at the same temperature.

The assumptions of this method are that the nucleation rate is negligible at the growth temperature and that a negligible number of the crystals nucleated at the nucleation temperature dissolve at the growth temperature (the critical radius being larger at the larger growth temperature). James' experiments show that the nucleation rate at the growth temperature is much lower than in the nucleation temperature range and that the time and temperature of the growth treatment does not influence the number of crystals in samples given similar nucleation treatments. This method is probably satisfactory for most glass forming compositions.

Counting the number of crystals per unit volume is difficult. The method James and most other authors use is to count the number of crystals per unit area in a polished section and convert this number to number per unit volume using the method developed by Dehoff and Rhines (1961) for ellipsoidal-shaped particles.

James' plot of the number of crystals per unit volume versus time at three different temperatures for nominally stoichiometric $Li_2O \cdot 2SiO_2$ are shown in Figure 6. At the higher temperature the relationship is linear at all times with no apparent incubation period. This would seem to be entirely steady state nucleation. At the lower temperatures there is an incubation time before any nuclei form, and the nucleation rate (slope) is small at short times and increases until it reaches a constant value. This would appear to be an initial transient time,

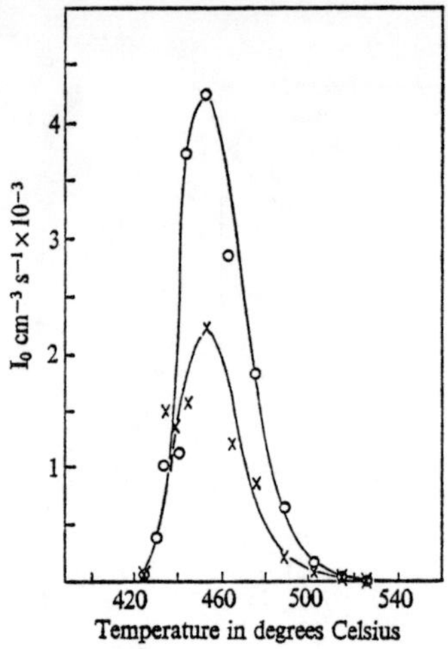

Figure 7. Steady state nucleation rate versus temperature for lithium disilicate (circles) and a composition with slightly more lithium (35.5% Li_2O curves) (From James, 1974).

during which an equilibrium distribution of clusters is being established, followed by a period of steady state nucleation. Figure 7 presents the steady state nucleation rates plotted against temperature.

James' interpretation is that his data represent homogeneous nucleation, all the crystals nucleating on the surface of the sample having been ground off. It certainly represents *internal nucleation*, which will be defined here as nucleation inside the sample not associated with the surface. Whether internal nucleation represents true homogeneous nucleation or heterogeneous nucleation on microscopic impurity particles in the melt is difficult to determine. One argument for it being true homogeneous nucleation is that other alkali-disilicate glasses produced by essentially the same methods and that contain similar impurity levels do not show internal nucleation. It is difficult to prove conclusively that no heterogeneous agent is involved. We will treat the data as if it does represent true homogeneous nucleation.

The relationships observed in these experiments are qualitatively as predicted by the theory discussed above. The steady state nucleation rate is low at relatively high temperatures (low undercoolings), increases to a maximum, and then decreases with falling temperature. The liquidus temperature for the composition is 1234°C, putting the maximum nucleation rate at an undercooling of about 740°C.

These data have been used by Rowlands and James (1979a,b) and Neilson and Weinberg (1979) to test the quantitative applicability of steady state nucleation theory and the assumptions used to make it treatable [equations (1-18)]. Both papers indicate that there are serious qualitative and quantitative difficulties with the classical theory.

Neilson and Weinberg use the Stokes-Einstein viscosity substitution method to approximate the activation energy term, but do have

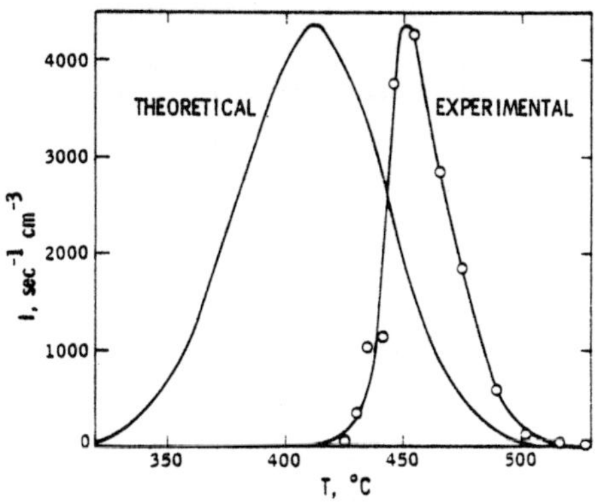

Figure 8. Comparison of experimental and calculated nucleation rates for lithium disilicate. Note that the experimental peak is narrower and at a higher temperature. Theoretical peak normalized to the same maximum value as the experimental peak (after Neilson and Weinberg, 1979).

thermochemical data, so that the latent heat approximation for the chemical driving force is not necessary. A value of 201 erg/cm^2 was obtained for σ from the log ηI versus $1/T\Delta G_v^2$ plot of the data. Using this value they calculated the temperature dependence of the nucleation rate. Figure 8 shows both the calculated and observed rates, with the calculated values normalized to give the same maximum nucleation rate. There are two significant qualitative differences between the observed and calculated values. The calculated temperature of maximum nucleation rate is lower than the observed temperature, and the width of the calculated peak is broader than the observed peak.

Quantitatively, the comparison is even poorer. Calculations using various formulations of the pre-exponential term in the nucleation rate equation and various methods of approximating ΔG_v give rates at least 10^{14} cm^{-3}sec^{-1} lower than the experimental values.

The most obvious explanation for this difference is that nucleation is taking place heterogeneously. As discussed above, however, James (1974) has shown that the nucleation rates are independent of starting material and preparation technique. In addition, the data of Matusita and Tashiro (1972) for the same composition prepared by different methods are only about an order of magnitude different than James'. In any case, classical heterogeneous effects have already been taken into account in the Neilson and Weinberg analysis because the value of the surface energy has been obtained from the data itself, and this is the only place that heterogeneous effects enter the classical theory [equation (25)].

Rowlands and James (1979a) analyzed the same data using slightly different assumptions, but also came to the conclusion that calculations using the classical theory give results orders of magnitude lower than the observed rates.

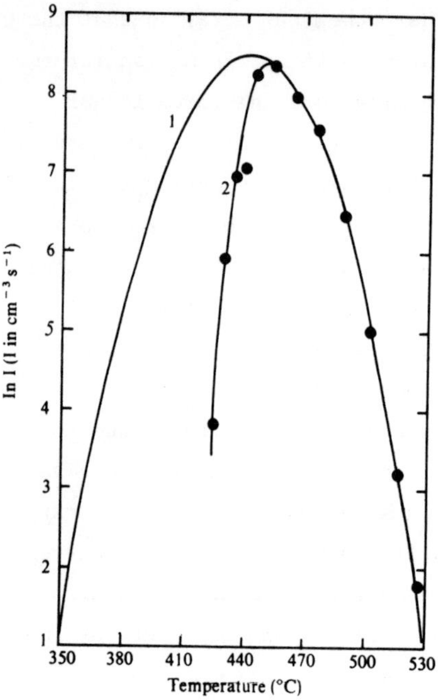

Figure 9. Comparison of experimental and calculated steady state nucleation rates for lithium disilicate after using the experimental data to determine the pre-exponential factor. Note that the calculated peak is still too broad (after Rowlands and James, 1979).

Disagreements of this magnitude have also been observed in other compositions (for instance, mercury crystals nucleating in droplets of liquid mercury, Turnbull, 1952), and have usually been ascribed to a poor theoretical understanding of the pre-exponential term. The results for lithium disilicate show that there is not only a problem with the magnitude of the nucleation rates but also with the qualitative form of the relationship. There seems to be no way that the classical theory can produce a peak as narrow as observed in the nucleation rate versus temperature plot, or put the maximum nucleation rate at the correct temperature. The best that can be said for the classical theory is that in broad qualitative terms it gives the correct form of this relationship.

The results for the sodium disilicate composition are not quite as bad. Klein and Uhlamnn (1977) have obtained internal nucleation rate data for this composition and, after fitting the data to obtain ΔG^*, calculated pre-exponentials only about 10^4 cm^{-3}sec^{-1} less than the observed values. They did not obtain rates at temperatures below the maximum rate, however, and this limits the comparisons that can be made.

It would appear that the classical theory needs significant revision if it is to be quantitatively applicable to silicate melts. One possibility discussed by Rowlands and James (1979b) is to allow the surface energy to decrease with decreasing temperature. They have shown (Fig. 9) that the data above the temperature of maximum rate can be fit to give a pre-exponential factor close to the theoretical values. At temperatures below the maximum, however, the rates are still too high, giving a peak that is still too broad.

Although adjustments to the classical theory may apparently be made to make the theory fit the data *post hoc*, it seems likely to me

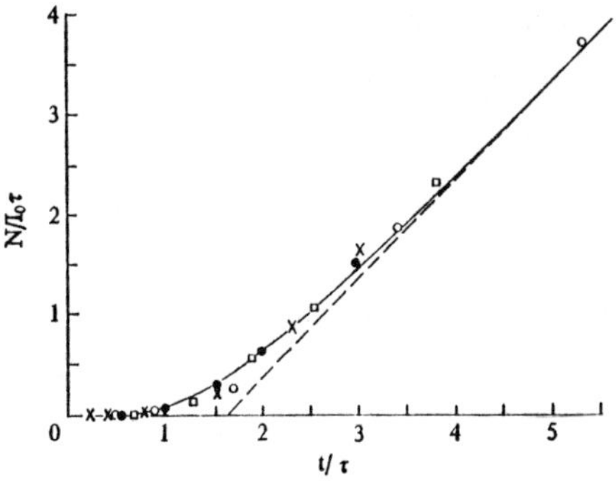

Figure 10. Comparison of experimental and theoretically predicted number of crystals versus time for lithium disilicate. Different symbols refer to different compositions and temperatures. All data should fall on the same curve (after James, 1974).

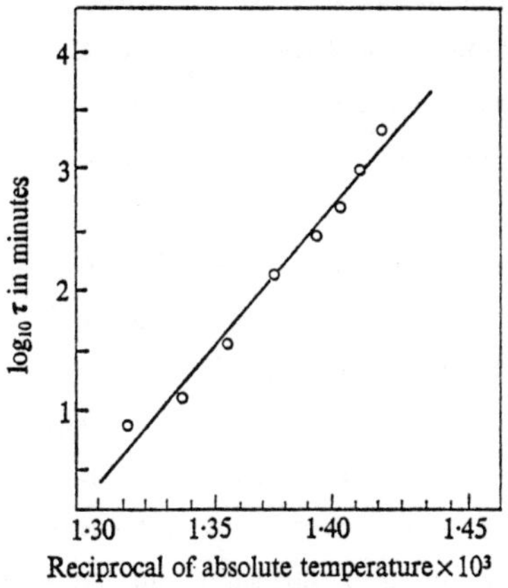

Figure 11. Arrhenius plot of the nucleation incubation time (τ) versus 1/T for lithium disilicate. The apparent activation energy is about 105 kcal/mole (after James, 1974).

that the difficulties with the fundamental assumptions of the theory are too great to be overcome, and that a totally different viewpoint is needed. Unfortunately, as far as I know, no other experimentally testable nucleation theory is available. This one will probably have to do for the time being.

James (1974) has also used the lithium disilicate data to investigate the qualitative validity of the theories of transient nucleation. Figure 10, comparable to Figure 3, is a plot of $N/I_o\tau$ versus t/τ for the lithium disilicate data. The solid line is the relationship predicted by the theory of Kashchiev (1969) and the dashed line is the long time ($t > 5\tau$) approximation. It is clear that Kashchiev's equation for the rate of transient nucleation is in good agreement with the data. Detailed analysis of the short time data indicate that the agreement is less good in that range, but James (1974) indicates that the experimental results are also less accurate in this range.

Kashchiev's model also predicts that the incubation time should decrease with decreasing temperature according to equation (23a). This expression indicates that plots of log τ versus 1/T can yield values of the apparent activation energy for nucleation. Figure 11 shows that this plot for the lithium disilicate data is quite linear over the temperature interval investigated. The value of the activation energy is about 105 Kcal/mole, which is essentially identical to the apparent activation energy for viscous flow of this composition in this temperature range. This would seem to indicate that the molecular processes occurring during nucleation are comparable to those occurring during viscous flow.

The System Na_2O-CaO-SiO_2. A number of investigators have studied nucleation in various compositions in the soda-lime-silica system. We will discuss the work of Kalinina *et al.* (1980) and Strnad and Douglas (1977). This work has been done in the large undercooling range around the nucleation rate maximum using essentially the same techniques as James (1974).

Kalinina *et al.* (1980) investigated the nucleation behavior of the composition $2Na_2O$-CaO-$3SiO_2$, which is a stoichiometric crystalline phase that melts incongruently with a peritectic point at 1141°C and a liquidus temperature at 1205°C. Figure 12 shows their steady state internal (homogeneous?) nucleation rate data plotted versus temperature. The maximum nucleation rate is at an undercooling of about 700°. At the lower temperatures there is a significant incubation time. Figure 13 shows a typical plot of the number of crystals per unit volume versus time. Their plot of log τ versus 1/T is a straight line. The apparent activation energy [equation (23a)] is about 146 kcal/mole. Unlike lithium disilicate, however, this value is significantly less than the activation energy of 197 kcal/mole for viscous flow for this composition in this temperature range. The apparent activation energy for crystal growth, which might also be expected to involve the same kinds of molecular processes, is only about 102 kcal/mole. It would seem that the Stokes-Einstein approximation is not always a good substitution to make when investigating thermally activated processes (see Hofmann and Magaritz, 1977).

Kalinina *et al.* (1980) also investigated the effects of previous thermal history on the nucleation rates for this composition. To do this they pretreated samples at 400°, 460°, and 470° for times slightly longer than those needed to reach a steady state distribution of clusters, and then heated the samples at 480° for various lengths of time. Figure 14 shows the number of crystals per unit volume versus time for the various sets of samples. Curve 1 is for no pretreatment, and others are for various pretreatments. Figure 15 shows the nucleation rates obtained from these curves; the intercept at t = 0 represents the number

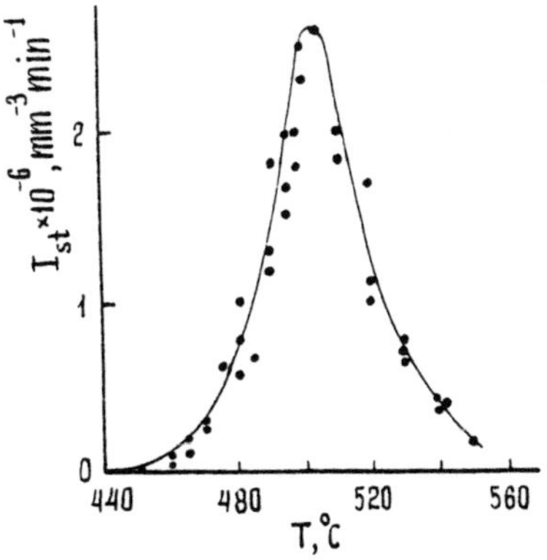

Figure 12. Steady state nucleation rate for the composition $2Na_2O$-CaO-$3SiO_2$ versus temperature (after Kalinina *et al.*, 1980).

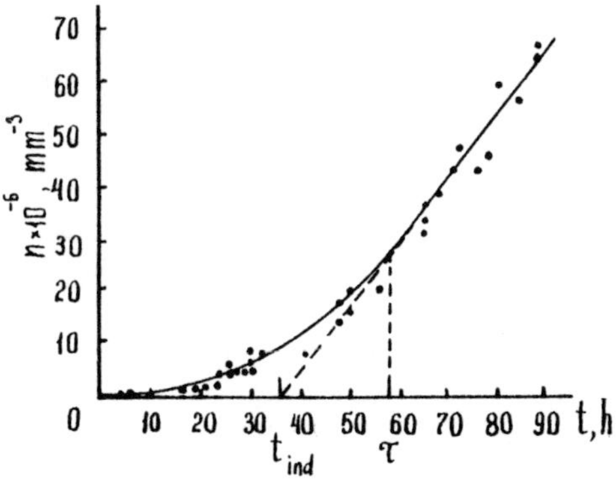

Figure 13. Comparison of experimental and theoretical predicted number of crystals versus time for the composition $2Na_2O$-CaO-$3SiO_2$ (after Kalinina *et al.*, 1980).

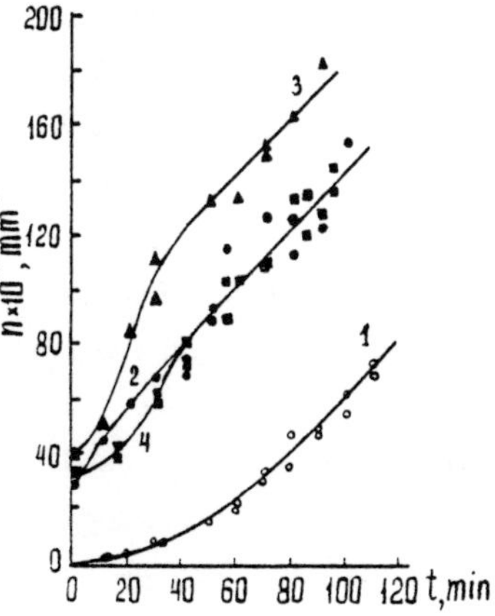

Figure 14. Dependence of the total number of nucleated crystals on time for preheated samples of $2Na_2O$-CaO-$3SiO_2$ composition nucleated at 480°C. Curve 1 - no pretreatment; Curve 2 - pretreatment of 4 hours at 470°C; Curve 3 - pretreatment of 17 hours at 460°C; Curve 4 - pretreatment of 65 hours at 450°C (after Kalinina *et al.*, 1980).

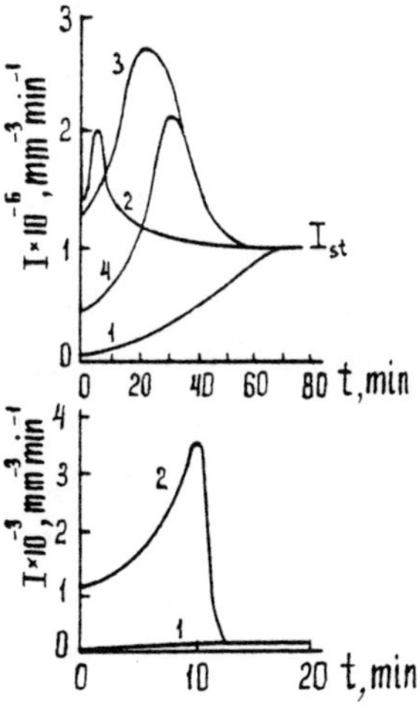

Figure 15. Nucleation rates calculated from curves shown in Figure 14. Note that the nucleation rate for the pretreated samples is large at first and then decreases to the steady state value (after Kalinina *et al.*, 1980, with permission).

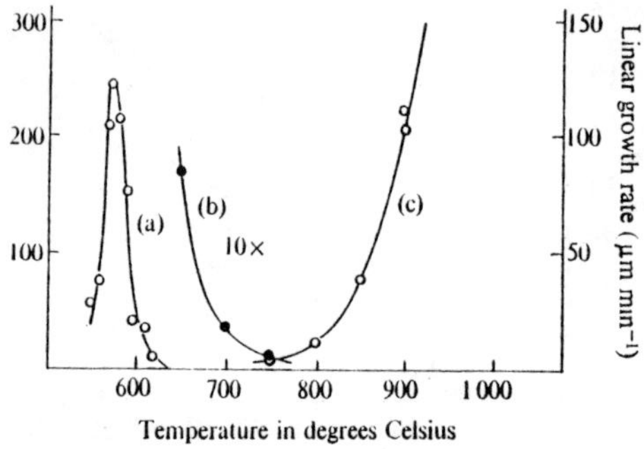

Figure 16. Experimental steady state internal nucleation rates (a), steady state surface nucleation rates of platinum (b), and linear growth rates (c) versus temperature for a soda-lime-silicate composition (after Strnad and Douglas, 1973).

of crystals nucleated during the pretreatment. The nucleation rate with pretreatment is initially higher than for no pretreatment, but decays to the steady state value as time proceeds. These initially higher nucleation rates are due to the presence of many clusters with just sub-critical radii at 480°, some of which quickly form crystals at this temperature. As time goes on the distribution of cluster sizes reaches the steady state appropriate for 480°, and the nucleation rate approaches its steady state value.

Strnad and Douglas (1973) have investigated both the steady state internal (presumably homogeneous) nucleation and the heterogeneous nucleation on the surface of the samples of several compositions in the soda-lime-silica system. Their samples were apparently cut from blocks of glass with the surface left untreated after cutting. Figure 16 shows their data. Although a full range of temperatures was not investigated, it is clear that significant surface nucleation occurs at temperatures higher than the temperature range in which significant internal nucleation takes place, and that many more crystals nucleate on 1 cm^2 of surface than nucleate within 1 cm^3 of volume. They have not interpreted their results in any more detail.

In a similar study in the soda-baria-silica system Burnett and Douglas (1971) have shown that for samples cast in a platinum-rhodium crucible, fire polishing the surface of the crucible reduces the surface nucleation rate relative to an untreated crucible, that scratching the inside of the crucible increases the surface nucleation rate, and that no matter what the treatment of the surface, surface nucleation is always more effective at temperatures higher than those at which internal nucleation occurs. These results are in good agreement with the theory for the effect of irregular surfaces, discussed in the section on heterogeneous nucleation.

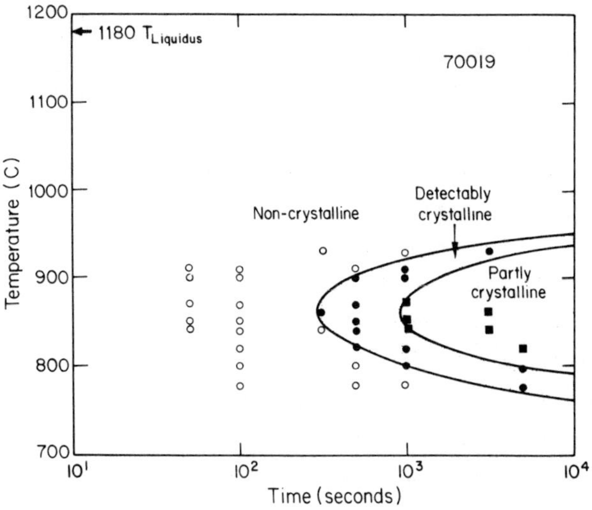

Figure 17. Time-Temperature-Transformation curves for Lunar Composition 70019. Open circles non-crystalline, filled circles detectably crystalline, filled squares partly crystalline, as determined by x-ray diffraction. Curves are loci of approximately equal volume fractions crystallized of 10^{-3} and greater than 10^{-2} (after Klein and Uhlmann, 1976).

More Complex Compositions

Nucleation rates have recently been measured in several more complex compositions of geologic interest. There are, however, no data as complete as those just discussed, and our understanding of nucleation in these complex compositions is at a far less detailed level.

Klein and Uhlamnn (1976) have calculated internal (presumably homogeneous) nucleation rates for Lunar composition 70019, a soil breccia, using a more complex method than used in the experiments just discussed. Samples were run isothermally at undercoolings of from about 250° to 400°C for known times. The layer of surface crystals was ground off, and the internal portion of the samples x-rayed. They classified the x-ray patterns into three categories: those with diffuse peaks, those with somewhat sharpened peaks, and those with sharp peaks. The results are shown in Figure 17. They then, apparently arbitrarily, assigned the onset of somewhat sharp peaks as representing a fraction crystallized of 10^{-3}. To calculate the nucleation rate they then assumed that the fraction crystallized, V_c/V, is given by the relationship

$$\frac{V_c}{V} = 1 - \exp(-\frac{\pi}{3} I_v Y^3 t^4) \tag{28}$$

where I_v is the internal nucleation rate, Y is the growth rate, and t is run time. This is the standard form of the isothermal Johnson-Mehl-Avrami relationship for nucleation and growth controlled reactions, which is derived in Appendix 1. Using experimentally determined growth rates and the known run times, they calculated the nucleation rate for the time needed to produce $V_c/V = 10^{-3}$.

Figure 18 is a plot of $\ln(\eta I)$ versus $1/\Delta T_r^2 T_4^3$ [equation (16)]. The fit to the expected straight line relationship is good. The calculated barrier to nucleation is about 55kT at $T_r = 0.2$, where k is Boltzmann's constant. The intercept at $1/\Delta T_r^2 T_r^3 = 0$ is 5×10^{33} $cm^3 sec^{-1}$ poise, which, they claim is within order of magnitude agreement of theoretical prediction. One significant assumption of the calculation is that the period of transient nucleation is much shorter than the run times, which are hundreds to thousands of seconds. This assumption is not discussed. If it is not valid, it seems likely that the interpretation of the data will have to be modified. Similar results were obtained for sodium disilicate by Klein *et al.* (1977) using the same method.

Nabelek *et al.* (1978) have attempted to measure nucleation rates using isothermal experiments for lunar composition 14310, a plagioclase-rich basalt. Figure 19 is their plot of temperature versus time showing the temperature variation of the incubation time before isothermal nucleation occurs. Figure 20 shows the variation of the number of crystals per unit volume versus time for several temperatures. The number of crystals per unit volume was calculated by counting the number per unit area in thin section and raising this to the 3/2 power. At the two highest temperatures the number of crystals first increases and then decreases with increasing time. At the lowest temperature the number decreases continually after about one hour, which is the shortest run. They attribute the initial increase in the number of crystals to nucleation and the decrease at longer times to recrystallization (Ostwald ripening). In this process the smaller grains, which have a relatively large surface area per unit volume (and therefore a large surface energy), shrink, while the larger grains grow. Although no rates were calculated, the initial slopes, which represent nucleation, are relatively low at high temperatures and much steeper at low temperatures.

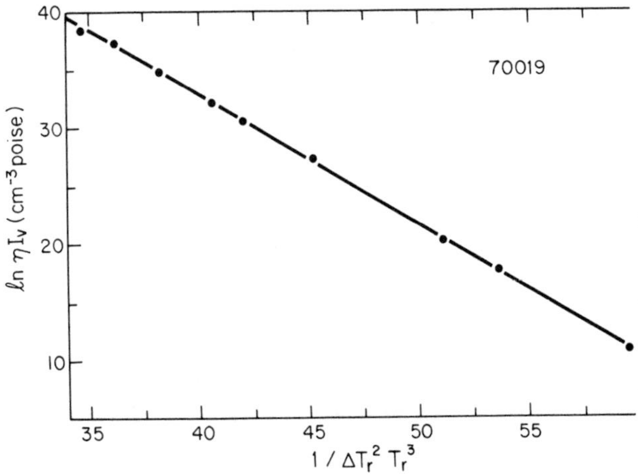

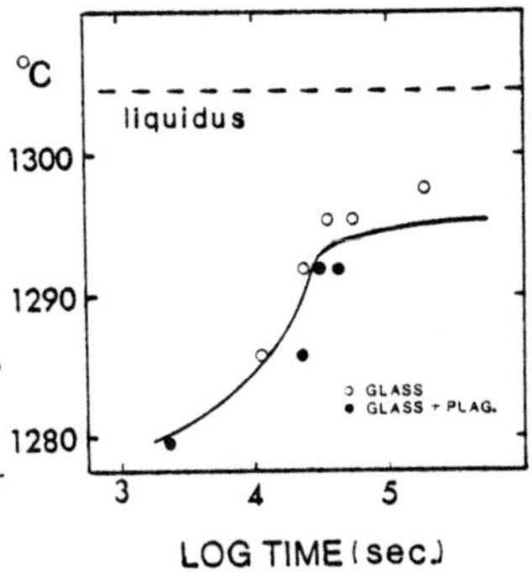

Figure 18. ln (nucleation rate x viscosity) versus $1/\Delta T_r^2 T_r^3$ for lunar composition 70019 (after Klein and Uhlmann, 1976).

Figure 19. Incubation period for plagioclase during isothermal crystallization of Lunar Composition 14310 (after Nabelek *et al.*, 1978).

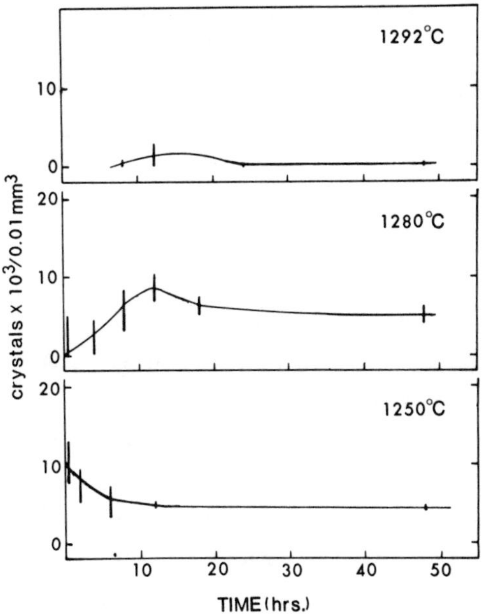

Figure 20. Number of plagioclase crystals per unit volume (nucleation density) versus time for Lunar Composition 14310 (after Nabelek *et al.*, 1978).

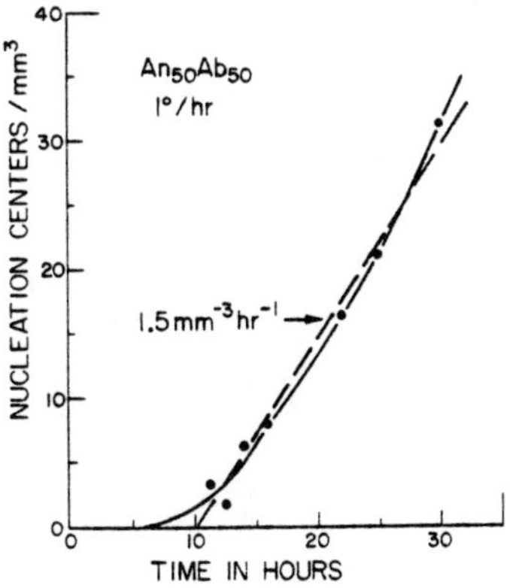

Figure 21. Number of plagioclase nucleation centers versus time in synthetic $An_{50}Ab_{50}$ (mole %) melt in programmed cooling experiments at 1°C/hr. The samples reached the liquidus at a run time of about 5 hours. The straight line fit to the data gives a nucleation rate of about 1.5 events per mm^3 per hour, far lower than in experiments at larger undercoolings.

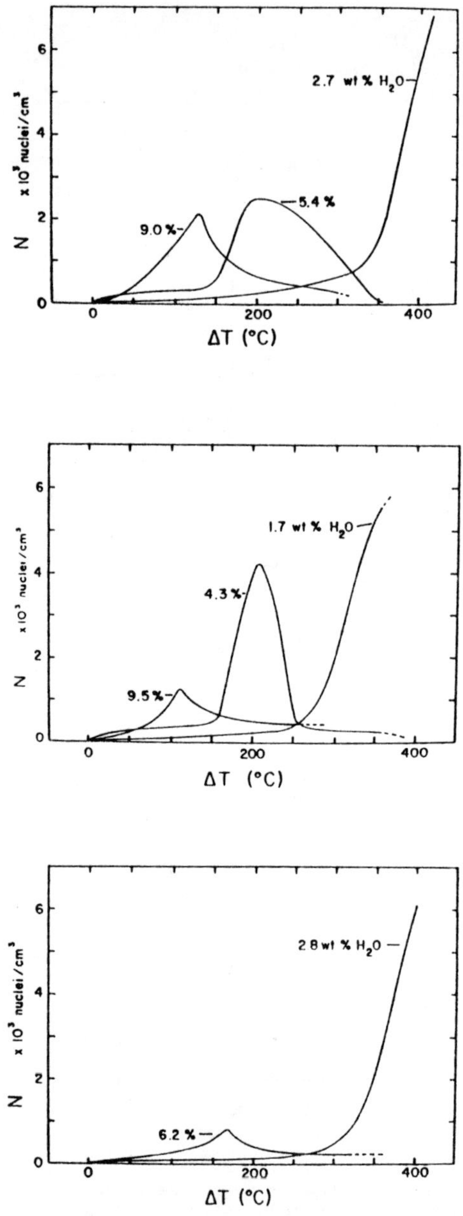

Figure 22. Number of crystals per unit volume versus system undercooling for experiments at 2.5 kbar water pressure for three compositions along the albite - K-feldspar join. (a) $Ab_{90}Or_{10}$, (b) $Ab_{70}Or_{30}$, (c) $Ab_{50}Or_{50}$ (after Fenn, 1977).

The rate of nucleation of internally nucleated clusters of plagioclase crystals from a synthetic albite-anorthite 50-50 mole percent melt in continuously cooled (programmed) experiments has been measured by Kirkpatrick *et al.* (1978). Figure 21 plots the number of clusters per unit volume (obtained by direct counting in thick sections) versus run time for the samples cooled at 1°/hour. The slope of the linear fit to the data yields a rate of 1.5 $cm^{-3}hr^{-1}$. In these experiments the fraction crystallized increases with increasing run time. This rate is orders of magnitude lower than the rates obtained for any of the compositions previously discussed, but were also obtained at much smaller undercoolings. These data are all that are available at such small undercoolings.

Fenn (1977) and Swanson (1977) have measured nucleation density, defined as the number of crystals per cm^3, for isothermal runs at high water pressures for haplogranitic compositions. Because of the difficulty of the experiments, which were done in an internally heated pressure vessel, absolute nucleation rates could not be obtained. The nucleation density in equal length runs should be related directly to the nucleation rate, barring extensive incubation periods.

Fenn (1977) examined several compositions in the system $NaAlSi_3O_8$-$KAlSi_3O_8$-H_2O. Figure 22 shows his results. For all compositions the nucleation density is low at small undercoolings and increases with increasing undercooling. For the more hydrous compositions it then decreases with even greater increases in undercooling. The same behavior would, presumably, be observed for the less hydrous compositions if larger undercoolings were examined. There is no significant variation in the nucleation density with Na/K ratio. For a given Na/K ratio the nucleation density and the temperature of maximum nucleation density decrease with increasing water content.

The temperature variation of the nucleation density in these experiments is that expected from classical nucleation theory. The decrease in the nucleation density with increasing water content is opposite that usually ascribed to "mineralizers" such as water. The reason for this variation is unexplained.

Swanson (1977) performed similar experiments to Fenn for water saturated and water undersaturated synthetic haplogranitic and haplogranodioritic compositions at 8 kb. Figure 23 presents his data. The nucleation densities are much higher in these experiments than in the alkali feldspars, but the qualitative temperature variation is the same. There are fewer data points, so that the variation with water content is not entirely clear.

Crystal growth rates obtained from both of these sets of experiments will be discussed in the section on crystal growth below.

Nucleation Delay in Undercooled Samples

Many investigators of nucleation in geologically important silicate melt compositions have concentrated not on determining quantitative nucleation rates but on examining the time and undercooling variation of the onset of nucleation after a sample has been cooled from a higher temperature (usually from above the liquidus). This variation is usually discussed in terms of the theory of transient nucleation and any scatter in the data to the random nature of nucleation.

Gibb (1974) has investigated this variation for a Columbia River basalt with plagioclase on the liquidus quickly cooled from above the liquidus and then allowed to crystallize isothermally below the liquidus. Figure 24 shows his results. Above the upper dashed line no

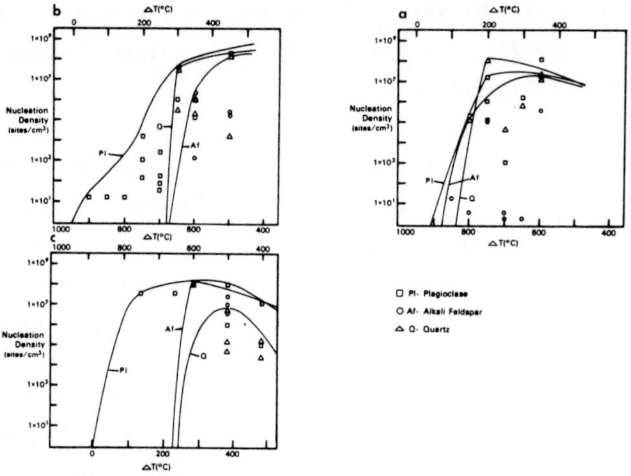

Figure 23. Number of crystals per unit volume versus system undercooling for (a) synthetic granite plus 3% water, (b) synthetic granodiorite plus 6.5% water, (c) synthetic granodiorite plus 12% water, all at 8 kbar (after Swanson, 1977).

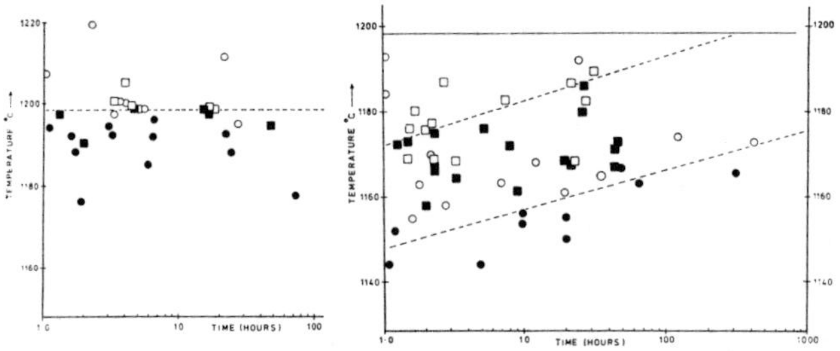

Figure 24. Results of (a) heating and (b) cooling experiments with a Columbia River basalt (after Gibb, 1974).

nucleation occurs. Between the dashed lines some samples contain crystals and some are still all glass. Below the lower dashed line all samples contain crystals. This variation is similar to that observed in most experiments of this type.

This data is in qualitative agreement with the theory of transient nucleations. At small undercoolings the size of the critical nucleus is large, and it takes a long time for the small heterophase clusters present above the liquidus to grow to the size of the critical radius after the sample has been cooled to below the liquidus. With increasing undercooling the size of the critical nucleus decreases, and it takes less time for the clusters to grow to this size. At much lower temperatures (larger undercoolings) the rate of atomic motion slows sufficiently that the incubation time once again increases with decreasing temperature (as seen for lithium disilicate), but this temperature range was not reached in Gibb's experiments. No attempt has been made to analyze these data quantitatively.

The temperature-time field in which an experiment may or may not contain crystals is large in these experiments. The simplest explanation of this is that at these small undercoolings there are few crystals nucleating, and in some runs the random attachment of atoms onto the cluster surfaces leading to a critical nucleus happens to occur quickly and in other samples relatively slowly. At much larger undercoolings, where the number of crystals is larger and the rates are higher, this randomness is probably not important.

Gibb heated his samples to as much as 100°C above the liquidus for as long as 20 hours before cooling them to the run temperature below the liquidus and found no effect of temperature or time of super-liquidus heating on plagioclase nucleation.

Donaldson (1979) has investigated the effect of superheating on the delay in nucleation of olivine (the liquidus phase) in several basalts. We will review the data for an alkali basalt from Skye. Donaldson used both isothermal and programmed cooling techniques. In both techniques samples were held in platinum-rhodium wire loops above the liquidus temperature for from one to 35 hours. In the isothermal experiments the samples were then brought rapidly (about 2500°C/hour) from above the liquidus to the run temeprature, where they were held for known times and then quenched. In the programmed experiments the samples were cooled from the initial heating temperature at known rates to known temperatures below the liquidus and then quenched. Donaldson's results show that for this composition the time of superheating has no effect of the nucleation delay, but that the temperature of superheating has a significant effect. Figure 25 shows the time of nucleation versus undercooling for the isothermal runs initially superheated to temperatures of from 4°C to 168°C. The general form of the relationship is similar to that found by Gibb, except that there is no interval where nucleation may or may not occur. For a given superheating temperature the delay is large at small undercoolings and becomes smaller with increasing undercooling. The length of the delay at a given undercooling, unlike in Gibb's experiments, increases with superheating temperatures.

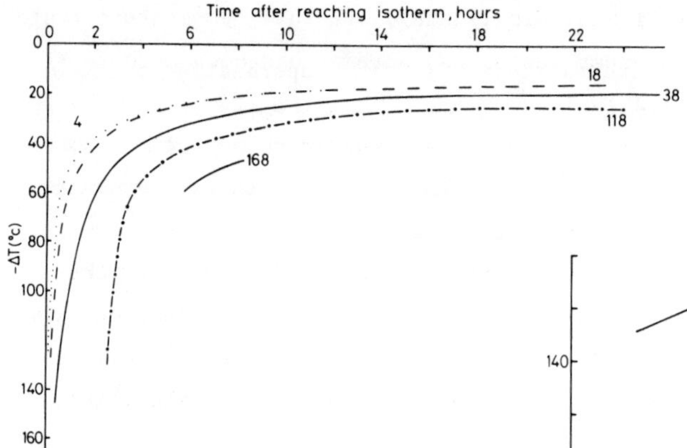

Figure 25 (above). Nucleation delay of olivine versus undercooling for isothermal experiments with a Skye alkali basalt (time for first crystal to form). Numbers by curves indicate amount of superheating (°C) to which the samples were heated before cooling (after Donaldson, 1979).

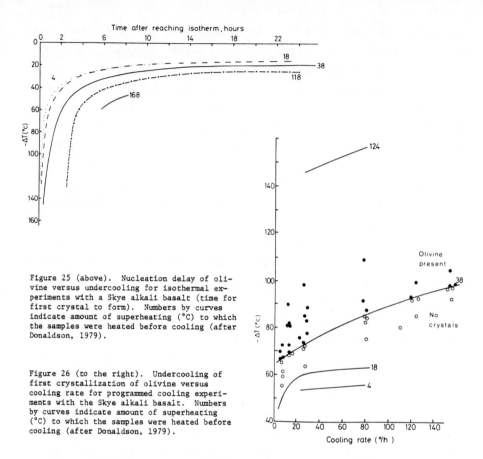

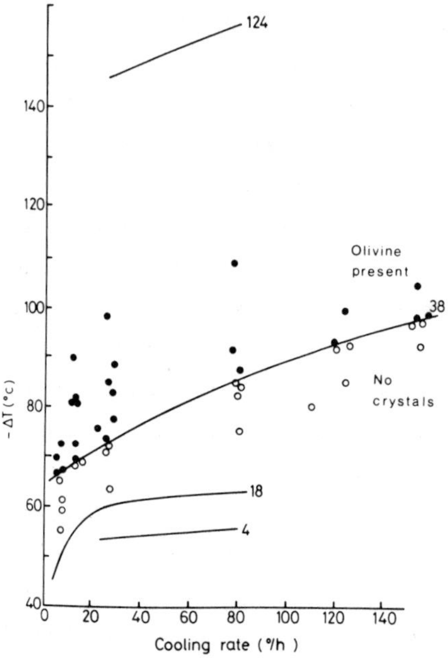

Figure 26 (to the right). Undercooling of first crystallization of olivine versus cooling rate for programmed cooling experiments with the Skye alkali basalt. Numbers by curves indicate amount of superheating (°C) to which the samples were heated before cooling (after Donaldson, 1979).

Figure 26 shows Donaldson's results for the temperature of olivine nucleation as a function of cooling rate in the programmed cooling experiments for initial superheatings of from 4°C to 124°C. For a given superheating the undercooling of nucleation increases with increasing cooling rate. For a given cooling rate the undercooling of nucleation increases with increasing superheating.

Donaldson's explanation of these variations is essentially transient nucleation. At any temperature below the liquidus there is a critical size nucleus with radius r*. This critical radius decreases with increasing undercooling. When a sample is brought rapidly to some undercooling the largest heterophase cluster which is present above the liquidus is much smaller than this critical radius, and it takes a while for one of these clusters to reach r*. Since the critical radius decreases with increasing undercooling it takes less time for a small cluster to reach r* at large undercoolings and the delay time is reduced. The delay time increases with increasing initial superheating because the size of the heterophase clusters present in the melt decreases with increasing superheating.

CRYSTAL GROWTH

Once a heterophase cluster exceeds the size of the critical nucleus, it generally continues to increase in size to become a visible crystal. Primarily because it can be studied by ordinary optical microscopy, far more work has been done on this crystal growth process than on nucleation. This is true for materials of both general and specifically geological interest.

Theory

The theory of crystal growth as applied to igneous petrology has recently been reviewed by Kirkpatrick (1975) and Tiller (1977), and this chapter will not discuss all the aspects of crystal growth presented in those papers. It will, however, briefly discuss some of the fundamentals of crystal growth theory, review some of the recent data on the rates of crystal growth and crystal morphology in compositions of interest to igneous petrology, and discuss some of the more recent contributions to crystal growth theory that may be of some geological significance.

Rate Controlling Processes. When a crystal grows into a melt there are at least four processes which must be considered in a complete theoretical description: the reaction at the crystal-melt interface, diffusion of components in the melt to and from the interface, removal of the latent heat of crystallization generated at the interface, and bulk flow of the melt. For silicate materials the interface process and diffusion are usually considered the most important. Heat flow is probably of little significance in most cases because the thermal diffusivities are several orders of magnitude larger than even the largest diffusion coefficients (10^{-2} to 10^{-3} cm^2/sec for heat flow relative to, at most, 10^{-6} cm^2/sec for diffusion). Thus, the magnitude of the temperature gradients present are likely to be much smaller than composition gradients. Bulk flow (convection) is usually ignored in most theoretical treatments of crystal growth. To my knowledge there has been no attempt to examine the effects of convection on the kinetics of crystal growth in magmatic situations. It is possible that density or surface tension driven flow could be important under certain circumstances, and more thought should be given to the problem. We will be concerned here primarily with the interface process and diffusion in the melt.

Considering only these two processes, there are two ideal situations which can be analyzed relatively easily. (1) When diffusion of the components in the melt to and from the interface is very fast relative to the rate of uptake and rejection of components at the interface, the interface composition remains essentially constant at the bulk liquid composition, and the rate of crystal growth is controlled by the interface attachment process. (2) When the rate of uptake or rejection of components is much faster than diffusion in the melt, the interface liquid composition remains fixed at essentially the equilibrium composition, and the rate of crystal growth is controlled entirely by the rate of diffusion in the melt. If the crystal and the liquid have the same composition, no diffusion is needed, and the interface attachment process is probably the rate controlling process, although in this case heat flow may be important.

In general, we would expect the interface process to be rate controlling at small undercoolings because the growth rate, and therefore the rate of uptake of components, is small and the rate of diffusion is large. At large undercoolings we would expect diffusion to be rate controlling because the growth rate is apt to be large, and since the temperature is lower, the rate of diffusion should be slower. In many intermediate situations both may be significant. As we shall see, this combined interface and diffusion control is difficult to treat theoretically but has been analyzed recently by Lasaga (1981).

We will first look at interface controlled growth, then at diffusion controlled or influenced growth, and finally at the problem of the stability of planar crystal-melt interfaces.

Interface Controlled Growth.

Rate laws. If the interface reaction is the rate controlling process, a rate law may be derived in the following way (Turnbull and Cohen, 1960). The rate of attachment of atoms to the interface, r_a, can be written

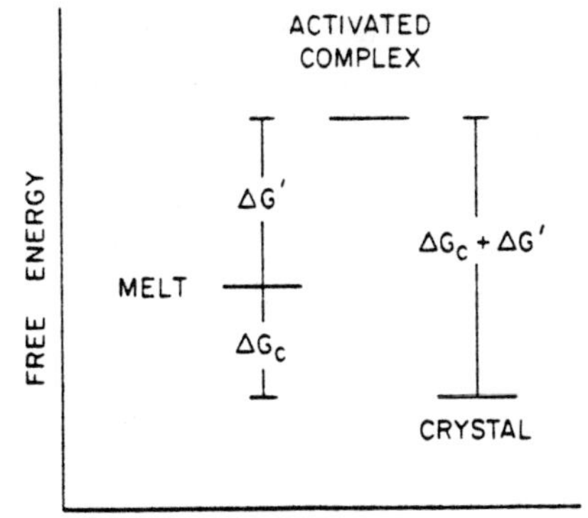

Figure 27. Free energies involved in reactions at crystal-melt interfaces.

$$r_a = \nu \exp(-\Delta G'/RT) \qquad (29)$$

where ν is an atomic vibration frequency, and $\Delta G'$ is the activation energy for attachment (see Fig. 27 for the energies involved). The rate of detachment can be written

$$r_d = \nu \exp[-(\Delta G_c + \Delta G')/RT)] \qquad (30)$$

where ΔG_c is the chemical driving force. The growth rate, Y, is then the difference between these two rates times the thickness per layer, a_o, and the fraction of sites on the surface on which atoms may successfully attach, f. Rearranging, this may be written

$$Y = f a_o \nu \exp(-\Delta G'/RT)[1 - \exp(-\Delta G_c/RT)] \ . \qquad (31)$$

This relationship has the same qualitative variation versus undercooling as the nucleation rate equation. The rate is zero at the liquidus because the chemical driving force is zero. The rate then increases with increasing undercooling because the chemical driving force is increasing. At lower temperatures (larger undercoolings) the activation energy term for attachment becomes important, reflecting the decrease in the rate of atomic motion. Because of this the growth rate passes through a maximum and finally decreases as the undercooling becomes very large. All observed growth rate versus undercooling relationships for silicate materials are of this form.

Growth mechanisms. Various simple physical models have been proposed for the nature of crystal surfaces. Each of these predicts a different variation in the way the fraction of sites available for attachment varies with undercooling.

In what is called the continuous model, the interface is assumed to be atomically rough. This allows atoms to attach at essentially all sites on the surface. In this case f equals one for all undercoolings, and the growth rate should be relatively fast. This appears to be the mechanism by which materials with low latent heats of fusion, such as metals, grow.

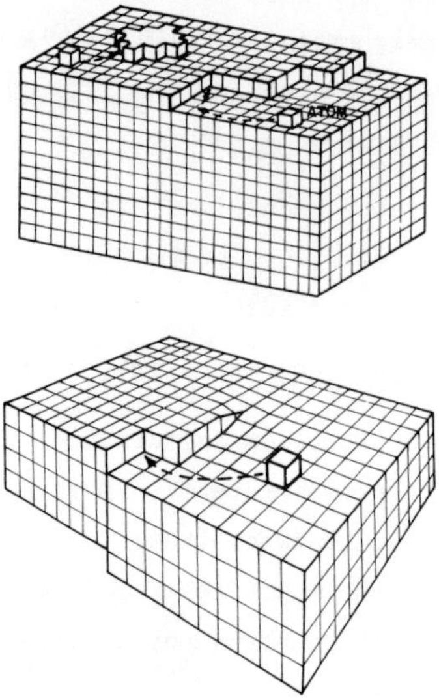

Figure 28. Diagramatic representation of atoms attaching on steps on crystal surfaces. (a) Surface nucleation controlled growth, (b) screw dislocation controlled growth.

In a general class of mechanisms called layer spreading mechanisms, the crystal surface is considered to be atomically flat except at steps, which are the only places atoms may attach onto the surface. We would expect attachment of atoms to be more likely at step sites because the atoms are bonded to more nearest neighbor atoms there than at a site in the middle of a flat surface (general adatom site). There are two generally recognized layer spreading growth mechanisms: surface nucleation and screw dislocation (Fig. 28). In the surface nucleation mechanism steps form by the nucleation of one atom thick caps which then spread over the surface as atoms attach to the edges. In the screw dislocation mechanism, a screw dislocation (see Chapter 7 by Lasaga) is assumed to emerge on the surface causing a spiral-shaped step on the surface that does not have to be renucleated for each layer. Each of these two mechanisms has a characteristic f versus ΔT relationship. Uhlmann (1972) and Kirkpatrick (1975) discuss the details of these mechanisms and how to determine from experimental data which one is operating. As in the nucleation rate equation [equation (10)], in the growth rate equation [equation (31)] it is possible to substitute a viscosity term for the activation energy term and approximate the thermodynamic driving force by a latent heat of fusion term. Thus, plots of $Y\eta/\Delta T$ versus ΔT can be used to determine the growth mechanism (Kirkpatrick, 1975).

Computer Simulation of Crystal Growth. Theoretical analysis of the atomic roughness of crystal surfaces and the computer simulation of crystal growth that followed this work have been very useful in understanding the nature of the interface, the mechanisms by which different kinds of materials grow, and the conditions under which different mechanisms can operate. This work is now finding application

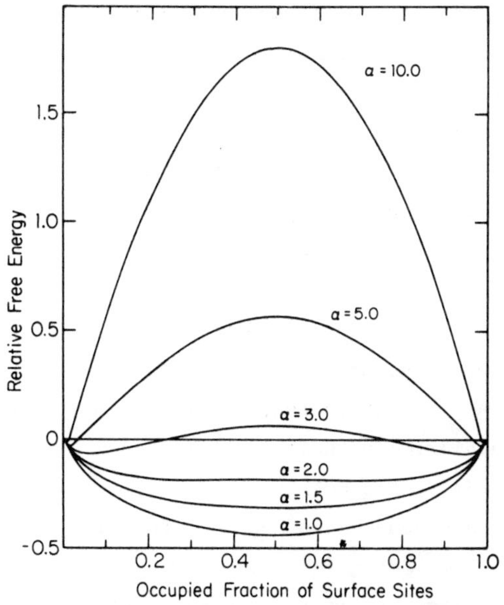

Figure 29. Relative free energies of surfaces versus occupied fraction of sites. α is the fraction of the binding energy which operates in the plane of the interface. Large α values represent densely packed planes. Small α values represent loosely packed planes (after Jackson, 1958).

in understanding the stability of planar interfaces and the origin of the various morphologies crystals may take on. It is based in large part on Jackson's (1958a,b) analysis of interface roughness. Interface roughness may be defined as the topographic relief on the crystal surface. In this model all the atoms are considered to be in either the liquid or the crystal. No transitional states are allowed. Jackson calculated the free energy of an interface as a function of the fraction of sites occupied by atoms and the parameter α, which is defined by

$$\alpha = \frac{L}{RT_L} \xi , \qquad (32)$$

where L is the latent heat of crystallization and ξ is the fraction of the total binding energy per atom which operates in the plane of the surface. ξ varies from one to zero and is larger for more closely packed planes than for less closely packed planes. Figure 29 shows Jackson's results for a simple cubic crystal. For α greater than about 2.0 the minimum free energy configuration of the surface is either mostly filled or mostly empty, with either a few surface vacancies or a few adatoms. For α less than about 2.0 the minimum energy configuration is a half-filled plane. The mostly filled or empty configuration is considered to correspond to an atomically flat surface which is likely to grow by a layer spreading mechanism. The half-filled configuration is considered to correspond to an atomically rough surface which is likely to grow by the continuous mechanism. Thus, densely packed planes of large latent heat materials, like most silicates, should be molecularly flat and grow by a layer spreading mechanism. Less densely packed planes of large latent heat materials and all planes of low heat materials should be atomically rough and grow by a continuous mechanism.

The more recent computer simulations have confirmed and extended Jackson's analysis. These simulations assume a simple cubic array of atoms and consider the attachment and detachment of atoms from a surface. The arrays considered may be as large as 100 atoms on a side, but there seems to be no significant effect of array size on the results. The two most important parameters are the bond strength (related to the latent heat) and the chemical potential difference between the crystal and the growth medium (i.e., the thermodynamic driving force). No overhanging configurations are allowed. Diffusion of adsorbed atoms, the effects of an emergent screw dislocation, and the growth of crystallographically different faces can all be considered. Monte Carlo simultaion techniques are the most popular. The details of the method are discussed by Gilmer and Bennema (1972), Binsberger (1972), Kohli and Ives (1972), Gilmer (1976, 1977), and Bennema and van der Eerdgen (1977) and will not be presented here.

While these simulations have been carried out for a very simple crystal using a simple picture of the bonding forces, it appears that the results may be useful in understanding, at least qualitatively, various aspects of the growth process in silicate melts.

The simulations confirm Jackson's conclusion that materials with large latent heats of crystallization should have atomically flat interfaces, while those with small latent heats should be atomically rough.

The Monte Carlo simulations of Gilmer (1976, 1977) have shown the importance of the latent heat, the magnitude of the thermodynamic driving force, and crystal perfection on the growth rate. Figure 30 shows Gilmer's (1976) results for the growth rate divided by the rate of atomic impingement on the surface and the step height versus the reduced thermodynamic driving force, $\Delta\mu/kT$ (proportional to ΔT), for various values of the reduced latent heat, L/kT, for perfect crystals. For $L/kT = 4.5$ (a low latent heat) the surface is rough and the growth rate is large even at small driving forces. For materials with larger latent heats the surface is smooth, and because there is no dislocation present, growth occurs by the surface nucleation mechanism. In this case there is a range of driving forces (relatively small for $L/kT = 6$, relatively large for $L/kT = 12$) where the rate of surface nucleation is so low that the growth rate is effectively zero.

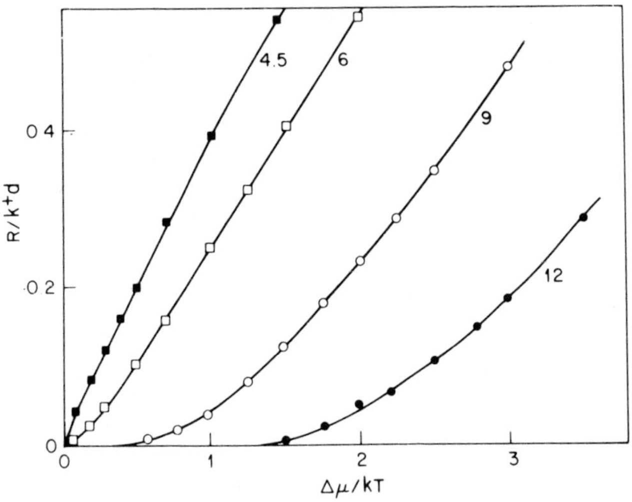

Figure 30. Calculated growth rates (normalized by the step height and rate of atomic impingement) versus thermodynamic driving force (μ/kT) of perfect (001) faces of a simple cubic crystal for materials with different values of the latent heat/kT. Crystals with low latent heats have low α values and grow by surface nucleation (no dislocations are present in the calculations) (after Gilmer, 1976).

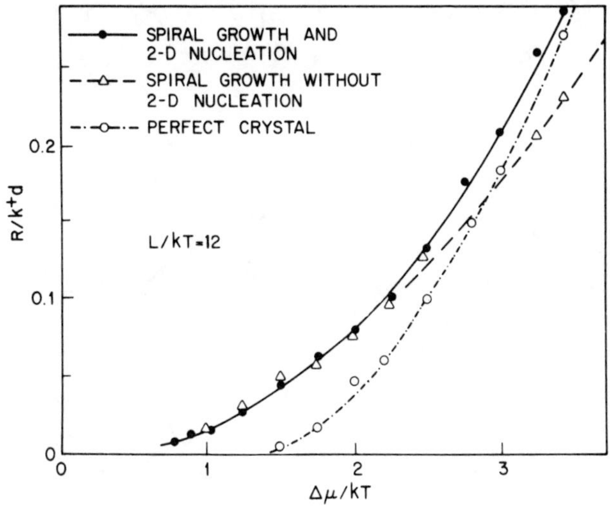

Figure 31. Calculated growth rates normalized as in Figure 30 versus thermodynamic driving force for a large latent heat of fusion material with and without a screw dislocation present. Note that the growth rate with both mechanisms allowed to operate approximates the screw dislocation-controlled rate at small driving forces and the surface nucleation-controlled rate at large driving forces (after Gilmer, 1976).

Figure legend on the next page →

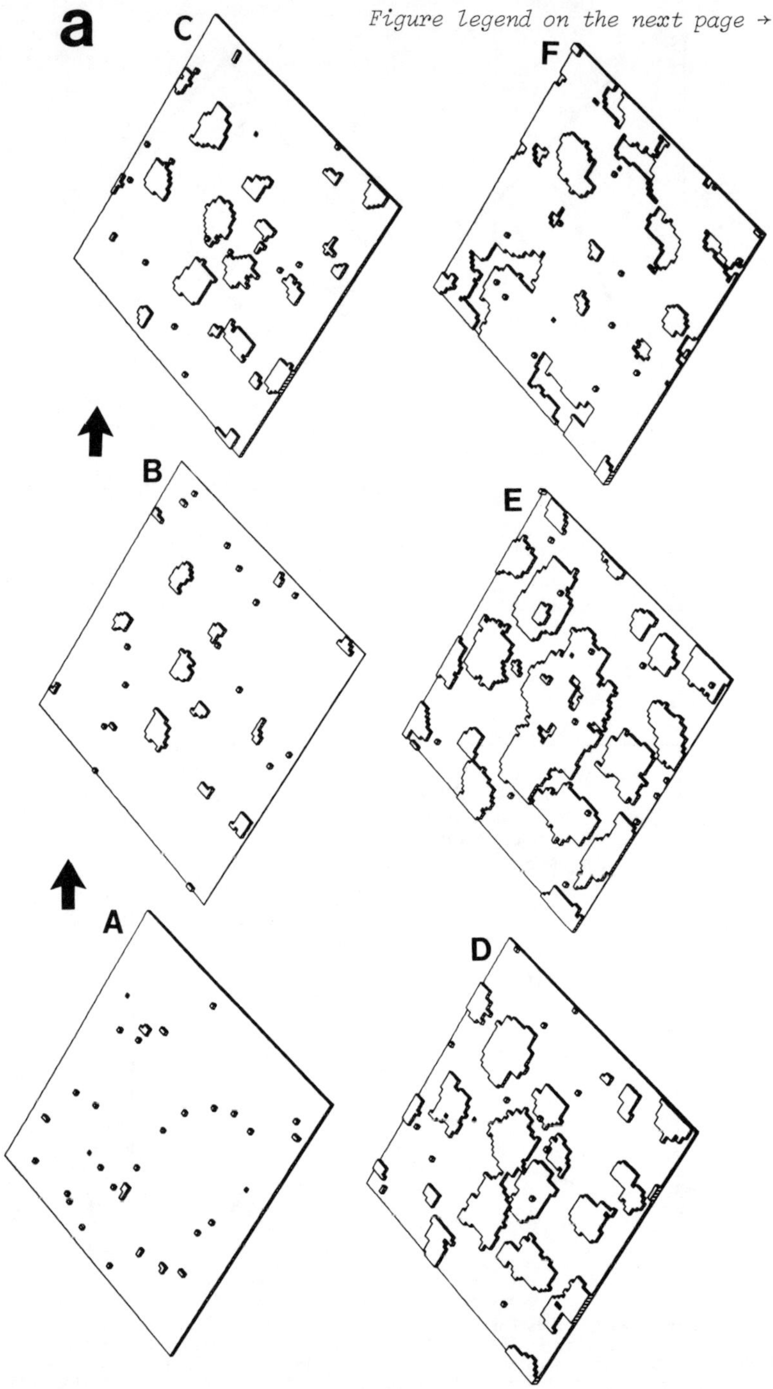

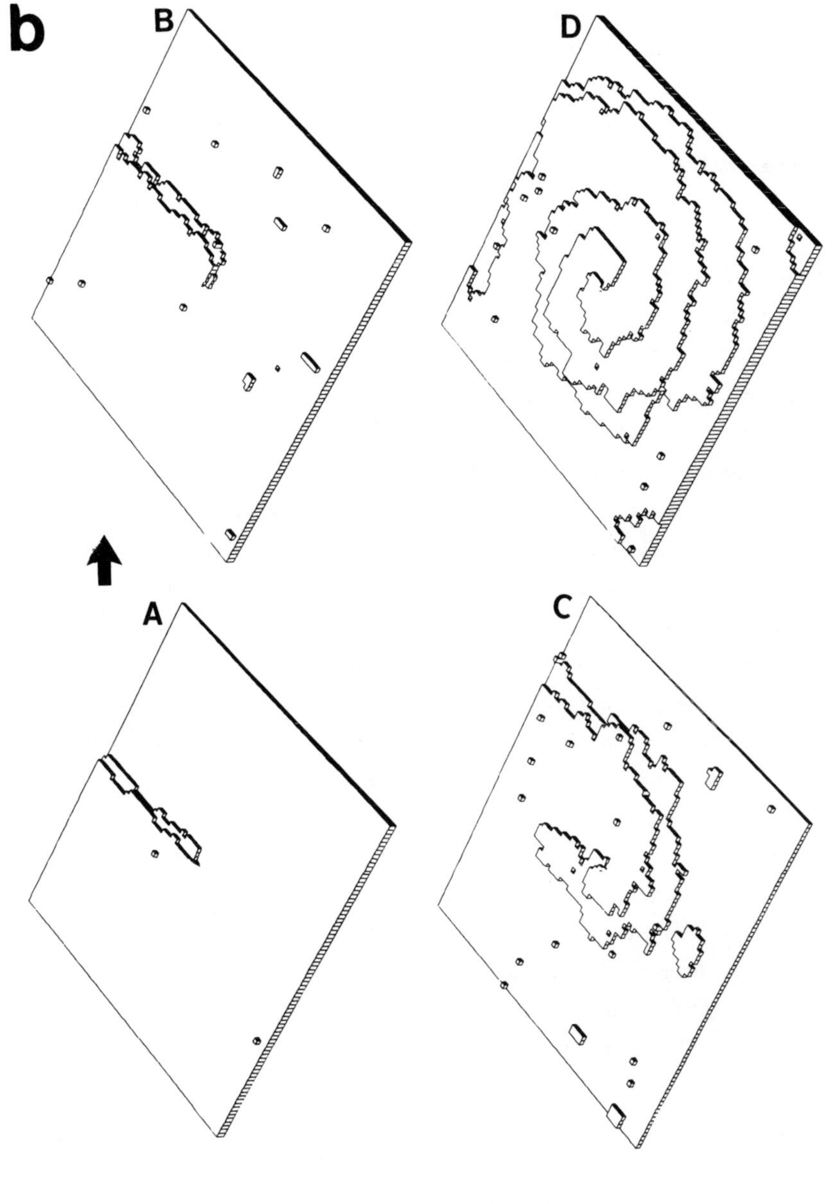

Figure 32 (a) Computer simulated configurations of a simple cubic (100) crystal face during a growth period after instantaneous supercooling with no dislocation present (A to F is a time sequence). Note the formation and growth of a number of surface nuclei which grow together to form the first new layer and the formation of surface nuclei in the second new layer before the first layer is completely filled. ($\Delta\mu/kT = 2$, $\phi = 4kT$). (b) Computer simulated development of a spiral step pattern on a simple cubic (100) face with a double Burger's vector screw dislocation emergent at the center (A to D is a time sequence). Note the minor contribution of surface nuclei to the overall growth. Figures courtesy of Dr. G. H. Gilmer.

The presence of an emergent screw dislocation greatly affects the growth rate of large latent heat materials in the small undercooling range. Figure 31 is a plot of the normalized growth rate versus normalized driving force for L/kT = 12. It shows curves for the total growth rate including the effects of surface nucleation and the screw dislocation, the effect of surface nucleation alone, and the effect of the dislocation alone. In the small driving force range dislocation controlled growth is much faster than surface nucleation controlled growth, and the total growth rate is essentially the screw dislocation controlled rate. With increasing driving force (increasing deviation from equilibrium) there is a transition region in which the dislocation and nucleation controlled rates are approximately equal. At large driving forces surface nucleation becomes dominant, with the total growth rate approximately equal to the nucleation controlled rate. Figure 32 shows the simulated surfaces of a large latent heat material with and without a screw dislocation present (Gilmer, 1976). Sunagawa (1977) has used these results to explain the morphological development of natural crystals. Sunagawa's results will be discussed in detail when we take up morphological stability.

Gilmer (1977) has simulated the anisotropy of the growth rate of a dislocation free simple cubic crystal. Figure 33 shows his results for the (111), (100), and (110) faces of a face-centered cubic crystal with L/kT = 12. For the most densely packed (111) plane the growth rate behaves as predicted for surface nucleation controlled growth, it is very low (essentially zero) at low driving forces but increases rapidly as the driving force increases. The less densely packed (100) face has a small driving force region of low growth rate, and the least densely packed (110) face shows no nucleation barrier at all. The result of this is that at small deviations from equilibrium the anisotropy of the growth rate is large, i.e., the (110) face grows much faster than the (111) face, while at the large deviations from equilibrium all faces grow at nearly the same rate. The reason the (110) face grows so rapidly at small driving forces is that it is atomically rough under all conditions, so that no nucleation is needed to initiate new layers, while the (111) face is smooth at small deviations but becomes rougher at large deviations as the rate of

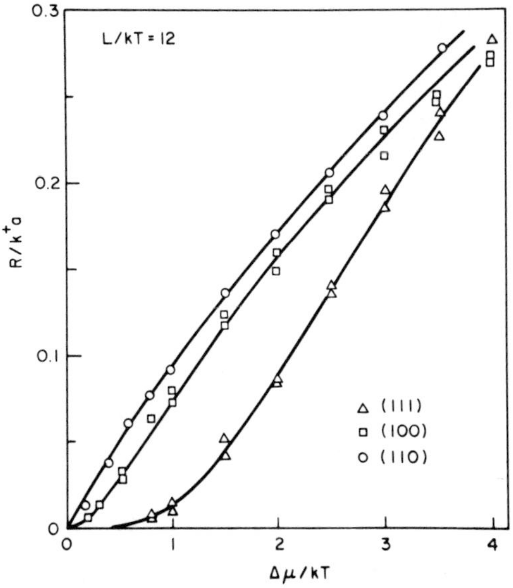

Figure 33. Calculated growth rates normalized as in Figure 30 versus thermodynamic driving force for different faces of a perfect simple cubic crystal. The densely packed (111) faces grow by surface nucleation control, while the loosely packed (110) faces do not require nucleation for growth to occur (after Gilmer, 1977).

addition of atoms to the general adatom sites on the surface approaches the rate at which they are removed.

Because of this, a crystal of a large latent heat material should be faceted at small deviations from equilibrium. All the non-planar faces should grow rapidly and disappear, so that only the slowly growing surfaces will remain as the planar crystal boundaries. At large undercoolings, on the other hand, the crystal should not be faceted, since the growth rate of all faces is about the same. Miller (1977) has used this theory to explain the transition from faceted crystallites in open spherulites at relatively small undercoolings to non-faceted crystallites in compact spherulites at larger undercoolings for several organic compounds grown from the melt. In general, the transition in morphology takes place at undercoolings larger than the undercooling of maximum growth rate. This kind of morphologic transition has not been investigated in silicate materials, but may prove to be a useful tool in determining the undercooling conditions under which crystallization has occurred in igneous rocks.

Diffusion Controlled Growth. When a crystal grows from a melt of a different composition, redistribution of the components in the melt must take place. Components rejected by the crystal build up at the interface and try to diffuse away, resulting in a composition gradient qualitatively similar to Curve 1 in Figure 34. Components preferentially incorporated into the crystal will be depleted in the melt at the interface, and will diffuse towards it, resulting in a composition gradient qualitatively similar to Curve 2 in Figure 34.

Under the conditions of diffusion controlled free growth (i.e., the growth rate not constrained in some way to be a fixed value), the

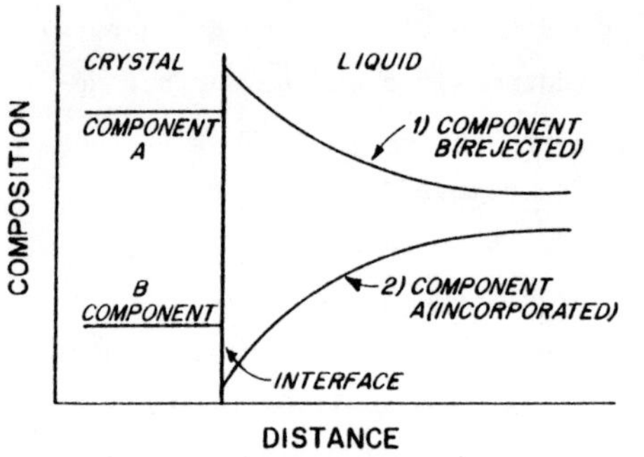

Figure 34. Schematic composition gradients in the melt near a growing crystal face when diffusion is not fast enough to eliminate the gradients.

growth rate is controlled by how rapidly components can diffuse to and from the interface, and the growth rate will decrease with increasing time. This is because as time proceeds the composition gradients must extend further and further out into the melt. In general, the growth rate will be proportional to $(D/t)^{\frac{1}{2}}$, where D is an exchange diffusion coefficient and t is time.

A solution to this problem for a two component system at arbitrary concentration has been given by Müller-Krumbhaar (1975). He assumes a one-dimensional system, constant temperature throughout, no diffusion in the solid, no volume change on crystallization, and no volume change in the melt with change in composition. The bulk liquid composition is C_o and is supersaturated with respect to the equilibrium composition, C_1. The crystal composition is C_s. The diffusion equation in the melt is

$$\frac{\partial C}{\partial t} = D \frac{\partial^2 C}{\partial u^2} \tag{33}$$

where C is concentration and u is distance in the laboratory fixed-reference frame.

The initial boundary conditions are

$$C(u,o) = C_o;\ C(\text{interface},\ t) = C_1;\ C(u,t) = C_o \quad [u \to \infty]\ . \tag{34}$$

The laboratory fixed-reference frame can be transformed into a reference frame fixed on the interface by the transformation of coordinates

$$X \leftarrow u - \int_o^t Y(t)dt\ , \tag{35}$$

where X is distance ahead of the interface in the melt and Y(t) is the crystal growth rate in the X-direction at time t. The diffusion equation is then

$$\frac{\partial C}{\partial t} = D \frac{\partial^2 C}{\partial X^2} + Y(t) \frac{\partial C}{\partial X}\ . \tag{36}$$

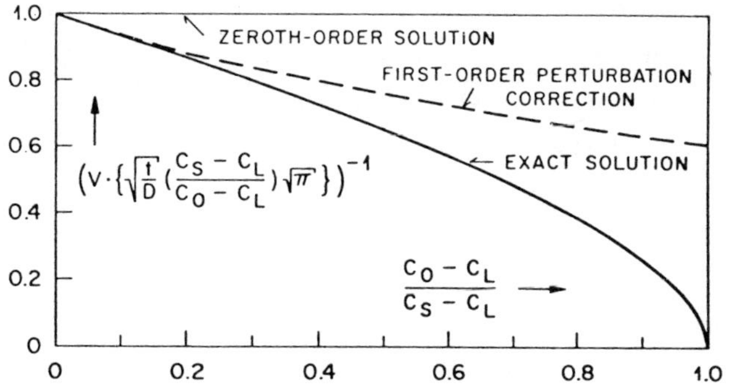

Figure 35. Normalized growth rate versus relative solute concentration for the condition of a constant interface concentration resulting from Equations 40 (after Müller-Krumbhaar, 1975).

The second term on the right-hand side accounts for the effects of the crystal acting as a source or sink of components. The initial and boundary conditions are now

$$C(X,0) = C_o; \; C(0,t) = C_1; \; C(\infty,t) = C_o \; . \tag{37}$$

Since matter must be conserved at the interface, the flux of atoms reaching the crystal by diffusion must equal the number being incorporated. This conservation condition can be written

$$-D \frac{\partial C}{\partial X} \Big|_{X=0} = Y(t)[C(0,t) - C_s] \; . \tag{38}$$

Equation (36) becomes

$$\frac{\partial C}{\partial t} = D \frac{\partial^2 C}{\partial X^2} + \left[\frac{1}{C_s - C_1} D \frac{\partial C}{\partial X} \Big|_{X=0}\right] \frac{\partial C}{\partial X} \; . \tag{39}$$

The solution to this equation is

$$Y(t) = \tilde{C}(D/t)^{\frac{1}{2}}$$

where $\tilde{C}$ is a constant, obtained from the following equation:

$$\tilde{C} = \frac{C_o - C_1}{(C_s - C_1)} \frac{1}{\sqrt{\pi}} \frac{e^{-\tilde{C}^2}}{\text{erfc}\cdot(\tilde{C})} \; . \tag{40}$$

Müller-Krumbhaar's exact solution for this is shown in Figure 35.

An important feature of this solution, as in all purely diffusion controlled processes, is that there is no possibility of reaching a steady state. Steady state solutions for planar interfaces exist only when the growth rate is constrained to be a constant value. Solutions for this condition have been discussed by Tiller *et al.* (1953) and Burton *et al.* (1953), among others. These solutions, while important in many industrial and experimental situations, would seem to be inappropriate for petrologically interesting situations. In plutons and lava flows the growth rates are controlled by the properties of the material and the local physical conditions. A constant growth rate would not be expected and is not observed in the limited results available (Kirkpatrick, 1977).

Another important feature of the Müller-Krumbhaar solution, and indeed most published solutions to diffusion controlled growth problems, is that at the interface the liquid and solid compositions are assumed to be the equilibrium values. This can then be used as a boundary condition in solving the diffusion equation. For many materials, including most silicates, a significant undercooling (or supersaturation) at the interface is needed to produce significant growth rates. Incorporation of interface kinetic effects into the diffusion problem has proven to be quite difficult. Hooper and Uhlmann (1977) have given a solution for isothermal conditions assuming a constant growth rate (i.e., an analog to the Tiller *et al.*, 1953, solution) and have also discussed the thermodynamic problem of assigning the interface conditions. Müller-Krumbhaar (1975) has circumvented the thermodynamic problem by using an expression for the growth rate that involves activation energy and thermodynamic driving force terms. He assumes an exponentially-shaped liquidus surface and a constant rate of cooling and uses finite difference methods. His results indicate that with increasing time at constant cooling rate the growth rate should first increase, pass through a maximum, and then decrease. The supersaturation at the interface should increase continually. Figure 36 shows these results for a situation where a large driving force is needed for growth. He also discusses how the temperature dependence of the diffusion coefficient can be taken into account. A more recent analysis of this problem, which includes a general solution for time-dependent and temperature-dependent growth rates, has been developed (Lasaga and Muncill, 1980; Lasaga, 1981, unpublished results).

While diffusion is clearly an important process that must occur in many situations, in actual fact the predicted $t^{-\frac{1}{2}}$ dependence of the

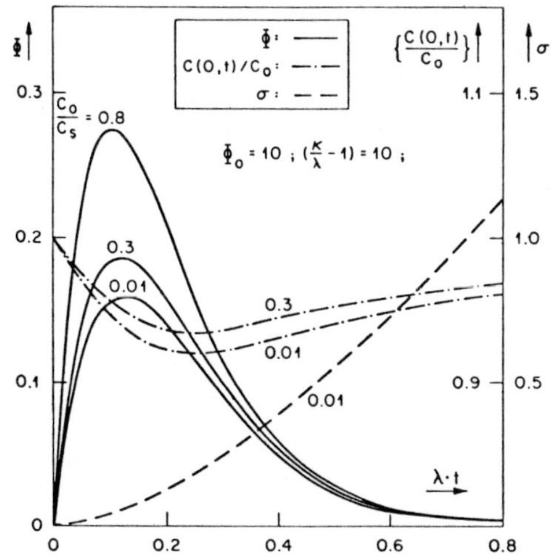

Figure 36. Renormalized growth rate, ϕ relative concentration, $C(0,t)/C_0$, and supersaturation at the interface, σ, versus normalized time assuming a large kinetic barrier to crystal growth (after Müller-Krumbhaar, 1975).

growth rate is rarely observed. In most experiments isothermal growth rates are time independent no matter what the composition of the crystal or melt. At small undercoolings this can be explained by having the interface controlled growth rate being much slower than the diffusion controlled growth rate. In general, though, other processes may allow a steady state to develop. Convection in the melt may disrupt the diffusion field around the crystal, causing an essentially constant thickness boundary layer to develop. Diffusion across this layer may then reach a steady state (Crank, 1956). Perhaps more important, however, is the breakdown of the planar interface to a skeletal, dendritic, or spherulitic morphology. These morphologies allow exchange of material parallel to the overall trend of the interface as well as perpendicular to it, allowing a steady state to develop. In effect these morphologies allow the individual parts of the crystal to grow out into undepleted melt and the rejected components to be trapped between the crystal segments or crystallites. Convection associated with freely growing crystals is poorly understood (Carruthers, 1979) and will not be treated here.

Stability of Planar Crystal Surfaces. When a crystal is growing, there is often a tendency for planar, crystallographically controlled interfaces to break down into a non-planar morphology. Common non-planar morphologies include hopper and skeletal shapes, dendrites, and spherulites. Less commonly discussed interface features associated with interface instability include steps and waves that move across the surface. The cause of interface instability is always non-uniform physical conditions near the interface. These non-uniformities may involve thermal gradients, compositional gradients, or both.

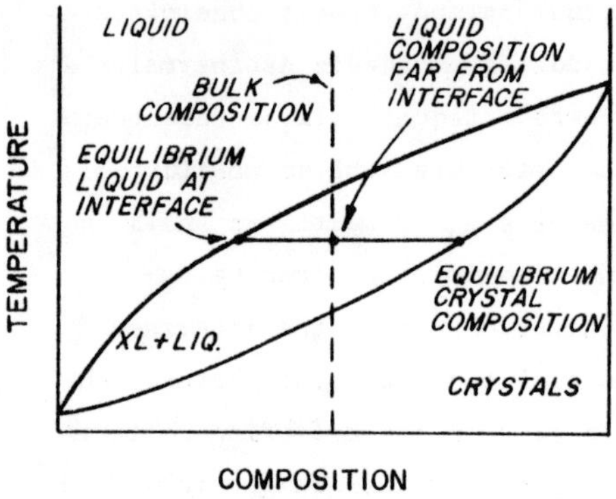

Figure 37. Important compositions involved in diffusion-controlled growth of crystals.

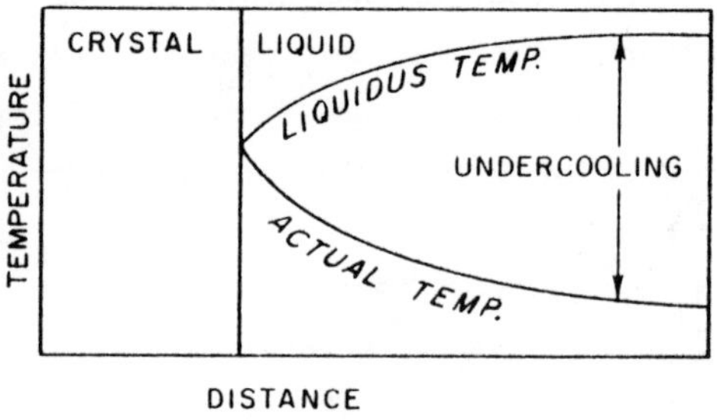

Figure 38. Generalized undercooling distribution in the melt in the vicinity of a growing crystal which is rejecting impurity.

Theoretical treatment of interface stability began with the analysis of the stability of materials for which interface kinetics is rapid and can be ignored relative to the diffusion (i.e., a continuous growth mechanism). Quantitative theories for this situation are available. Incorporation of layer spreading growth kinetics into the stability analysis has proven to be very difficult, however, and no quantitative stability criteria are available for materials with large latent heat of fusion.

From a geological standpoint one of the objectives of studying interface stability is to determine the conditions under which an interface will be planar and when it will take on one or another of the non-planar morphologies. It is then possible to determine by looking at the crystals the conditions under which a particular rock crystallized.

The first interface stability criterion to be developed was constitutional supercooling (Tiller *et al.*, 1953; Elbaum *et al.*, 1959). This criterion says that if the undercooling is least at the crystal-melt interface and increases with increasing distance from the interface into the melt, the interface will be unstable. For almost all crystals growing in such a way that the latent heat of crystallization generated at the interface must be lost through the melt, this criterion predicts interface instability. In most incongruently melting compositions the crystals are richer in the high melting temperature components (Fig. 37), and the interface liquid at or close to the equilibrium value will have the lowest liquidus temperature of any liquid composition in the diffusion field surrounding the crystal. In addition, because of the release of latent heat at the interface, the interface liquid will be the highest temperature of any melt near the crystal. These variations in liquidus temperature and actual temperature surrounding the crystal are shown in Figure 38. Under these circumstances, the tip of any protrusion which happens to develop on the crystal surface will be at a larger undercooling than the rest of the interface and will, therefore, grow faster. The interface will then break down into a periodic array of growth cells which may, for instance, take on a dendritic morphology.

This simple model has been extended to include such factors as surface energy, diffusion parallel to the interface, and isotropic growth kinetics (i.e., a continuous growth mechanism). The work is based on the initial analysis of Mullins and Sekerka (1964) and has been reviewed by Delves (1975). The details will not be reviewed here. This method assumes an initially steady state diffusion field in the melt ahead of a planar interface, allows the interface to be perturbed, expresses the shape of the perturbed (i.e., non-planar) interface as a Fourier series, and solves the diffusion equation in the melt. Analysis of the solution may indicate that perturbations with certain wavelengths tend to grow rather than decay into a planar interface. In this case, a crystal larger than any such wavelength is unstable. For crystals which must lose their latent heat of crystallization through the melt, surface energy effects allow stabilization of planar interfaces only if the dimensions are of the order of a micron. Larger planar crystal surfaces appear to be unstable.

This is clearly at variance with observation. Centimeter-sized, fully-faceted crystals of many materials can be grown by free growth methods, and millimeter-sized, fully-faceted crystals are common in rocks.

The Mullins- and Sekerka-type stability theories fail for materials with high latent heat of fusion because they neglect the stabilizing effect of layer spreading growth mechanisms (in the terminology of stability theory-anisotropic growth kinetics). For materials with low latent heats of fusion and for high-Miller-index planes of materials with large latent heats of fusion, however, these stability theories seem to be sufficient.

In simple mechanistic terms, layer spreading growth mechanisms promote interface stability for two reasons: (1) It is difficult for bumps to develop on surfaces that have a tendency to remain atomically flat, and (2) if a bump does develop it grows far more rapidly sideways than it does forward and it simply grows sideways to become part of a flat interface. These effects have proven difficult to quantify, and no fully satisfactory theory is available.

The observation for many silicate materials is that interface breakdown first takes place not by bumps developing all over a planar interface, but only at corners where two or three faces meet. The qualitative explanation of this seems to be that, in cases where convection is not important, the interface melt composition is closer to the bulk melt composition at corners than in the center of a face. This is because there is a larger melt volume to surface area ratio near the corners, and rejected components can diffuse more rapidly away from the interface at the corners than in the center. This results in a larger, local undercooling at the corners than at the face centers and in diffusion parallel to the interface. This diffusion parallel to the interface reduces the composition difference between the melt at corners and at face centers but cannot eliminate it completely.

There is presently no quantitative model for when breakdown at corners should occur, but consideration of the recent computer modeling of crystal growth has lead to a better understanding of the conditions that should promote it. The following discussion will follow the ideas of Wilcox (1977) and Sunagawa (1977).

The results of Gilmer's (1977) computer simulations shown in Figure 32 indicate that at small undercoolings the rate of surface nucleation is vanishingly slow and that growth in this range should be controlled by the screw dislocation mechanism if dislocations with a screw component emerge on the surface. At larger undercoolings, however, surface nucleation becomes dominant. For a crystal growing from a melt with a low bulk undercooling both crystal corners and face centers will be in the dislocation controlled range. Because dislocations are more likely to emerge in the center of the face (this part is directly over the oldest part of the crystal), steps will be generated in the center and move towards the edges. In this case the vicinal

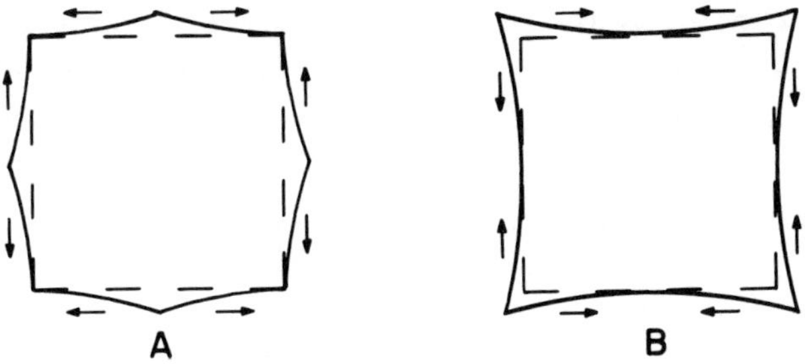

Figure 39. Vicinal interfaces (exaggerated) associated with (A) growth-controlled by screw dislocations emergent at the center of the face at low undercoolings, and (B) surface nucleation-controlled growth at relatively large undercoolings. Arrows indicate direction of step motion across the surfaces.

(actual) interface will slope slightly from the center to the edges (Fig. 39). Under these conditions steps move into regions of progressively larger undercooling. Because of this, they grow progressively more rapidly until they reach the corner and can never pile up to form a step or wave on the surface.

At larger bulk undercoolings the crystal will be in the surface nucleation controlled range, and the behavior will be quite different. Because the rate of generation of steps increases rapidly with increasing undercooling in this regime, more layers will be generated at corners or edges, which are at larger undercoolings, than at face centers. These layers will spread towards the center, and the vicinal interface will slope towards the center (Fig. 39). These steps will move into regions of progressively lower undercooling and will slow as they approach the face center. There will be a tendency for the steps to bunch together to form steps or waves on the surface. These steps may trap inclusions, and eventually the growth in the center of the face may stop altogether (as in the case of a hopper crystal). Kuroda *et al.* (1977) have shown that growth by the screw dislocation mechanism leads to stable planar interfaces, while for surface nucleation dominated growth there is a transition regime in which the interface is stable and then a regime at larger undercoolings in which planar interfaces are not stable.

Wilcox (1977) has recognized at least four stages of interface breakdown that develop with increasing undercooling. These are: (1) stable planar growth, (2) macroscopically (i.e., in an optical microscope) planar interfaces with inward moving microsteps that may trap tiny inclusions, (3) development of waves or macrosteps on the interface that move from the edges to the face centers, and (4) growth of dendrites from the corners. Hopper crystals (which he did not discuss) probably develop between Stages 3 and 4. Examples of some of these morphologies are illustrated in Figure 40.

Although no quantitative criteria are available to determine the conditions under which these various morphologies develop, Wilcox (1977) has analyzed the diffusion field around a growing cube using finite difference methods and has concluded that surface nucleation dominated growth from corners, and therefore interface breakdown of this type is promoted by a large bulk undercooling, large crystal size (because composition gradients parallel to the interface cannot readily be eliminated), and a low concentration of the crystal components in the melt. Different materials should behave differently because the more slowly a phase grows at a particular undercooling, the more likely it is to grow by surface nucleation at corners, but the less likely it is to develop waves or macrosteps. It would seem, then, that the maximum size a given face of a crystal of a particular material can reach before becoming unstable decreases with increasing undercooling. Thus, although it is not possible to determine quantitatively the undercooling or range of undercoolings a particular rock crystallized under, it is possible to use the maximum size of a particular planar face of the same phase to determine relative undercoolings within a magma body or between magma bodies of approximately the same composition. Application of this idea is discussed below.

Crystal Growth Data

The amount of data available on the rates of crystal growth and the morphological development in geologically interesting compositions has increased substantially in the past decade. Much, but certainly not all, of this work has been stimulated by the lunar science program.

Rates of Crystal Growth. The rates of crystal growth from the melt have been collected for a number of geologically important compositions at both one atmosphere and elevated water pressure. These results qualitatively confirm the theoretically predicted undercooling variation (low at small undercoolings, increasing to a maximum, and then decreasing with increasing undercooling) and the idea that low viscosity

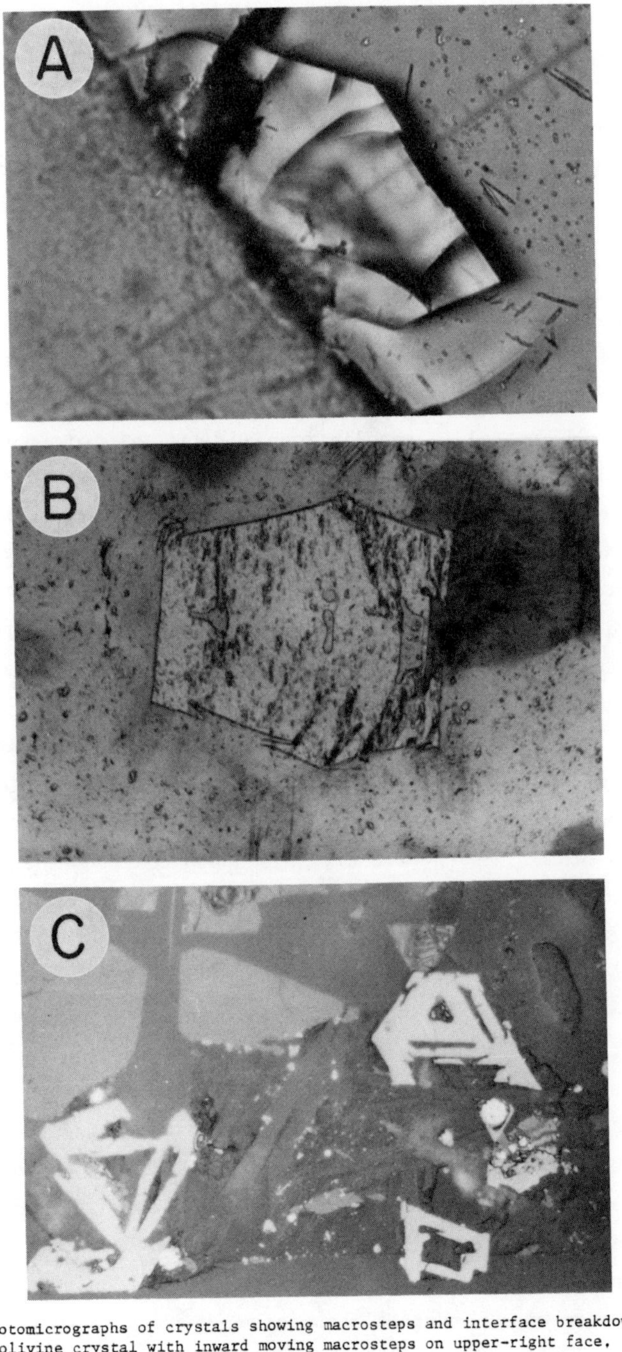

Figure 40. Photomicrographs of crystals showing macrosteps and interface breakdown at corners. (a) Synthetic olivine crystal with inward moving macrosteps on upper-right face. The crystal is ∿ 0.2 mm long. (b) Olivine microphenocryst in a basalt pillow with dendrite arms growing from corners. Inclusions are trapped when dendrite arms from opposite corners meet. The crystal is ∿ 0.25 mm long. (c) Reflected light photomicrograph of titanomagnetite grains in a basalt with dendrites growing from corners and dendrite entrapment of inclusions. Magnetites are ∿ 50 μm across.

compositions have higher growth rates than high viscosity compositions at the same undercooling. The long held idea that increasing water content increases the growth rate does not seem to be true.

Data for synthetic diopside ($CaMgSi_2O_6$) and for compositions in the system albite ($NaAlSi_3O_8$)-anorthite ($CaAl_2Si_2O_8$) have been measured by Kirkpatrick (1974), Klein and Uhlmann (1974), and Kirkpatrick *et al.* (1977, 1979). These data were obtained by both microscope heating stage techniques and by the quenched sample method. In the microscope heating stage technique the melt is held in a small platinum crucible in the heating stage. To measure the growth rate the sample is heated above the liquidus and homogenized and is then cooled to the desired undercooling. A seed crystal is then introduced, and the growth of the crystals on the seed photographed with a motion picture camera. Rates are obtained from the movie by plotting the crystal lengths versus time, the rate being the slope of the plot. In the quenched sample method several samples of glass are heated at the same temperature for different lengths of time. The length of the crystals are measured in each sample (the crystals usually nucleate on the surface), and this length is then plotted versus run time. Again, the rate is the slope of this plot. In all cases, including the incongruently melting compositions in which the crystals should not have the composition of the melt, the growth rates are independent of time, except when two crystals approach each other. As the two crystals approach each other the growth rate slows, presumably because of the increase in temperature or concentration of rejected components between the crystals.

Figure 41 shows the data for the rate of growth of clinopyroxene at small undercoolings from a diopside composition melt. The maximum growth rate is about 2×10^{-2} cm/sec, which is comparable to the largest growth rates yet measured in geologically interesting compositions. A plot of $Y\eta/\Delta T$ versus ΔT (Fig. 42), which can be used to determine the growth mechanism (see Uhlmann, 1972, or Kirkpatrick, 1975, and above) for the diopside data at small undercoolings shows that growth is controlled by the screw dislocation mechanism. Fitting this data with a straight line forced to pass through the point 0,0 allows calculation of the growth rates in the very small undercooling region, where data cannot be obtained experimentally (Fig. 43). This calculation indicates that at an undercooling of 1°C the growth rate is about 10^{-6} cm/sec. At this rate

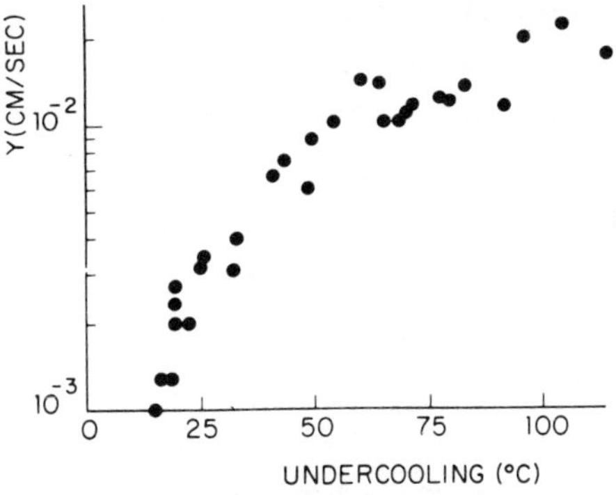

Figure 41. Observed growth rate versus undercooling relationship for clinopyroxene growing from diopside-composition melt (after Kirkpatrick *et al.*, 1976).

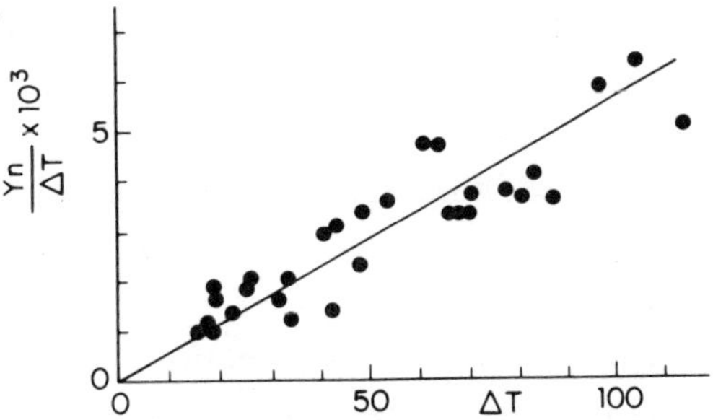

Figure 42. Observed reduced growth rate versus undercooling relationship for clinopyroxene growing from diopside composition melt. The straight line relationship indicates that the screw dislocation mechanism is dominant (after Kirkpatrick *et al.*, 1976).

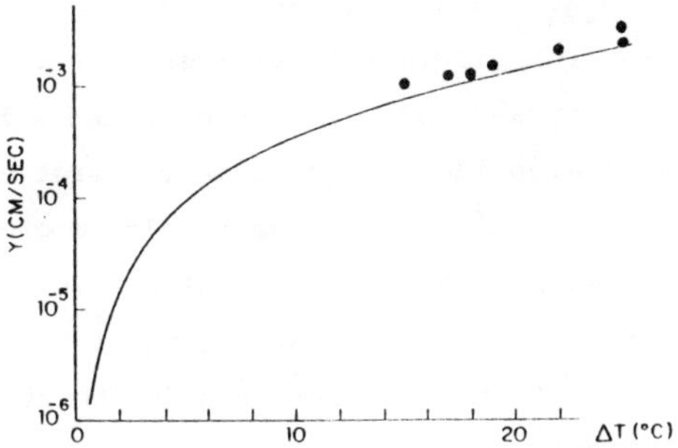

Figure 43. Calculated growth rate versus undercooling relationship in the small undercooling region for clinopyroxene growing from a diopside-composition melt (after Kirkpatrick *et al.*, 1976).

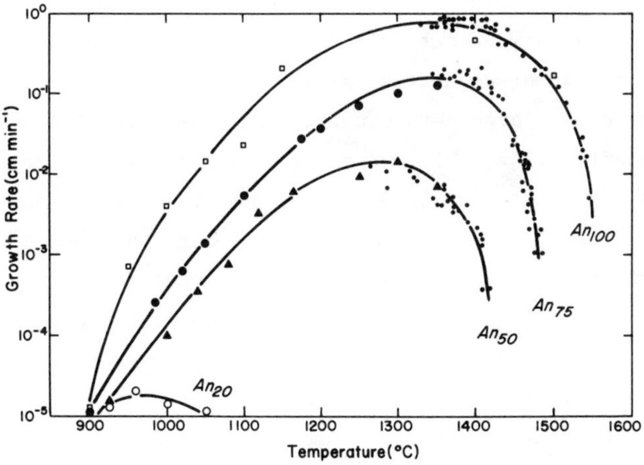

Figure 44. Variation of growth rate with temperature for compositions in the join albite-anorthite. High-temperature data by microscope heating stage, low temperature data by quenching (after Kirkpatrick *et al.*, 1979).

it would take about 10 days to grow a one-centimeter crystal. This rate is considerably higher than the rates expected for most rocks because the temperature (about 1390°C) is higher than the liquidus for most common rock types, because the viscosity is lower than for most rocks, and because diffusion in the melt is not a significant problem.

The growth rate data for compositions in the system albite-anorthite are shown in Figure 44 (Kirkpatrick *et al.*, 1979). The growth rates at a given undercooling and at a given temperature, as well as the maximum growth rate all decrease with increasing albite content. This correlates well with the increase in viscosity and decrease in liquidus temperature which occurs with increasing albite content.

Although the maximum growth rate of anorthite from the An_{100} composition is comparable to the maximum growth rate of diopside, in the small undercooling region the growth rate of anorthite is lower at small undercoolings and increases much more rapidly with increasing undercooling than that of diopside. Figure 46 is a plot of $Y\eta/\Delta T$ versus ΔT for the An_{100} data (implying a surface nucleation mechanism for this composition), and Figure 46 is a plot of the calculated growth rates in the small undercooling region. At an undercooling of 1°C the growth rate is about 10^{-8} cm/sec. At this rate it would take about three years to grow a one-centimeter crystal. Again, this rate is probably larger than the rates for common rock compositions at the same undercooling because of the high liquidus temperature (about 1550°C), and low melt viscosity, and because diffusion is not important.

As part of a comprehensive study of lunar glasses one atmosphere growth rate data have been obtained for a wide variety of lunar basalts and soil compositions by D. R. Uhlmann and his co-workers (Scherer *et al.*, 1972; Cukierman *et al.*, 1973; Klein *et al.*, 1975; Uhlmann *et al.*, 1976, 1977). These data have been collected using the quenched sample technique. Viscosity data have also been collected for most of the compositions. The data are far too extensive to reproduce entirely, but Figure 47 does present the observed growth rate versus undercooling relationships for a variety of compositions covering the range from the most slowly to the most rapidly growing. The composition with the slowest growth rate (14310) is very anorthite rich. Most of the medium growth rate compositions are basalts or breccia glasses. The most rapidly growing composition is the Apollo 17 Green Glass, which is a

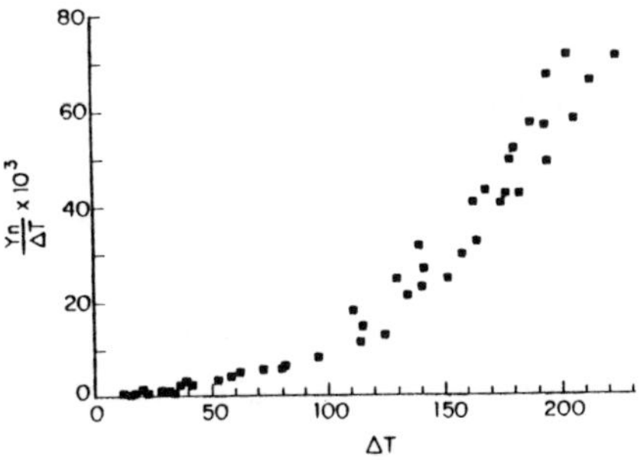

Figure 45. Observed reduced growth rate versus undercooling relationship for anorthite growing from its own melt (after Kirkpatrick *et al.*, 1976). Curved relationship indicates that surface nucleation is the dominant growth mechanism.

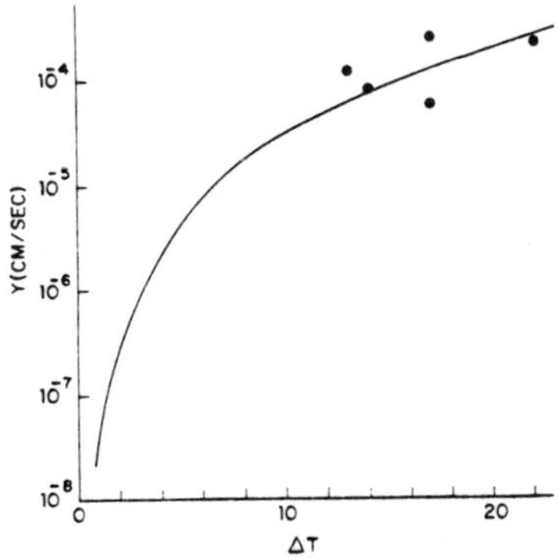

Figure 46. Calculated growth rates for anorthite growing from its own melt in the small undercooling region (after Kirkpatrick *et al.*, 1976).

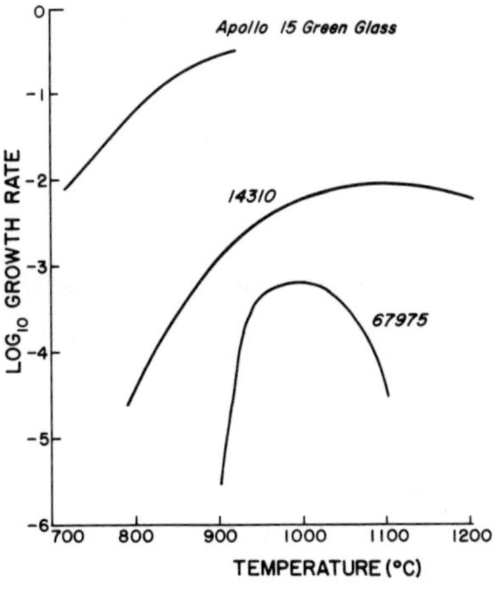

Figure 47. Growth rate versus temperature relationship for Lunar Compositions. The three data sets cover the observed range of rates. The Apollo 15 Green Glass is a picrite, 14310 is an anorthosite, and 67975 is a vitric breccia.

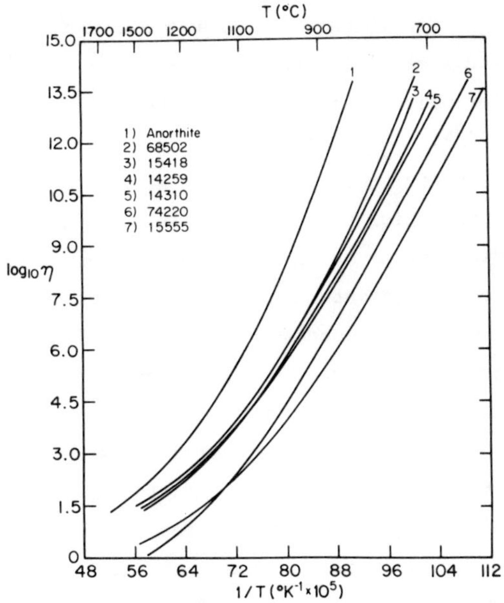

Figure 48. Viscosity versus temperature relationships for selected lunar compositions and anorthite (after Cukierman *et al.*, 1973).

very olivine-rich basalt composition. This increase in growth rate correlates well with the observed decrease in viscosity (Fig. 48). The difference in growth rate between the slowly growing 14310 compoition, which is mostly anorthite, and the synthetic anorthite composition, both of which have about the same viscosity, is probably related to the low liquidus temperature of 1310°C for 14310 relative to the high liquidus of 1550°C for pure anorthite.

The purpose of collecting the data for the lunar compositions has been to analyze the formation of lunar glasses and to estimate cooling times, cooling rates, and cooling unit sizes. The data collected are the data needed for this analysis. For the most part, however, rates have been measured at large undercoolings where the crystal morphology is spherulitic. In addition, in most cases the phases crystallizing have not been identified. They are, therefore, of little use in elucidating the growth of the large, more or less equant, crystals which occur in most igneous rocks. These, presumably, grow at relatively small undercoolings, in the range where no growth rate data are presently available for basaltic compositions.

Growth rates at elevated water pressures have been estimated for the synthetic alkali feldspar, granite, and granodiorite compositions for which nucleation rates have been determined (Fenn, 1977; Swanson, 1977). The major limitation of these data is that they are calculated using only one run time at each temperature. The rate is obtained by determining the length of the longest internally nucleated crystal and dividing it by the run time. This procedure assumes that nucleation occurs instantaneously and that the growth rate is constant throughout the run. It seems likely that this method underestimates the actual growth rate because nucleation is probably delayed to some extent. The data appear to be internally consistent, however, and relative to each other probably do demonstrate the form of the variation. Figure 49 presents Fenn's (1977) data for the alkali feldspar compositions at 2.5 kb P_{H_2O}. Figure 50 presents Swanson's (1977) data for the granite and granodiorite compositions at 8 kb P_{H_2O}. The forms of the relationships are as expected. The maximum growth rate observed in all these experiments is about 7 x 10^{-6} cm/sec, which is more than three orders of magnitude less than the maximum growth rate for the most slowly

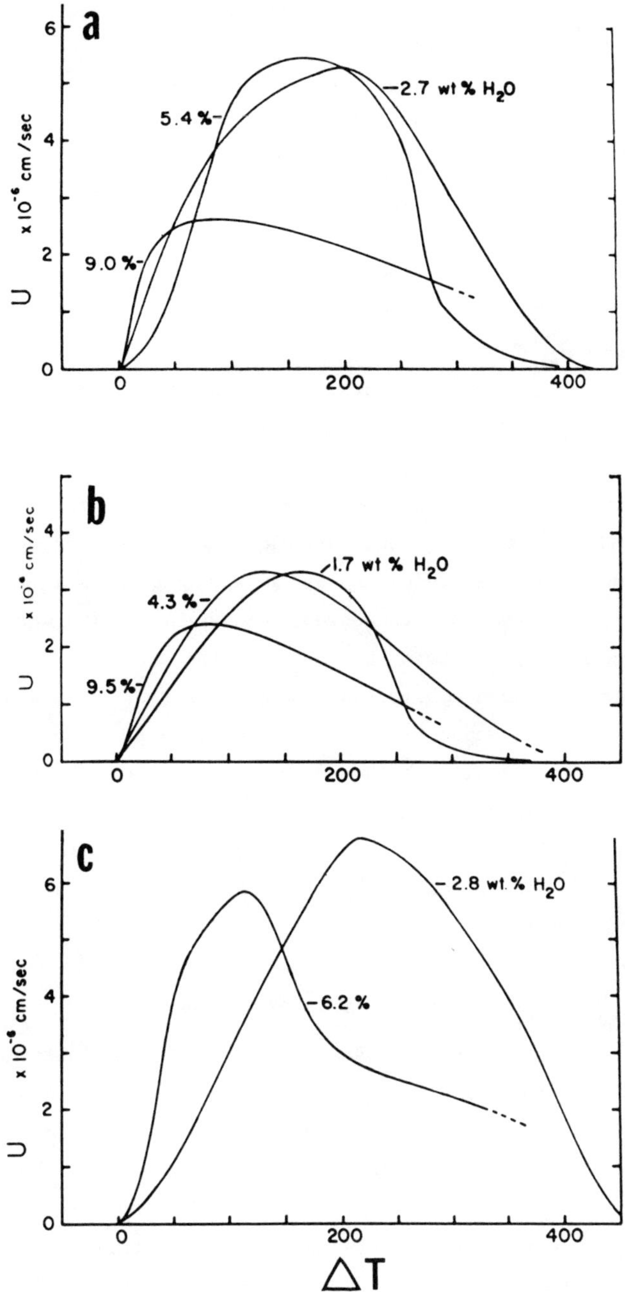

Figure 49. Growth rate versus undercooling relationships for alkali feldspar crystals from Ab-Or melts at 2.5 kbar (after Fenn, 1977). (a) $Ab_{90}Or_{10}$; (b) $Ab_{70}Or_{30}$; (c) $Ab_{50}Or_{50}$.

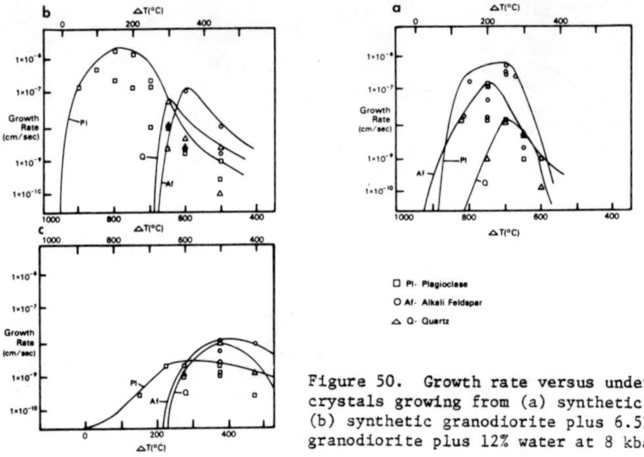

Figure 50. Growth rate versus undercooling relationships for crystals growing from (a) synthetic granite plus 3.5% water, (b) synthetic granodiorite plus 6.5% water, and (c) synthetic granodiorite plus 12% water at 8 kbars (after Swanson, 1977).

growing basalt composition. This is clearly the result of the higher melt viscosity and the lower liquidus temperatures for the haplogranitic compositions. The second major feature of these experiments is that for a given composition the growth rate decreases with increasing water content. This variation has not been fully explained.

Although there is now a considerable body of growth rate data for geologically interesting compositions, there is still not enough data or sufficient theory to predict the growth rates for real rock compositions. This kind of predictive ability is essential to a quantitative understanding of the kinetics of crystallization of igneous rocks. What is clearly needed are more extensive data for rock compositions in the small undercooling region.

BULK CRYSTALLIZATION OF IGNEOUS ROCKS

Holocrystalline igneous rocks consist of a large but finite number of variously sized crystals and have formed by the nucleation and growth of those crystals. Figure 51 shows a sequence of photomicrographs of samples taken from the zone of crystallization at the base of the crust on Makaopuhi lava lake, Kilauea volcano, Hawaii. The increase in size and number of the crystals with increasing percent crystallized is clearly visible. The discussion above has shown that the finite rates of nucleation and growth needed to produce these crystals require some deviation from equilibrium. In order to address the question of bulk

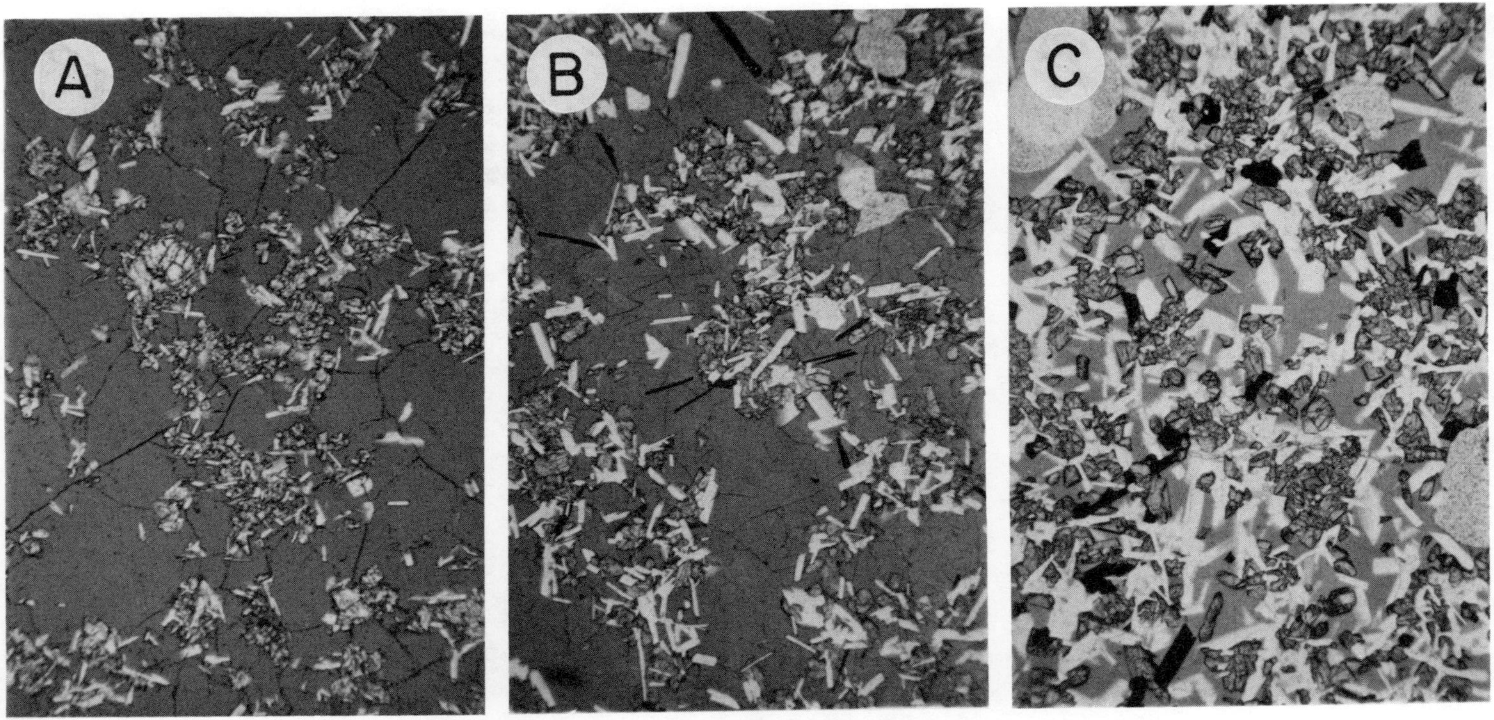

Figure 51. Photomicrographs of core samples from the zone of crystallization at the base of the crust on Makaopuhi lava lake, Kilauea volcano, Hawaii. The crystalline phases are olivine, clinopyroxene, plagioclase, and ilmenite. (a) About 33% crystallized, (b) about 55% crystallized, and (c) about 70% crystallized. The increase in size and number of crystals as crystallization proceeds implies simultaneous nucleation and growth. Note that most of the crystals form in clusters, indicating that nucleation is mostly heterogeneous.

crystallization kinetics of igneous rocks, it is necessary to determine what controls the deviation from equilibrium in igneous environments. Since, in the simplest case, crystallization is due to loss of heat from the melt, this deviation must in simple cases be related to the thermal regime.

In the simplest possible igneous event, magma or lava is quickly emplaced, cools, and crystallizes. If no crystallization were to occur, the body would cool to a glass with the undercooling increasing continually as the temperature falls. In situations where thermal conduction is the dominant heat transport mechanism the thermal regime in a body can be found by solving the heat flow equation, which in one dimension can be written

$$\frac{\partial T}{\partial t} = \kappa \frac{\partial^2 T}{\partial X^2} \tag{41}$$

where κ is the thermal diffusivity, and X is position. If crystallization does occur, this equation does not fully describe the thermal conditions because of the release of the latent heat of crystallization inside the body. This requires an additional term involving the rate of release of latent heat. With this term included, the heat flow equation can be written

$$\frac{\partial T}{\partial t} = \kappa \frac{\partial^2 T}{\partial X^2} + \frac{1}{\rho C_p} \frac{dQ}{dt} , \tag{42}$$

where ρ is density (of crystals + melt), C_p is the specific heat capacity of crystals + melt, and dQ/dt is in general the rate of extra heat generation or consumption per unit volume (of crystals + melt). In our case dQ/dt is the rate of generation of latent heat of crystallization. This can be written

$$\frac{dQ}{dt} = \frac{dV}{dt} L_v , \tag{43}$$

where dV/dt is the rate of change of the volume of crystals (rate of bulk crystallization) per unit volume and L_v is the latent heat per unit volume of crystal formed. Because crystallization adds heat to the system, it reduces the rate at which the temperature falls.

Since the rate of bulk crystallization must be related in some way to the rates of nucleation and growth, equation (42) can be written in general as

$$\frac{\partial T}{\partial t} = \kappa \frac{\partial^2 T}{\partial X^2} + \frac{1}{\rho C_p} L_v \, f(I_t, Y_t) , \tag{44}$$

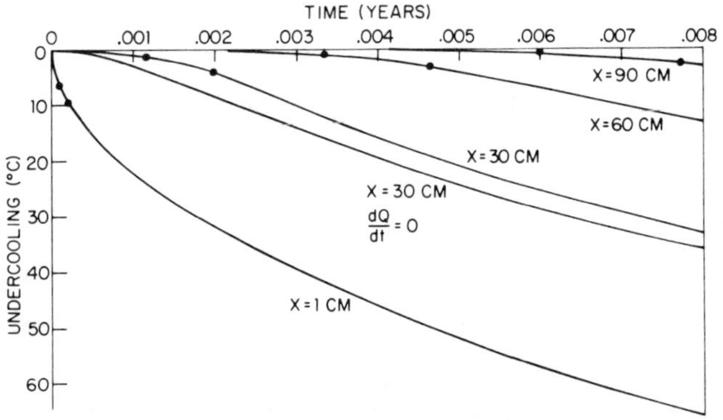

Figure 52 (above). Calculated undercooling versus time relationships for various distances from the margin of a 2 m dike. Dots on each curve indicate the beginning and ending of crystallization at each point. Curve without dots is relationship for 30 cm from the margin assuming no heat production due to crystallization (after Kirkpatrick, 1976).

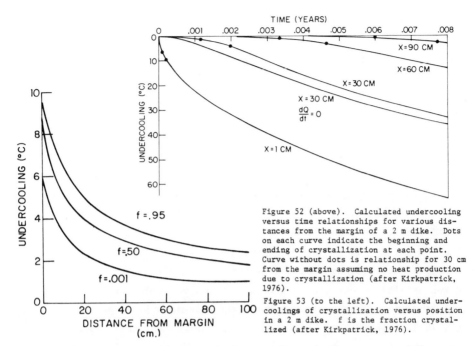

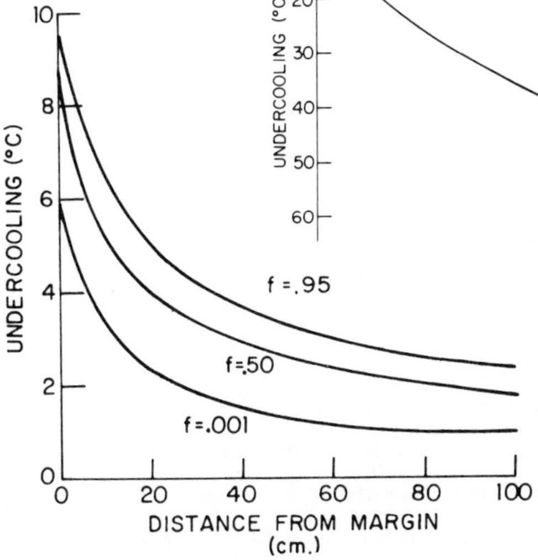

Figure 53 (to the left). Calculated undercoolings of crystallization versus position in a 2 m dike. f is the fraction crystallized (after Kirkpatrick, 1976).

where the nucleation rate, I, and the growth rate, Y, are certainly functions of temperature, melt composition, and melt liquidus temperature. Thus, there is a feedback relationship between the rate of temperature change and the rates of nucleation and growth. Qualitatively, this relationship may be thought of in the following way. As the temperature falls the undercooling increases. This causes the rates of nucleation and growth, and therefore the overall rate of crystallization, and therefore the rate of release of latent heat, to increase. This slows the rate at which the temperature falls. For any given physical and chemical situation there will be one temperature and crystallization path that will be followed.

If an expression for the rate of bulk crystallization in terms of the rates of nucleation and growth and the dependences of the nucleation and growth rates on temperature and liquid composition are known, it is possible to solve equation (44) for a given set of boundary and initial condition by numerical methods. This has been done for an extremely idealized situation using reasonable, but nontheless assumed, rates of nucleation and growth (Kirkpatrick, 1976). The results of these calculations are not quantitatively applicable to any actual geological situation, but do illustrate how the calculations are done and the form of the results. These results are qualitatively in agreement with observations in igneous rocks. The expression used for the relationship between the fraction crystallized is the simplest form of the Johnson-Mehl-Avrami theory [equation (A-15), which is derived in the Appendix], and is the same expression used by Klein and Uhlmann (1976) to calculate their nucleation rates. To make calculations simple the material is assumed to be congruently melting. The time-temperature relationships for a 2 m thick dike are shown in Figure 52. Figure 53 shows the calculated undercooling versus position relationships for different fractions crystallized.

The most important features of Figure 53 are that, for the same fraction crystallized, the undercooling decreases with distance from the surface of cooling, and that at the same position the undercooling increases with fraction crystallized. Thus, we would expect to see indicators of larger undercoolings (extreme morphologies and non-equilibrium mineral compositions, for instance) near the margins of bodies and increases in the rates of nucleation and growth with fraction crystallized at any given position in a body. In these calculations all crystallization occurs at relatively small undercoolings, and the nucleation and growth rates increase with falling temperature. In situations where glass is actually found in the rock, the rate of crystallization, and therefore the rates of nucleation and growth, must go to zero at the time crystallization stops, and the undercooling must reach very large values.

Figures 54 and 55 show the calculated nucleation and growth rates versus fraction crystallized for various distances from the surface of cooling. At all positions the rates increase monotonically, but there is an inflection point at about 50% crystallized. The rates are higher near the surface of cooling than in the center because the undercooling is larger near the margin.

Figure 56 shows the calculated relationship between fraction crystallized and time for various distances from the surface of cooling. The outer part cools most rapidly and is fully crystallized before significant crystallization occurs in the central part. The curves have the characteristic sigmoidal shape expected for nucleation and growth controlled reactions.

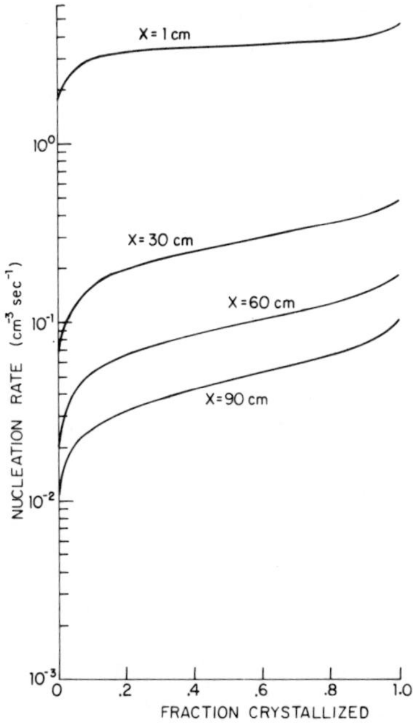

Figure 54. Calculated nucleation rates versus fraction crystallized for various distances from the margin of a 2 m dike (after Kirkpatrick, 1976).

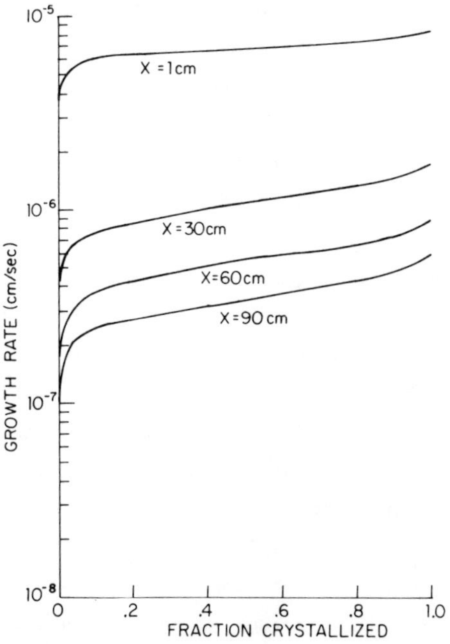

Figure 55. Calculated growth rate versus fraction crystallized for various distances from the margin of a 2 m dike (after Kirkpatrick, 1976).

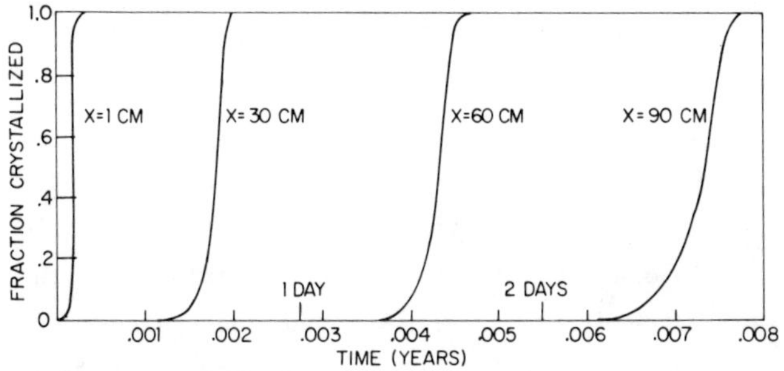

Figure 56. Caculated fraction crystallized versus time relationships for various distances from the margin of a 2 m dike. Note the characteristic sigmoidal shape of the curves (after Kirkpatrick, 1976).

These calculations greatly oversimplify the cooling and crystallization of igneous bodies. Factors neglected include incongruent behavior, variation of the nucleation and growth rates with changing melt composition, multiple phase crystallization, convective heat transport, crystal settling and flotation, and multiple intrusion. Incongruent behavior by itself is not expected to cause significant qualitative changes, because there must still be an undercooling for crystallization to occur. The changing melt composition as crystallization proceeds could be significant. The magnitude and direction of these changes are unknown, except that the rates will likely be lower for silica-rich liquids. Under many circumstances convection is likely to be the dominant heat transport mechanism when the fraction crystallized is low. The primary effects of convection are likely to be to reduce the temperature variation in the body, to increase the cooling rate in the mostly liquid portion of the body, and decrease the cooling rate in the mostly crystallized portions of the body. The results from the Kilauea lava lakes (Wright *et al.*, 1976) show that convection becomes unimportant when the fraction crystallized reaches about 40%. The effects of crystal settling and flotation and multiple intrusion on the crystallization are likely to be complex.

Even though these calculations are very simplified, it appears that they do qualitatively describe many kinetically controlled features of igneous rocks. Quantitative comparisons are not possible because of the lack of nucleation and growth rate data.

To determine whether the calculations are useful in describing the kinetics of igneous crystallization, it is necessary to find igneous bodies that cooled in a simple way and then to examine features which are expected to be kinetically controlled. Quantitative data as functions of distance from the surface of cooling and time at one position are needed. A few examples of both are available.

Because crystal morphology varies greatly with undercooling and is relatively easily examined, its variation within lava flows is an important tool in examining the undercooling variation. The chilled margins of submarine pillow basalts are perhaps the easiest places to look for this kind of variation becuase they have cooled in a simple manner and show a wide range of crystal morphologies. Figure 57 presents photomicrographs of the textural variations in a basalt pillow from the mid-Atlantic ridge. The very margin is all glass except for a few microphenocrysts. Inward olivine dendrites and complex plagioclase/olivine spherulites/dendrites have crystallized. Further in, high-calcium clinopyroxene, magnetite, and ilmenite also appear. Kirkpatrick (1978) has divided this textural variation into six zones based on the phases present and their morphology.

A

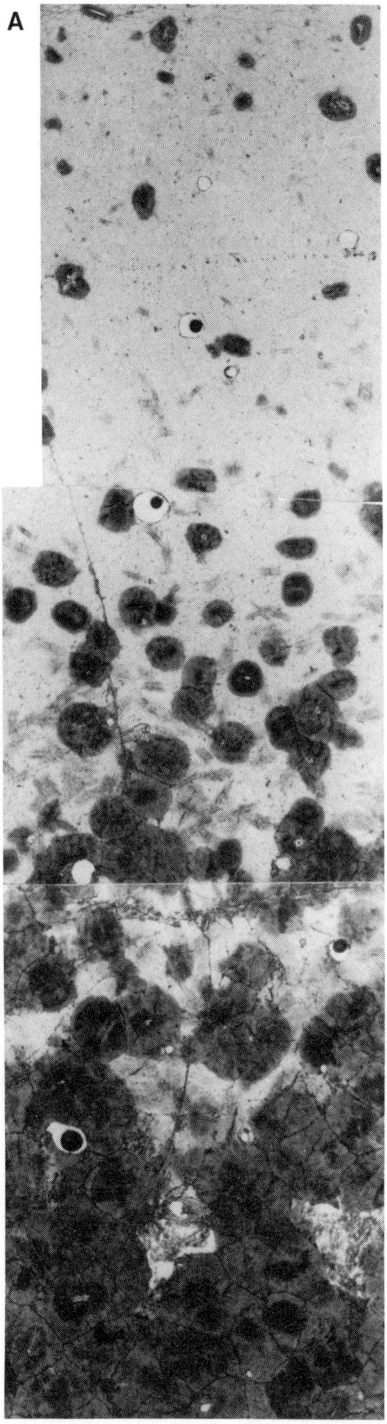

B▲ ▼C

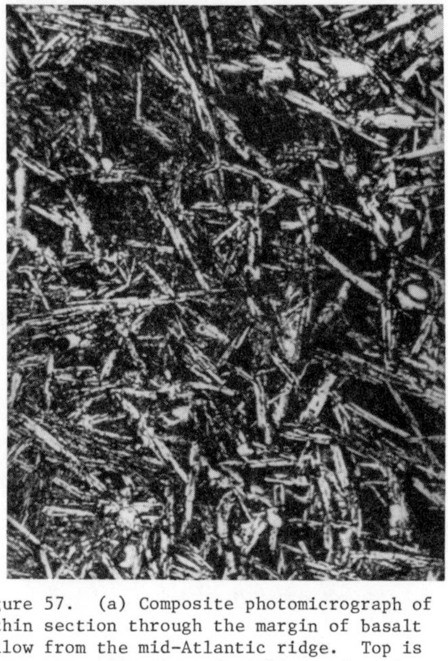

Figure 57. (a) Composite photomicrograph of a thin section through the margin of basalt pillow from the mid-Atlantic ridge. Top is glassy rim with only a few olivine dendrites and olivine + plagioclase spherulites. In the middle dendrites and spherulites become more abundant. At the bottom the spherulites have coalesced to fill the entire volume. Photo width at base is 4.25 mm. (b) Bow-tie spherulites form a few cm further into the same pillow. Photo width is ∿ 1.5 mm. (c) Microlitic center of the same pillow. Scale same as (b).

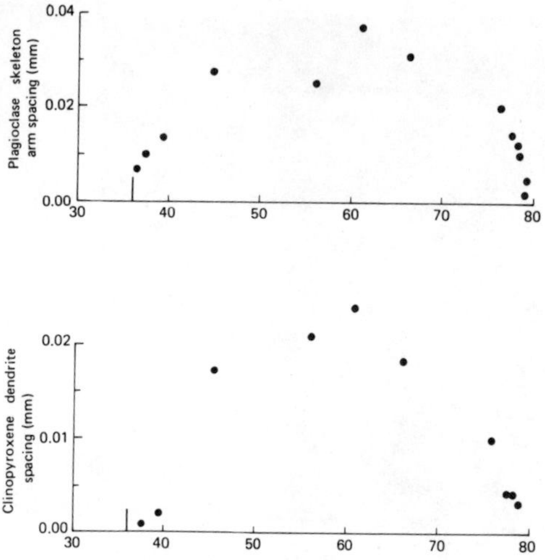

Figure 58. Plagioclase skeleton spacing and pyroxene dendrite arm spacing in a basalt pillow from the mid-Atlantic ridge. The rapid increase in spacing near the margins implies a rapid decrease in the undercooling of crystallization with increasing distance from the margin (surface of cooling).

As discussed above, the spacing between the arms of skeletal or dendritic crystal should decrease with increasing undercooling, and the average undercooling should decrease with increasing distance from a margin of cooling. Figure 58 shows the observed spacing for skeletal plagioclase and dendritic pyroxene crystals from the MAR pillow. Both increase with increasing distance from the rim, as predicted by the calculations.

The calculated increase in undercooling with increasing time at one place in a cooling body implies that the arm spacing should decrease from center to rim within one crystal or spherulite. Figure 59, which presents SEM photomicrographs of plagioclase spherulites from the mostly glassy pillow rim, shows that this is the case. The crystal fibers are significantly larger in the center of the spherulite than near the margin. Thus, it would appear that the petrographic features of pillow basalts are consistent with the calculated time and distance variations in undercooling within a cooling body.

Within the limits of present knowledge, the only way to determine the variations of the growth and nucleation rates in real magma bodies is to examine one in the act of crystallizing. Observable bodies like this are rare. The only ones presently available are the lava lakes from Kilauea volcano, Hawaii (Wright *et al.*, 1976). Drilling through the solidified crusts of these lava lakes has produced samples from the zones of crystallization at the bases of the crusts which have as little as 33% crystals. From these samples it has been possible to estimate the rates of nucleation and growth of plagioclase in these zones of crystallization (Kirkpatrick, 1976).

Figure 60 shows the observed fraction crystallized versus time curves for two sets of data from Alae and Makaopuhi lava lakes.

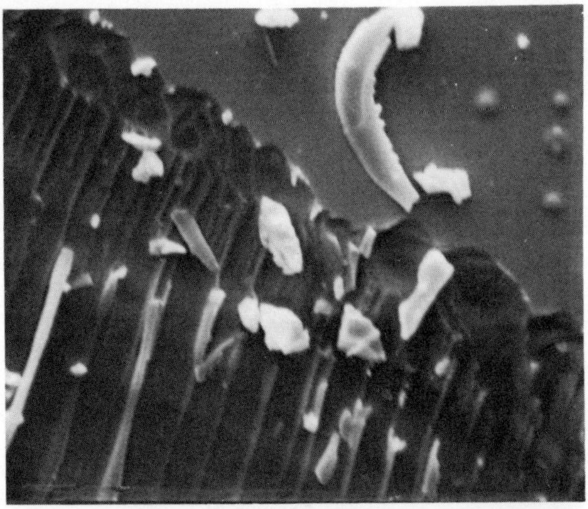

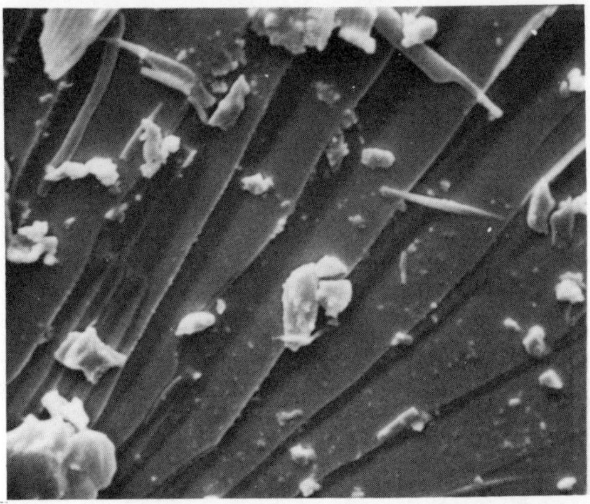

Figure 59. Scanning electron photomicrograph of a fractured surface of a plagioclase spherulite from the margin of a basalt pillow from the mid-Atlantic ridge. (a) Margin of the spherulite is contact with glass (b) spherulite center. The decrease in spacing of the crystal fibers towards the outside of the spherulite implies an increase in the undercooling as crystallization proceeds. Note the faceted tips of the fibers (width of field is about 40 μm).

Both curves, especially that for Makaopuhi, which are the best data available, are consistent with the sigmoidal relationship predicted by the calculations. This curve cannot be the equilibrium relationship. It must be kinetic. If it were the equilibrium relationship, the percent crystallized would eventually reach 100%, but all of the cold, "fully crystallized," lava lake samples contain about 5% glass. The undercooling when crystallization stopped must have been quite large.

Figure 61 shows the observed growth rate of plagioclase perpendicular to (010) versus fraction crystallized for the same samples.

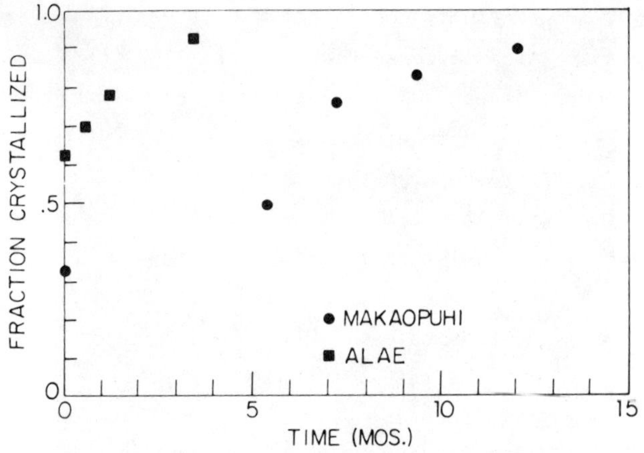

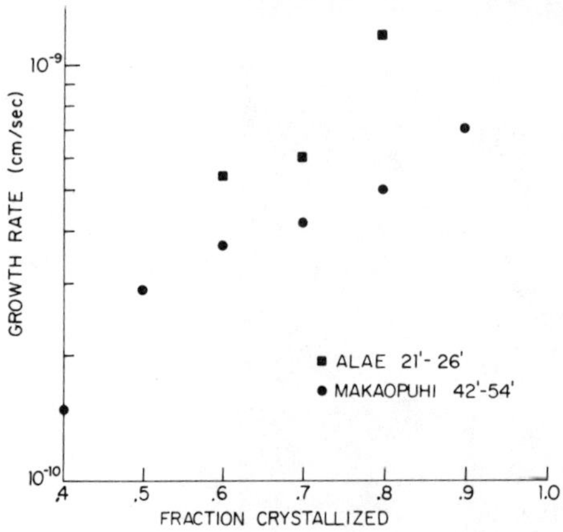

Figure 60 (above). Observed fraction crystallized versus time relationships for samples taken from the Hawaiian lava lakes Makaopuhi and Alae. The zero of time is taken arbitrarily to be the deepest sample.

Figure 61 (to the left). Observed plagioclase growth rates perpendicular to (010) versus fraction crystallized for samples from the Hawaiian lava lakes Makaopuhi and Alae.

Figure 62 (below). Observed plagioclase nucleation rates versus fraction crystallized for samples from the Hawaiian lava lakes Makaopuhi and Alae.

[Figures 60-62 after Kirkpatrick (1976).]

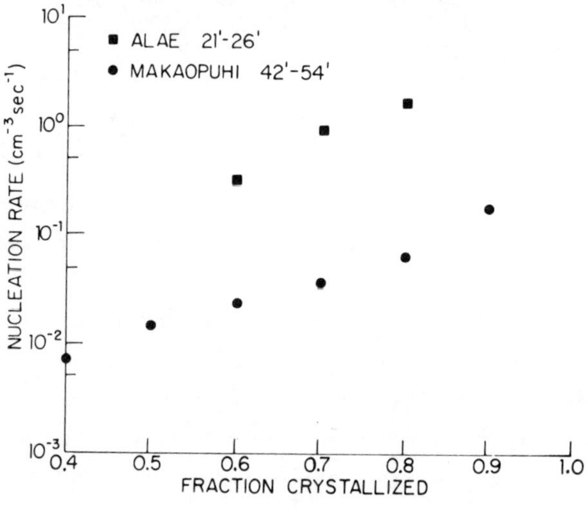

These results are again fully consistent with the predicted relationships. The growth rates increase monotonically with time, but there is an inflection point near 50% crystallized. This increase is consistent with an increasing undercooling as crystallization occurs. In addition, the rates for the Alae samples, which are from closer to the surface of cooling, are larger than those for the Makaopuhi samples. Both lakes have similar bulk compositions, and this difference is probably due to a larger undercooling for the Alae samples. This, too, is consistent with the results of the calculations.

Figure 62 presents the rates of nucleation of plagioclase for the samples. Although these data are less compelling than the growth rate data and have been challenged by Gray (1978), they too are consistent with the predictions of the calculations. As for the growth rates, the nucleation rates increase monotonically with increasing fraction crystallized, there is an inflection point in the relationships, and the rates are larger for Alae than for Makaopuhi.

It seems, then, that the magma body kinetic model discussed above can qualitatively predict some kinetically controlled features of real magma bodies. A fuller understanding of the crystallization processes in igneous bodies, however, must await further theory, experiment, and observation. The greatest need is for more extensive data for the rates of nucleation and growth for real rock compositions at small undercoolings and a way to quantitatively understand these rates as functions of composition. Nucleation is by far the less understood process. A better understanding of when nucleation will occur heterogeneously and when it will occur homogeneously in natural compositions as well as a better theoretical treatment of heterogeneous nucleation dominated bulk crystallization processes are also major concerns. In addition, far more work is needed in examining both plutonic and volcanic magma bodies in terms of the kinetics of the crystallization processes, if for no other reason than to focus the experimental and theoretical work on problems related to real rock bodies.

PROGRAMMED COOLING EXPERIMENTS

One of the major findings of the recent kinetic studies of igneous rocks is that the temperatures of crystallization and the sequences of phase appearance in a continuously cooled environment can be much different than predicted by the equilibrium relationships. Reversals of phase appearance have been shown to occur in lunar volcanic rocks. Whether it occurs commonly in plutonic rocks is unknown.

These results have been obtained primarily through programmed cooling experiments. In these experiments a sample is heated for some time at a high temperature above or just below the liquidus, cooled at a known rate, quenched, and examined. Most experiments have been done using lunar basalt compositions because the lunar rocks are dry, and dry, controlled-oxygen-fugacity conditions are relatively easy to generate. Terrestrial basalts always contain at least a small amount of water and other volatiles, and controlling the amount of water in experimental changes at low pressure is difficult. Many investigators are reluctant to extend the results of dry experiments to rocks with a significant volatile content.

The literature on programmed cooling experiments is extensive, and a complete review here is not possible. Lofgren (1980) has given a thorough review of this work. I do, however, want to review two well-established results of these experiments: estimates of absolute cooling and rates of rocks and the effect of cooling rate on temperature and order of phase appearance. Most of the papers discussing programmed cooling experiments have been published in the *Proceedings of the Lunar Science Conferences* beginning in 1974. Papers not referenced here are readily available there.

The first application of programmed cooling techniques to the study of absolute cooling rates of rocks was by Lofgren *et al.* (1974, 1975) on the Apollo 15 quartz normative basalts. In these experiments the samples were melted about 50°C above the liquidus temperature and then cooled to temperatures well below the solidus at constant rates of from 1.2°C to 126°C/hr. One important result of these experiments is that a porphyritic texture (in this case with olivine and pyroxene phenocrysts) can be generated at constant cooling rates and does not require a two-stage cooling history. To determine the cooling rates they compared the textures and crystal morphologies in the experimental charges with those in the basalt samples. Table 1 presents the variations of crystal morphology with cooling rate for the phases present in the experiments.

TABLE 1

Variation of crystal morphology with cooling rate in programmed cooling experiments with a synthetic Apollo 15 quartz-normative basalt composition. After Lofgren *et al.* (1974).

Cooling rate °C/hr	Olivine	Pyroxene Phenocryst	Pyroxene Groundmass	Plagioclase	Opaques
1260	Dendrites with 1Y, 2Y 2Y, and 4Y branching	none			
430	coarser dendrites with less complex branching	acicular skeletons with internal dendritic structure and complex dendrites with 1Y, 2Y, and 3Y branches, some of which are curved		None	chromite dendrites
220	thinly tabular parallel growth				
115	acicular skeletons				
60	elongate to subequant skeletons	elongate skeletons more complete and larger than above	fan spherulites preferentially nucleated at phenocryst margins	acicular, intergrown with fan spherulites of pyroxene; few larger crystals near phenocryst margins	subequant euhedral chromite
30					subequant subhedral chromite mantled by ulvöspinel
20	subequant skeletons		fan spherulites that coarsens with decreasing cooling rate and rare small subequant crystals		
10		elongate to subequant skeletons		acicular, intergrown with fan spherulites of pyroxene; skeletal laths increasingly common with decreasing cooling rate	subequant subhedral chromite mantled by ulvöspinel and feathery ilmenite
5					
2.5	none	subequant euhedral to subhedral crystals with a central cavity	subequant to equant crystals with few coarse, open fan spherulites	skeletal laths and few acicular crystals in pyroxene fan spherulites	equant subhedral chromite mantled by ulvöspinel; euhedral ulvöspinel; and blocky ilmenite
1.2					

TABLE 2

Estimated cooling rates of Apollo 15 quartz-normative basalt samples based on programmed cooling experiments. Sample numbers are condensed to the last three digits; e.g., 485 = 15485. After Lofgren *et al.* (1975).

Pyroxene phenocrysts		Matrix	
485	↑	597	↑
486	5–20°C/hr	(125)	> 30°C/hr
597		485	f_{O_2} change?
499	↓	486	↓
(125)	↑	595	↑
595		499	
596	2–5°C/hr	(666)	10–30°C/hr
(666)		596	↓
(682)	↓	(682)	↑
(118)	↑	476	
476		495	1–5°C/hr
495		(118)	
475		(684)	↓
(684)		475	↑
058	< 1°C/hr	058	
075		076	
076		075	< 1°C/hr
(116)		(116)	
065		065	
085	↓	085	↓

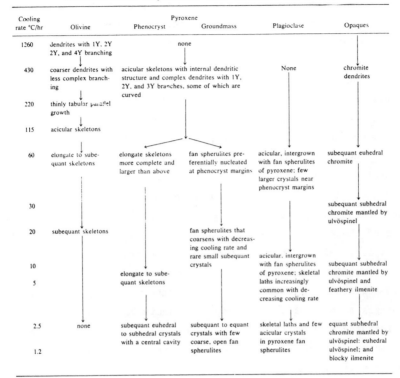

In general, most phases are still skeletal (indicating significant interface instability) even at 1.2°C/hr. Table 2 presents the cooling rates estimated for each sample using both the phenocryst and groundmass morphologies. In general, the estimated cooling rates using the phenocrysts are higher than those using the groundmass. Except for the vitrophyres, this is consistent with the increase in cooling rate with increasing time experienced by a point in the interior of lava flow. These cooling rates are similar to estimates made using Zr partitioning between ilmenite and ülvospinel (Taylor *et al.*, 1975).

Following the lead of this work, most investigators using programmed cooling methods have attempted to estimate the cooling rates of the samples they are studying. In more recent work non-equilibrium mineral compositions have also been used as an estimator of cooling rate.

The first investigation of the effect of cooling rate on temperature and order of phase appearance used the Apollo 12 picritic basalt 12002 (Walker *et al.*, 1976). In these experiments several charges were run at the same cooling rate, but quenched from different temperatures within the crystallization range to examine the mineralogical and textural changes as crystallization occurs. Figure 63 shows their results. Several features of this diagram are important: All phases except the liquidus phase (olivine) begin crystallization at temperatures well below their equilibrium temperature of appearance; the amount of temperature suppression increases with increasing cooling rate; even at cooling rates as low as about 0.5°C/hr. significant crystallization continues at temperatures well below the solidus; and the order of appearance of ilmenite and plagioclase is reversed in the equilibrium and programmed experiments. The porphyritic textures in the programmed experiments are very similar to those in the rocks of the Apollo 12 picritic suite. The petrographically determined order of phase appearance in the 12002 sample is ilmenite before plagioclase, which is the order in the programmed experiments. In addition, the pyroxene compositions in the programmed experiments and the 12002 sample are significantly different from the compositions in the equilibrium experiments (Fig. 64). Taken together, these data strongly suggest that programmed experiments give a much truer picture of the crystallization of basalts than equilibrium experiments. Walker *et al.* estimate that

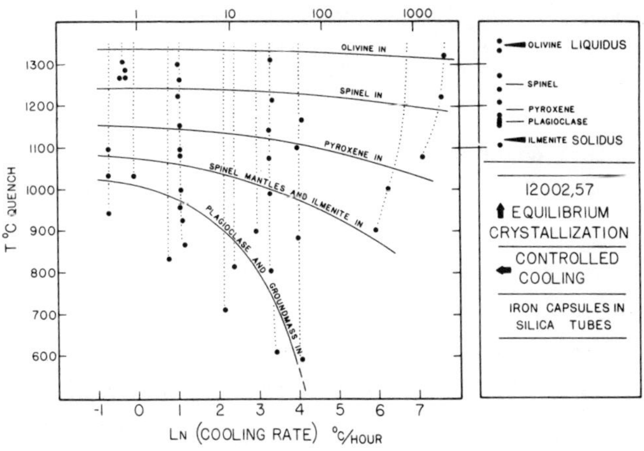

Figure 63. Results of programmed cooling experiments with lunar composition 12002. Dotted lines are cooling paths, dots indicate quench temperatures, and solid curves indicate temperature of first appearance of a phase. The equilibrium relationships are shown on the right (after Walker *et al.*, 1976).

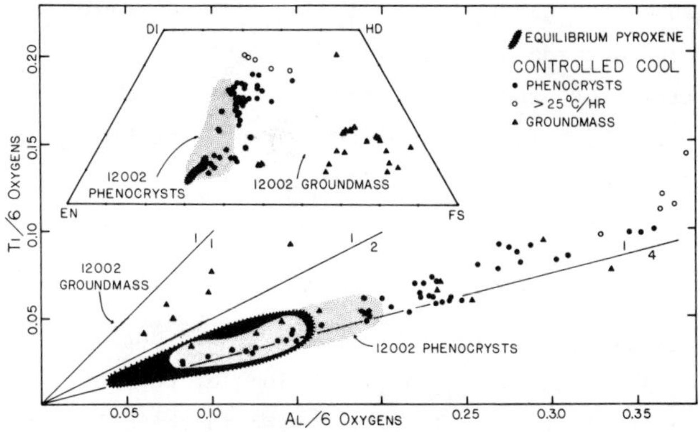

Figure 64. Compositions of pyroxenes in the 12002 sample, programmed cooling experiments with 12002, and equilibrium experiments with 12002 (after Walker *et al.*, 1976).

the actual cooling rate for 12002 decreased from about 1°C/hr. when olivine first crystallized to about 0.2°C/hr. during the crystallization of the pyroxene phenocryst. This decreasing cooling rate is consistent with crystallization in the outer part of a lava flow.

APPENDIX: THE RATE OF CRYSTALLIZATION OF NUCLEATION AND GROWTH CONTROLLED REACTIONS

The mathematical description of the overall rate of crystallization in situations where both nucleation and growth occur simultaneously was first described by Johnson and Mehl (1939) and Avrami (1940, 1941, 1942). Because their ideas are not well understood among geologists, it seems worthwhile to develop their ideas in some detail here and to extend their derivations to the general case of arbitrarily variable rates of nucleation and growth. The derivation given here will be for the general case. At the end it will be shown how this reduces to the traditional JMA form for constant rates.

The most important assumptions of the model are that there is a uniform parent phase with no crystals (volumes of the product phase), that nucleation occurs randomly, and that the crystals grow as spheres. This later could be changed to any desired shape, but would complicate the derivation and not add any additional insight.

Consider first the growth of a single crystal which nucleates at arbitrary time τ. The volume, V_1, of this crystal at time t later than τ is $V_1 = \frac{4\pi}{3} r^3$. The radius, r, is given by $r = \int_\tau^t Y_t \, dt$, where Y_t is the growth rate, which is allowed to vary with time. Therefore,

$$V_1 = \frac{4\pi}{3} \left(\int_\tau^t Y_t \, dt\right)^3 . \qquad \text{(A-1)}$$

The number of crystals, dn, nucleated in the liquid during a small time, $d\tau$, is $dn = V_{liq} I_\tau d\tau$, where V_{liq} is the volume of the melt. The total volume at time t of all the crystals nucleated in the melt during the time interval $d\tau$, V_2, is the product of V_1 and dn,

$$V_2 = \frac{4\pi}{3} V_{liq} \, I_\tau d\tau \left(\int_\tau^t Y_t \, dt\right)^3 . \qquad \text{(A-2)}$$

This assumes that all the crystals are still perfect spheres and that no impingement has occurred (i.e., the crystals have not run into each other at all). Consider now the number of crystals that could have nucleated during the same time interval, $d\tau$, in the volume occupied by previously formed crystals if the volume occupied by those crystals were occupied by melt. This number, dn', is $dn' = V_{x\ell} I_\tau d\tau$, where $V_{x\ell}$ is the volume of existing crystals. Avrami (1939) has called these imaginary nuclei phanton nuclei. The volume of the phantom crystals at time t, had they formed and not impinged in any way, V_2', is

$$V_2' = \frac{4\pi}{3} V_{x\ell} I_\tau d\tau \left(\int_\tau^t Y_t \, dt\right)^3 . \quad \text{(A-3)}$$

The sum of the volumes of the real crystals and the phantom crystals that formed or could have formed in the time interval $d\tau$ is called the instantaneous extended volume, V_2^{ext}, and is given by the relationship

$$V_2^{ext} = \frac{4\pi}{3} V I_\tau d\tau \left(\int_\tau^t Y_t \, dt\right)^3 , \quad \text{(A-4)}$$

where V is the total volume of the system. The total extended volume of crystals that formed or could have formed from the start of crystallization at time 0 to some later time t, V_{ext}, is then given by the relationship

$$V_{ext} = \frac{4\pi}{3} V \int_{\tau=0}^{\tau=t} I_\tau \left(\int_\tau^t Y_t \, dt\right)^3 d\tau . \quad \text{(A-5)}$$

This extended volume is larger than the true volume crystallized in two ways. It includes the volume of the phantom nuclei that could have nucleated in the already crystallized volume if it were not crystal, and it contains that part of the volumes of both the real and phantom crystlas that have not actually formed because the crystals have impinged on each other. In a sense the calculations have let the crystals nucleate in and grow through each other. The extended volume is useful in that it is directly related to the rates of nucleation and growth. We need only find a relationship between the extended volume crystallized and the true volume crystallized to develop an expression for the true relationship between volume crystallized and time.

To do this, consider a small randomly chosen region. The true volume fraction not crystallized at time t is by definition $(1 - \frac{Vx\ell}{V})$. During the next small increment of time, dt, the extended volume increases by dV_{ext} and the true volume crystallized by $dV_{x\ell}$. On the average, $(1 - \frac{Vx\ell}{V})$ of the new extended volume will be in uncrystallized volume. If this new material is randomly distributed,

$$dV_{x\ell} = \left(1 - \frac{Vx\ell}{V}\right) dV_{ext} . \quad \text{(A-6)}$$

This expression can be integrated to give

$$V_{ext} = -V \ln \left(1 - \frac{Vx\ell}{V}\right) . \quad \text{(A-7)}$$

Combining this with equation (A-5) gives

$$-V\ell n\ (1 - \frac{Vx\ell}{V}) = \frac{4\pi}{3} V \int_{\tau=0}^{\tau=t} I_\tau \ (\int_\tau^t Y_t \ dt)^3 \ d\tau \ . \qquad (A\text{-}8)$$

Defining $\frac{Vx\ell}{V} = \phi$,

$$\ell n\ (1-\phi) = - \frac{4\pi}{3} \int_{\tau=0}^{\tau=t} I_\tau \ (\int_\tau^t Y_t \ dt)^3 \ d\tau \ . \qquad (A\text{-}9)$$

Raising the base of the natural logarithms, e, to the power of both sides and rearranging gives

$$\phi = 1 - \exp[- \frac{4\pi}{3} \int_{\tau=0}^{\tau=t} I_\tau \ (\int_\tau^t Y_t \ dt)^3 \ d\tau] \ . \qquad (A\text{-}10)$$

This expression is generally valid for any variation of the rates of nucleation and growth with time and should be used in all cases where they vary. In most cases numerical methods will be needed to calculate the fraction crystallized.

In the case of constant rates of nucleation and growth, equation (A-10) reduces to give the classic JMA equation in the following way. If I and Y are constant, they can be removed from the integrals in equation (A-9) to give

$$\ell n\ (1-\phi) = - \frac{4\pi}{3} IY^3 \int_{\tau=0}^{\tau=t} (\int_\tau^t dt)^3 \ d\tau \ . \qquad (A\text{-}11)$$

$$(\int_\tau^t dt)^3 = (t-\tau)^3 \ , \qquad (A\text{-}12)$$

so that

$$\ell n\ (1-\phi) = - \frac{4\pi}{3} IY^3 \int_{\tau=0}^{\tau=t} (t-\tau)^3 \ d\tau \ . \qquad (A\text{-}13)$$

The integral in this expression is of the form $\int_0^t (a+bx)^n dx$, where b = -1. The value of the integral is then $t^4/4$. Equation (A-13) is then

$$\ell n\ (1-\phi) = - \frac{\pi}{3} IY^3 \ t^4 \ . \qquad (A\text{-}14)$$

Again raising e to both sides and rearranging gives

$$\phi = 1 - \exp(- \frac{\pi}{3} \ IY^3 \ t^4) \ , \qquad (A\text{-}15)$$

which is the classical JMA expression.

This expression, the general form given in equation (A-10) and all other expressions like them yield a sigmoidal shape of the fraction crystallized versus time curve. The rate of transformation is slow at first because there are few crystals onto which new material can be added and these crystals are small. With time the rate increases as more crystals are nucleated and those present become larger. Eventually, however, the rate begins to decrease as the crystals begin to impinge on each other to a significant extent and there is progressively less and less volume for new crystals to nucleate in. The model breaks down in the limit of long times because it never lets the volume crystallized reach 100%, and this clearly happens in many situations.

This kind of model is applicable not only to the crystallization of igneous rocks, but to any process which is controlled by the rates of simultaneous nucleation and growth. The original JMA work was done for recrystallization of metals, and it seems likely that this kind of thinking is also applicable to many metamorphic and sedimentary situations. One important assumption of this derivation is that nucleation occurs randomly. If heterogeneous nucleation on existing crystal surfaces is to be considered, the geometrical relationships are considerably more complex. In this case the final relationship will certainly be different, although the sigmoidal form of the relationship will likely be retained.

Avrami, M. (1939) Kinetics of phase change, I. J. Chem. Phys., 7, 1103-1112.

________ (1940) Kinetics of phase change, II. J. Chem. Phys., 8, 212-224.

________ (1941) Granulation, phase change, and microstructure. J. Chem. Phys., 9, 177-184.

Becker, R., and W. Döring (1935) Kinetische behandburg der Keim building in übersattigten dampfen. Ann. Phys., ser. 5, 24, 719-752.

Bennema, P., and J. P. van der Eerdgen (1977) Crystal growth from solution: development in computer simulation. J. Crystal Growth, 42, 201-213.

Binsberger, F. L. (1972) Computer simulation of nucleation and crystal growth. J. Crystal Growth, 16, 249-258.

Burnett, D. G. and R. W. Douglas (1971) Nucleation and crystallization in the soda-baria-silica system. Phys. Chem. Glasses, 12, 117-124.

Burton, J. A., R. C. Prim and W. P. Slichter (1953) The distribution of solute in crystals grown from the melt. Part I. theoretical. J. Chem. Phys., 21, 1987-199.

Carruthers, J. A. (1979) Dynamics of crystal growth. In *Crystal Growth: A Tutorial Approach.* W. Bardsley, D. T. J. Hurle and J. B. Mullin, eds., North-Holland Publishing Co., 157-188.

Christian, J. W. (1975) *The Theory of Transformations in Metals and Alloys, Part I,* 2nd Edition. Pergammon Press, Oxford.

Crank, J. (1956) *The Mathematics of Diffusion,* Oxford University Press, Oxford.

Cukierman, M., L. Klein, G. Scherer, R. W. Hopper, and D. R. Uhlmann (1973) Viscous flow and crystallization behavior of selected lunar compositions. Proc. 4th Lunar Sci. Conf., 2685-2696.

Dehoff, R. T. and R. N. Rhines (1961) Determination of number of particles per unit volume from measurements made on random plane sections: the general cylinder and the ellipsoid. Trans. Metal. Soc. AIME, 221, 975-982.

Delves, R. T. (1974) Theory of interface stability. In *Crystal Growth,* B. R. Pamplin, ed., Pergammon Press, Oxford.

Donaldson, C. H. (1976) An experimental investigation of olivine morphology. Contrib. Mineral. Petrol., 57, 187-213.

________ (1977) Laboratory duplication of comb layering in the Rhum pluton. Mineral. Mag., 41, 326-336.

________ (1979) An experimental investigation of the delay in nucleation of olivine in mafic magmas. Contrib. Mineral. Petrol., 69, 21-32.

Donaldson, C. H., T. M. Usselman, R. J. Williams and G. E. Lofgren (1975) Experimental modeling of the cooling history of Apollo 12 olivine basalt. Proc. 6th Lunar Sci. Conf., 843-869.

Dowty, E., K. Keil, and M. Prinz (1974) Lunar pyroxene-phyric basalts: crystallization under supercooled conditions, J. Petrol., 15, 419-453.

Elbaum, C. (1959) Substructures in crystals grown from the melt, in *Progress in Metal Physics,* v. 8, B. Chalmers, and R. King, eds., Pergammon Press, London.

Fenn, P. M. (1977) The nucleation and growth of alkali feldspars from hydrous melts, Canadian Min., 15, 135-161.

Fine, M. E. (1964) *Introduction to Phase Transformations in Condensed Systems,* MacMillan Co., New York.

Frenkel, J. I. (1946) *Kinetic Theory of Liquids.* Oxford University Press, Oxford.

Gibb, F. G. F. (1974) Supercooling and the crystallization of plagioclase from a basaltic magma. Mineral. Mag., 39, 641-653.

Gilmer, G. H. (1976) Growth on imperfect crystal faces: I. Monte-Carlo growth rates. J. Crystal Growth, 36, 15-28.

________ (1977) Computer simulation of crystal growth. J. Crystal Growth, 42, 3-10.

Gilmer, G. H. and P. Bennema (1972) Computer simulation of crystal surface structure and growth kinetics. J. Crystal Growth, 13/14, 148-153.

Grey, N., and R. J. Kirkpatrick (1978) Nucleation and growth of plagioclase Makaopuhi and Alae Lava Lakes, Kilauea Volcano, Hawaii: Discussion and Reply. Geol. Soc. Amer. Bull., 89, 797-830.

Hofmann, A. W. and Magaritz, M. (1977) Diffusion of Ca, Sr, Ba, and Co in a basalt melt: Implications for the geochemistry of the mantle. J. Geophys. Res., 82, 5432-5440.

Hopper, R. W. and D. R. Uhlmann (1974) Solute redistribution during crystallization at constant velocity and constant temperature. J. Crystal Growth, 21, 203-213.

Jackson, K. A. (1958a) Mechanism of growth. In *Growth and Perfection of Crystals,* R. H. Doremus, B. W. Roberts, and D. Turnbull, eds., John Wiley and Sons, New York.

———— (1958b) Interface structure. In *Growth and Perfection of Crystals*, R. H. Doremus, B. W. Roberts, and D. Turnbull, eds., John Wiley and Sons, New York.

James, P. F. (1974) Kinetics of crystal nucleation in lithium silicate glasses. Phys. Chem. Glasses, 15, 95-105.

Johnson, W. A. and Mehl, R. F. (1939) Reaction kinetics in processes of nucleation and growth. Amer. Inst. Mining Eng., Tech. Publ. 1089, 1-27.

Kalinina, A. M., V. N. Filipovich, V. M. Fokin (1980) Stationary and non-stationary crystal nucleation rate in a glass of $2Na_2O \cdot CaO \cdot 3SiO_2$ stochiometric composition. J. Non-Crystalline Solids, 38/39, 723-728.

Kaschiev, D. (1969) Solution of the non-steady state problem in nucleation kinetics. Surface Science, 14, 209-220.

Kirkpatrick, R. J. (1974) The kinetics of cyrstal growth in the system $CaMgSi_2O_6$-$CaAl_2SiO_6$. Amer. J. Sci., 273, 215-242.

———— (1975) Crystal growth from the melt: A review. Amer. Mineral., 60, 798-814.

———— (1976) Towards a kinetic model for the crystallization of magma bodies. J. Geophys. Res., 81, 2565-2571.

———— (1977) Nucleation and growth of plagioclase, Makaopuhi and Alae Lava Lakes, Kilaueau Volcano, Hawaii. Geol. Soc. Amer. Bull., 88, 78-84.

———— (1978) Processes of crystallization in pillow basalts, Hole 396B, DSDP Leg 46. In *Initial Reports of the Deep Sea Drilling Project*, 46, U. S. Government Printing Office, Washington, D. C.

Kirkpatrick, R. J., L. Klein, D. R. Uhlmann, J. F. Hays (1979) Rates and processes of crystal growth in the system anorthite-albite. J. Geophys. Res., 84, 3671-3676.

Kirkpatrick, R. J., L.-C Kuo, and J. Melchior (1978) Rates of nucleation and growth from programmed cooling experiments: $An_{50}Ab_{50}$. Abstracts of the 1978 Geol. Soc. Amer. Ann. Meeting, p. 435.

Kirkpatrick, R. J., G. R. Robinson, and J. F. Hayes (1976) Kinetics of crystal growth from silicate melts: anorthite and diopside. J. Geophys. Res., 81, 5715-5720.

Klein, L. C., P. I. K. Onorato, D. R. Uhlmann and R. W. Hopper (1975) Viscous flow, crystallization behavior, and thermal histories of lunar breccias 70019 and 79155, Proc. 6th Lunar Sci. Conf., 579-593.

Klein, L. C., C. A. Handwerker, and D. R. Uhlmann (1977) Nucleation kinetics of sodium disilicate. J. Crystal Growth, 42, 47-51.

Klein, L. C. and D. R. Uhlmann (1974) Crystallization behavior of anorthite. J. Geophys. Res., 79, 4869-4874.

———— (1976) The kinetics of lunar glass formation, revisited, Proc. 7th Lunar Sci. Conf., 1113-1121.

Kohli, C. S. and M. B. Ives (1972) Computer simulation of crystal dissolution morphology. J. Crystal Growth, 16, 123-130.

Kuroda, T., T. Irisawa, and A. Ookawa (1977) Growth of a polyhedral crystal from solution and its morphological stability. J. Crystal Growth, 42, 41-46.

Lasaga, A. C. (1981) Towards a master equation in crystal growth. Amer. J. Sci., in press.

Lasaga, A. C. and Muncill, G. E. (1980) A general solution to the composition of crystals grown from melts and applications. Geol. Soc. Amer. Abstracts, 12(7), 469.

Lofgren, G. E. (1977) Dynamic crystallization experiments bearing on the origin of textures in impact generated liquids. Proc. 8th Lunar Sci. Conf., 2079-2095.

———— (1980) Experimental studies of the dynamic crystallization of silicate melts. In *Physics of Magmatic Processes*, Hargraves, R. B., ed., 487-551, Princeton University Press, Princeton, N. J.

Lofgren, G. E., C. H. Donaldson, T. M. Usselman (1975) Geology, petrology and crystallization of Apollo 15 quartz-normative basalts. Proc. 6th Lunar Sci. Conf., 79-99.

Lofgren, G. E., C. H. Donaldson, R. J. Williams, O. Mullins, and T. M. Usselman (1974) Experimentally reproduced textures and mineral chemistry of Apollo 15 quartz normative basalts. Proc. 5th Lunar Sci. Conf., 549-567.

Lofgren, G. E., D. P. Smith, R. W. Brown (1978) Dynamic crystallization and kinetic melting of the lunar soil. Proc. 9th Lunar Planet. Sci. Conf., 959-975.

Maaløe, S. (1976) The zoned plagioclase of the Skaergaard Intrusion, East Greenland. J. Petrol. 17, 398-419.

Matusita, K., and M. Tashiro (1973) Rate of homogeneous nucleation in alkali disilicate glasses. J. Non-Crystalline Solids, 11, 471-484.

Miller, C. E. (1977) Faceting transition in melt grown crystals. J. Crystal Growth, 42, 357-363.

Müller-Krumbhaar, H. (1975) Diffusion theory for crystal growth at arbitrary solute concentration. J. Chem. Phys., 63, 5131-5138.

Mullins, W. W., and R. F. Sekerka (1964) Stability of a planar interface during solidification of a dilute binary alloy. J. Appl. Phys., 35, 444-451.

Nabelek, P. I., L. A. Taylor, and G. E. Lofgren (1978) Nucleation and growth of plagioclase and the development of textures in a high-alumina basaltic melt. Proc. 9th Lunar Planet. Sci. Conf., 725-741.

Neilson, G. F. and M. C. Weinberg (1979) A test of classical nucleation theory: crystal nucleation of lithium disilicate glass. J. Non-Crystalline Solids, 34, 137-142.

O'Hara, J., L. A. Tarshis, W. A. Tiller, and J. P. Hunt (1968) Discussion of interface stability of large facets on solution growth crystals. J. Crystal Growth, 3-4, 555-561.

Rowlands, E. G. and P. F. James (1979a) Analysis of steady state nucleation rates in glasses, Part I. Methods of analysis and application to lithium disilicate glass. Phys. Chem. Glasses, 20, 1-8.

Rowlands, E. G. and P. F. James (1979b) Analysis of steady state nucleation rate in glasses, Part II. Further comparison between theory and experiment for lithium disilicate glass. Phys. Chem. Glasses, 20, 9-14.

Scherer, G., R. W. Hopper, and D. R. Uhlmann (1972) Crystallization behavior and glass formation of selected lunar compositions. Proc. 3rd Lunar Sci. Conf., 2627-2637.